QUANTUM MECHANICS

QUANTUM MECHANICS
Concepts and Applications

John D. McGervey
Department of Physics
Case Western Reserve University
Cleveland, Ohio

ACADEMIC PRESS

San Diego   New York   Boston   London   Sydney   Tokyo   Toronto

Find Us on the Web! http://www.apnet.com

This book is printed on acid-free paper. ∞

Academic Press, Inc.
A Division of Harcourt Brace & Company
525 B Street, Suite 1900, San Diego, California 92101-4495

United Kingdom Edition published by
Academic Press Limited
24-28 Oval Road, London NW1 7DX

Library of Congress Cataloging-in-Publication Data

McGervey, John D., date.
    Quantum mechanics / by John D. McGervey.
        p.   cm.
    Includes index.
        ISBN 0-12-483545-7   ISBN 0-12-483546-5 (Diskette)
        1. Quantum Theory.   I. Title.
QC174.12.M378   1994
530.1'2--dc20   94-33103
                                                    CIP

PRINTED IN THE UNITED STATES OF AMERICA
96 97 98 99 00 MM 9 8 7 6 5 4 3 2 1

Ver. 2

CONTENTS

*Denotes chapters or chapter sections that can be omitted from a shorter course without loss of continuity.

CHAPTER 15   Molecular Structure and Spectra*

CHAPTER 16   Quantum Statistics*

APPENDIX A   Probability and Statistics

APPENDIX B   The Boltzmann Factor

APPENDIX C   Relativistic Dynamics

APPENDIX D   Derivation of the Eigenfunctions of the L^2 Operator

APPENDIX E   Solution of the Radial Equation for the Hydrogen Atom

PREFACE

This book has been designed for and tested in an upper-level one-semester course in quantum mechanics for students majoring in physics or related fields. It is an outgrowth of my book *Introduction to Modern Physics*, which has been used in quantum-mechanics courses at a number of schools since the first edition was published in 1971. Like that book, this book emphasizes concepts and connections with experimental results rather than sophisticated techniques that can be better learned at the graduate level. For example, the Heisenberg uncertainty relation is discussed at length from an experimental point of view, first in one dimension with reference to the Heisenberg microscope, and then later for angular momentum components with reference to the results of two consecutive Stern–Gerlach experiments. The latter results are then used in analyzing the Bell Inequality, which was treated for the first time in a book at this level in the second edition (1983) of *Introduction to Modern Physics*.

It is a well-established principle in education (which is, alas, not commonly applied in physics teaching) that people learn a generalization more readily if they see particular examples first, to motivate the generalization. Therefore the operator method is introduced only after the groundwork has been laid through analysis of wave behavior, and quantum-mechanical postulates are introduced by means of examples that are thoroughly discussed before generalizations are made. At each step in a discussion, difficulties are anticipated and addressed by explicit reference to questions commonly posed by students. Students are further aided by fully worked out examples in each chapter. Students here at Case Western Reserve University and at other schools have commented appreciatively on these rather unusual features of *Introduction to Modern Physics*.

However, this book differs from *Introduction to Modern Physics* in many important ways:

Background Information Crucial experiments are covered in depth, but other "modern physics" topics are treated only to the extent needed and are placed in appendices. Selected computational methods are treated in more depth to illustrate the power of the theory, and methods formerly given only in problem sets are integrated into the mainstream of the text.

Examples Illustrating Basic Concepts   Numerous fresh examples are given; these include the "quantum bouncer" (Frank Crawford, 1989, *American Journal of Physics* **57**, 261), oscillations in the two-level system (Quantum Pot Watching, 1989, *Science* **246**, 888), "Neutrino Oscillations in Matter, (1989, *Reviews of Modern Physics* **61**, 917), and "Experiments with an Individual Atomic Particle ... " (Hans Dehmelt, 1990, *American Journal of Physics* **58**, 17). Some examples are enriched by means of problems based on computer-generated wave functions.

Many Numerical Solutions   The student will be able to run the software that is provided with the book to see probability amplitudes and probability densities and find energy levels for single and multiple square wells, the quantum bouncer, the simple harmonic oscillator, the radial component for the hydrogen atom, and the radial component for the spherical square well. Some of these can be used to test perturbation calculations by slightly varying the energy, as in the square well with a step in the middle. Some exact solutions are used for reinforcement after the numerical integration has been done. Others are functions that are routinely omitted from this type of course, because of their mathematical difficulty (e.g., the quantum bouncer or a series of 22 consecutive square wells); for the computer, these are no more difficult than any other function. The software, which is included with the book, runs on IBM-PC compatible computers with VGA graphics. It has also been used to generate a number of figures for the book.

New Exercises   Many of the exercises involve explicit situations to be solved with the computer software. These have been assigned in my classes for several years and have led to improved comprehension of wave functions and probability densities.

The book has enough material for a two-semester course, but it can be used for a one-semester course by omitting some chapters and sections of other chapters. Chapter 1 in particular contains a large selection of background material, allowing the instructor to choose those portions that he or she prefers as an introduction. In the table of contents, chapters or chapter sections that can be omitted without loss of continuity are indicated by asterisks.

ACKNOWLEDGMENTS

I am grateful for the help of many people who have used *Introduction to Modern Physics* in their courses and have made helpful suggestions. These, and others who have helped, include Robert Brown, Arnold Dahm, Tom Eck, David Farrell, Leslie Foldy, Cyrus Taylor, and Bill Tobocman of CWRU, Karl Canter of Brandeis University, Mark Heald of Swarthmore College, and Eric Sheldon of the University of Lowell. I am also indebted to those who gave technical help during the production of this book, in particular Rich Sones, Charles Knox, and Zhibin Yu. Finally, I must acknowledge the hospitality of Professor Theodor Hehenkamp and his group at the University of Göttingen, with whom I worked for three months during the final stages of preparation of the manuscript.

JOHN D. McGERVEY

CHAPTER 1

The Quantum Concept

The concept of the quantum began with the discovery that electromagnetic radiation can gain or lose energy only in discrete quantities, or quanta. The credit for the discovery of quanta belongs to Max Planck, who postulated their existence in his analysis of blackbody radiation.[1] However, the most direct and easily explained evidence for this property of radiation is found in the photoelectric effect. After this effect was analyzed in Albert Einstein's 1905 paper, many other quantum effects were discovered. We shall discuss a few of these in this chapter, to lay the groundwork for the more general development of quantum mechanics.

1.1 THE PHOTOELECTRIC EFFECT: FAILURE OF CLASSICAL EXPLANATIONS

At a time when few people believed Planck's theory, Einstein made a sweeping generalization. He wrote:[2]

> According to the assumption considered here, the spreading of a light beam emanating from a point source does not cause the energy to be distributed continuously over larger and larger volumes, but rather the energy consists of a finite number of energy quanta, which move without breaking up and which can be absorbed or emitted only as wholes.

Einstein went on to suggest that consequences of this property of light could be observed in the *photoelectric effect*: the light-induced emission of

---

[1] For a treatment of blackbody radiation, see Section 3.1, J. D. McGervey, *Introduction to Modern Physics*, Academic Press, New York, (1983).

[2] A. Einstein, *Ann. Phys. (Leipzig)* **17**, 132–148 (1905). Translated by J. McGervey.

electrons from a metal. It was known at that time that light shining on a metal caused the metal to become positively charged (raising its electro-static potential). The potential would increase up to a definite limit, and this limit was higher when light of a shorter wavelength was used, but rigorous quantitative experiments had not been done.

Einstein said that an electron escaped from a metal after receiving energy from a single quantum of light (now called a *photon*). He also said that the photon's energy E is proportional to the light frequency ν, or

$$E = h\nu \tag{1.1}$$

where the proportionality constant h is *Planck's constant*, which Planck had introduced in his analysis of blackbody radiation.

In Einstein's view, an electron that absorbed a photon of energy $h\nu$ could escape the metal if $h\nu$ were greater than the minimum energy (called the *work function W*) required to escape from that metal. In that case its kinetic energy K after escaping could be no greater than

$$K_{\max} = h\nu - W \tag{1.2}$$

It is important to understand that electrons in a metal do not all have the same energy. For those with less energy, K is less than $K_{\max}$ after escaping.

Equation (1.2) was first tested by R. A. Millikan, who set up what he described as "a machine shop *in vacuo*" to obtain a clean alkali-metal surface, which he then irradiated with monochromatic light of various frequencies. He determined the value of $K_{\max}$ for each frequency by applying a retarding potential needed to stop the flow of electrons from the alkali metal plate to a collector (Figure 1.1) in agreement with Eq. (1.2), he found that the retarding potential V was linearly dependent on the frequency ν:

$$V = (h/e)\nu - W/e \tag{1.3}$$

By measuring the slope of the plot of V vs. ν (Figure 1.2) and using the value of the electron charge e (which he had measured in his famous oil-drop experiment), Millikan found h to be 6.56×10^{-7} erg-s, in excel-lent agreement with the value that Planck had found by analyzing the blackbody spectrum and only 1% below the currently accepted value.

It is often convenient to use the value of hc rather than h and to express it in units more suitable for atomic dimensions:

$$hc \approx 1240 \text{ eV-nm} \tag{1.4}$$

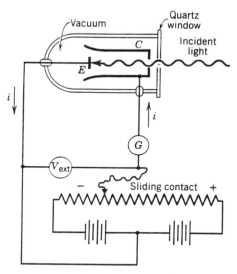

FIGURE 1.1 Photoelectric apparatus. The potential difference V_{ext} is varied until gal-vanometer G shows zero current. (Adapted from R. Resnick and D. Halliday, *Basic Concepts in Relativity and Early Quantum Theory*, Macmillan, 1992.)

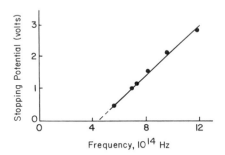

FIGURE 1.2 Plot of Millikan's measurements of stopping potential at various frequencies for a sodium emitter. (Adapted from R. Resnick and D. Halliday, *Basic Concepts in Relativity and Early Quantum Theory*, Macmillan, 1992.)

where the electron volt (eV) is the kinetic energy acquired by an electron when its electrical potential changes by one volt, and one nanometer (nm) is 10^{-9} meters (approximately ten times the diameter of an atom). Then, in terms of the wavelength λ, and using $\nu = c/\lambda$, we find the energy of a photon to be

$$E = hc/\lambda \approx 1240 \text{ eV-nm}/\lambda \qquad (1.5)$$

1.1.1 Significance

The premise leading to Eq. (1.2) is totally inconsistent with the original concept of light as a wave. Classical theory indicates that electrons are accelerated by the oscillating electric field of the light wave. To give the electrons greater energy, one should therefore increase the *intensity* of the light; that would increase the amplitude of the oscillations of the field and hence of the electrons, giving the electrons more energy. But it was shown experimentally, as early as 1902, that the energy of an ejected electron is independent of the light intensity. The total energy of all ejected electrons is larger, but this happens because there are more photons.

It is also significant that an electron can be ejected immediately when the light is turned on, even when the light intensity is very low. An electron can be ejected with an energy of several eV at a time when the wave would have deposited that much energy over an area of many square meters! This is the first known manifestation of what we now call wave–particle duality. In a way that we cannot describe adequately in the context of our limited senses, energy that clearly travels in a wave pattern is absorbed as if it were localized in a small region, like a particle.

Example Problem 1.1   The stopping potential V for electrons emitted from sodium is:

Incident wavelength λ (nm):	155	310	400
Stopping potential V (volts):	5.7	1.7	0.8

Find the work function of sodium and the value of hc from these data. Also verify that the data are consistent with Eq. (1.3).

Solution. The stopping potential in volts gives the value of $K_{\max}$ in electron volts. Substitution into Eq. (1.3) twice, using the first two sets of values, gives

$$5.7 \text{ eV} = hc/(155 \text{ nm}) - W$$
$$1.7 \text{ eV} = hc/(310 \text{ nm}) - W$$

Subtraction yields

$$4.0 \text{ eV} = hc/(310 \text{ nm}), \quad \text{or} \quad hc = 1240 \text{ eV-nm}$$

Thus

$$W = hc/(310 \text{ nm}) - 1.7, \quad \text{or} \quad W = 2.3 \text{ eV}$$

Verification. Using the third set of given values, with $hc = 1240$ eV-nm and $W = 2.3$ eV, yields

$$0.8 \text{ eV} = (1240/400) \text{ eV} - 2.3 \text{ eV} = 0.8 \text{ eV}$$

1.2 ATOMIC SPECTROSCOPY AND DISCRETE ENERGY LEVELS IN ATOMS

Quantum phenomena were observed long before they were recognized. In 1817 J. Fraunhofer reported his discovery of dark lines (now called Fraunhofer lines) in the spectrum of radiation from the sun.[3] Much later, Kirchhoff developed the theory that these lines result from selective absorption of light by gases in the sun's outer atmosphere,[4] and Foucault showed that absorption of light by sodium vapor produced two dark lines at the same position as two prominent lines in the sun's spectrum.[5] For an excellent and thorough treatment of these and other early developments, the book by Harvey White is recommended.[6]

Systematic studies then showed that each element emits and absorbs its own characteristic *line spectrum* consisting of a unique set of discrete frequencies. The lines can be grouped into *series* in which lines become more closely spaced with decreasing wavelength, approaching a *series limit* at which the line spacing converges to zero. (See Figure 1.3.)

After many unsuccessful attempts to derive an equation that would give the wavelengths of any series, J. J. Balmer constructed the formula for

[3]J. Fraunhofer, *Gilbert's Ann.* **56**, 264, (1817). If you go to Bavaria, you can still visit Fraunhofer's original laboratory and see his equipment. You simply knock on the door of the house across the road from the old building; a woman comes out and unlocks the door for you.
[4]G. Kirchhoff, *Monatsber. Berlin Akad. Wiss.*, p. 662 (1859).
[5]L. Foucault, *Ann. Chim. Phys.* **68**, 476 (1860).
[6]H. E. White, *Introduction to Atomic Spectra*, Chapter 1, McGraw-Hill, New York (1934).

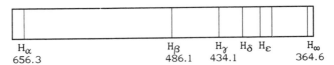

$$H_\alpha \qquad\qquad\qquad\qquad H_\beta \qquad H_\gamma \ \ H_\delta \ H_\epsilon \qquad H_\infty$$
$$656.3 \qquad\qquad\qquad\qquad 486.1 \qquad 434.1 \qquad\qquad 364.6$$

FIGURE 1.3 Part of the emission spectrum of hydrogen gas, showing the wavelengths, in nm, of the Balmer series lines. H_α shows the longest wavelength in this series; H_∞ shows the position of the series limit.

what is now called the *Balmer series*:

$$\lambda_n = 364.56[n^2/(n^2 - 4)] \text{ nm} \qquad n = 3, 4, 5, \ldots \qquad (1.6)$$

Balmer's formula gives all of the correct wavelengths for this one series (and equally important, gives no wavelengths that are *not* observed). The other series follow similar formulas; the general formula was developed by Rydberg in 1890. It can be written

$$k_n \equiv \lambda_n^{-1} = R(n^2 - n'^2)/n^2 n'^2 \qquad n' = 1, 2, 3, \ldots, n > n' \quad (1.7)$$

where R, called the Rydberg constant, equals 0.109720 mm^{-1}.

For the Balmer series, $n' = 2$. Other series (in Figure 1.4) are the Lyman series ($n' = 1$), the Paschen series ($n' = 3$), and the Brackett series ($n' = 4$).

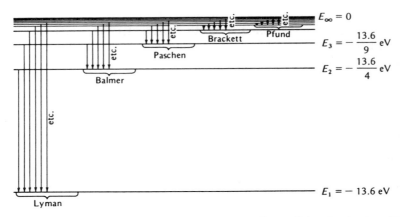

FIGURE 1.4 Energy levels of the hydrogen atom, according to Bohr, showing transitions that give rise to the various series of spectral lines.

1.3 THE BOHR MODEL OF THE ATOM

Many theorists, notably J. J. Thomson, tried to devise models of the hydrogen atom that would yield the Rydberg formula. The early models were scrapped when Rutherford's results with alpha particles showed that the atom contains a positive nucleus possessing almost all of the atom's mass in a tiny fraction of its volume [within a radius of a few femtometers (fm, 10^{-15} m)].[7]

This fact poses the question, "Why doesn't the electron fall into the nucleus?" One might suppose that the electron orbits around the nucleus, just as planets orbit around the sun without falling in. The problem is that the electric force is significantly different from the force of gravity. Accelerated charges radiate energy much more rapidly than do accelerated neutral masses.[8] Niels Bohr postulated this difficulty away by assuming:

1. Electrons move in circular orbits around the center of mass of the atom. (This assumption is incorrect, as we shall see in Sections 1.5 and 2.6.)
2. The only allowed orbits are such that the angular momentum of the atom about its center of mass is an integral multiple of $h/2\pi$.
3. Energy is radiated when an electron "falls" from an allowed orbit of energy E to one of energy $E - \Delta E$, and a photon of frequency $\nu = \Delta E/h$ is emitted, in accordance with Einstein's photoelectric equation.

Although Bohr's first postulate is incorrect and his work applies only to one-electron systems, his derivation of the Rydberg formula is noteworthy.

Following Bohr, we use assumptions 1 and 2 to find the radius of the nth allowed orbit and then determine its energy. Let M be the mass of the hydrogen nucleus and m the electron mass. The nucleus and the electron both revolve around the center of mass, at distances R_n and r_n, respectively; because they must always be on opposite sides of the center, their angular velocities ω_n are equal. From the definition of center of mass we have

$$MR_n = mr_n \quad \text{or} \quad R_n = r_n(m/M) \tag{1.8}$$

---

[7]E. Rutherford, *Phil. Mag.* **21**, Ser. 6, 669 (1911).

[8]Accelerated masses radiate gravitational waves, but the radiation rate is so small that we have not yet detected such waves. Energy loss from gravitational radiation causes planets to spiral toward the sun at an undetectably low rate.

The angular momentum is

$$m\omega_n r_n^2 + M\omega_n R_n^2 = n\hbar \tag{1.9}$$

where $\hbar$ is defined as $h/2\pi$, and n is an integer as prescribed by assumption 2. Another equation comes from setting the Coulomb force on the electron equal to the product of its mass and its centripetal acceleration:

$$m\omega_n^2 r_n = \frac{Ze^2}{4\pi\varepsilon_0(r_n + R_n)^2} \quad \text{(SI units)} \tag{1.10}$$

where the factor $4\pi\varepsilon_0$ is omitted when cgs units are used, and the numerator Ze^2 is the product of the nuclear charge Ze and the electron's charge e. For hydrogen Z is obviously equal to 1, but the same equations are valid for ionized atoms of higher Z having only one electron (for example, Li^{2+} or C^{5+}).

1.3.1 Results Based on the Bohr Model

We can evaluate a number of atomic parameters by means of calculations based on the Bohr model, as follows:

A. Calculation of the Orbit Radius We eliminate R_n from Eq. (1.9) by substituting from Eq. (1.8), obtaining

$$m\omega_n r_n^2 = n\hbar M/(M + m) \tag{1.11}$$

We can eliminate R_n from Eq. (1.10) via the same substitution, to obtain

$$m\omega_n^2 r_n = \frac{Ze^2 M^2}{4\pi\varepsilon_0 r_n^2 (M + m)^2} \tag{1.12}$$

Squaring Eq. (1.11) and dividing by Eq. (1.12) eliminates ω_n and yields

$$r_n = \frac{4\pi\varepsilon_0 n^2 \hbar^2}{mZe^2} \quad \text{(SI units)} \tag{1.13}$$

Again, deleting the factor $4\pi\varepsilon_0$ converts the equation to cgs units. Notice that r_n depends only on m, not on M; *the orbit radii are the same for all isotopes of hydrogen.* Substituting values of the constants into Eq. (1.13) gives (in nanometers)

$$r_n = 0.0529177n^2/Z = a_0 n^2/Z \tag{1.14}$$

where a_0, the orbit radius for the lowest-energy state of the hydrogen atom, is called the *first Bohr radius*.

B. Calculation of the Energy The total internal energy E_n of the atom in state n is the total kinetic energy of nucleus and electron plus the negative electrostatic potential energy

$$E_n = \frac{m\omega_n^2 r_n^2 + M\omega_n^2 R_n^2}{2} - \frac{Ze^2}{4\pi\varepsilon_0(r_n + R_n)} \tag{1.15}$$

If we again use Eq. (1.8) to eliminate R_n, this becomes

$$E_n = \frac{(m\omega_n^2 r_n^2)(M + m)}{2M} - \frac{Ze^2 M}{4\pi\varepsilon_0 r_n(m + M)} \tag{1.16}$$

Equation (1.12) shows that the kinetic energy is half of the magnitude of the potential energy, so we have

$$E_n = -\frac{Ze^2 M}{8\pi\varepsilon_0 r_n(m + M)} \tag{1.17}$$

Using the expression for r_n from Eq. (1.13), we finally find that

$$E_n = -\frac{Z^2 e^4 mM}{32\pi^2\varepsilon_0^2 n^2\hbar^2(m + M)} = -\frac{Z^2 e^4 m_r}{32\pi^2\varepsilon_0^2 n^2\hbar^2} \tag{1.18}$$

where m_r is defined as $mM/(m + M)$, a quantity called the *reduced mass*. Because the proton mass is $1836.15m$, the reduced mass for normal hydrogen (^{1}H) is $(1836.15/1837.15)m$; the use of m instead of m_r in Eq. (1.18) results in an error of less than 0.06%. The energy E_n is written in condensed form as

$$E_n = -m_r c^2\alpha^2 Z^2/2n^2 \tag{1.19}$$

where α is defined as $e^2/4\pi\varepsilon_0\hbar c$, a dimensionless number equal to about $1/137$ known as the *fine structure constant*. Introduction of this constant makes the expression for E_n independent of the units used. Numerically, we have

$$E_n = -13.5984/n^2 \text{ eV} \quad \text{for } {}^1\text{H} \qquad (1.20)$$

C. Calculation of the Frequencies of the Radiation   In a transition from level n to level n', the energy of the emitted photon is $h\nu_{nn'}$, where $\nu_{nn'}$ is the radiated frequency and

$$h\nu_{nn'} = E_n - E_{n'} = 13.5984\left(\frac{1}{n'^2} - \frac{1}{n^2}\right) \text{ eV} \quad \text{for } {}^1\text{H} \qquad (1.21)$$

Bohr calculated Rydberg's constant R, converting R from an empirical constant to a derived quantity; from Eqs. (1.7), (1.19), and (1.21), using current values for the constants, we have[9]

$$R = m_r c \alpha^2/2h = 0.010967758 \text{ nm}^{-1} \qquad (1.22)$$

Although these equations cannot be generalized to systems with more than one electron, they can be applied to any one-electron system, such as ${}^2\text{H}$, ${}^3\text{H}$, He^+, Li^{2+}, etc., by using the appropriate values for Z and m_r.

Example Problem 1.2   Write a formula like (1.20) for energy levels of ${}^6\text{Li}^{2+}$.

Solution. The nuclear mass of ${}^6\text{Li}$ is about $10962m$ and $Z = 3$, making $Z^2 m_r(\text{Li})$ equal to $9 \times (10962/10963)m$, while the value used in Eq. (1.20) for Zm_r is $Z = 1$, and $m_r(\text{H}) = (1836.15/1837.15)m$. Thus Eq. (1.20) becomes the equation for ${}^6\text{Li}$ if we divide by $m_r(\text{H})$ and multiply by $9m_r(\text{Li})$, with the result that

$$E_n = -122.44n^2 \text{ eV for } {}^6\text{Li}^{2+}.$$

Spectra of many one-electron ions have been observed and the results are in agreement with theory. In fact, deuterium (${}^2\text{H}$) was discovered in 1931 by observation of "satellite" lines in the spectrum of hydrogen.

[9]*Physics Today* **44**, August 1991, 9–13. Rydberg's value was slightly different from ours, but spectroscopic results agree well with the value derived from the Bohr formula. However, small corrections must be applied because of the spins of the electron and proton (Chapter 11).

1.4 ATOMIC COLLISIONS

It is significant that atomic energy levels can also be observed by means that do not involve emission or absorption of photons. In an *inelastic collision* between two atoms or between an electron and an atom, an atom can acquire enough energy to "jump" from its original energy level to a higher level. The energy lost by the electron was first measured in the *Franck-Hertz experiment* (1914), which showed that these losses are quantized, occurring only in discrete amounts, which are always the same for a given type of atom.

In this experiment, the collisions occurred between electrons and atoms of mercury vapor. Electrons were accelerated in a vapor-filled tube by an electric potential V_{acc} (Figure 1.5).

A meter showed the current of electrons reaching the anode. In a vacuum tube, increasing V_{acc} should always lead to increasing current. This is true when vapor is present, if collisions with atoms of the vapor are elastic. But Figure 1.6 shows that as V_{acc} increased beyond about 6.6 volts, the current fell, presumably because of inelastic collisions with Hg atoms. It then rose again, but fell when V_{acc} was large enough to permit *two* inelastic collisions for a single electron. Figure 1.6 shows a series of drops in current, spaced about 5 volts apart in V_{acc}.

Apparently, 5 eV is needed to excite an Hg atom from its lowest energy level to the next lowest level; a collision must be elastic if this energy is not available. This result agrees with other observations; there is

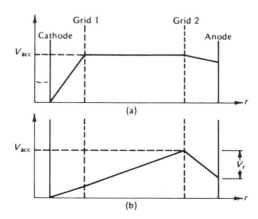

FIGURE 1.5 Two distributions of the accelerating potential V_{acc} in the mercury-vapor tube of Franck and Hertz.

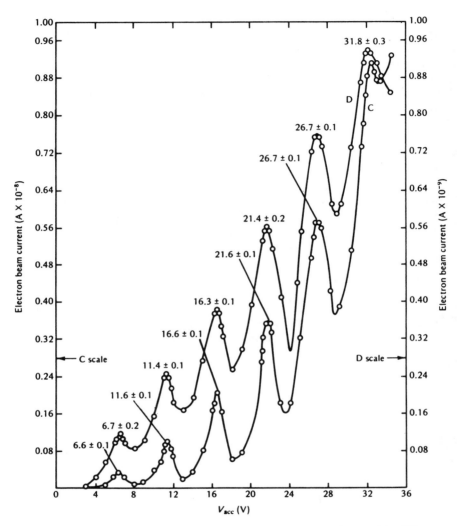

FIGURE 1.6 Dependence of current on potential V_{acc} in Hg vapor with potentials of Figure 1.5b. Curve C (left scale) obtained with filament at 2.5 volts, curve D with filament at 1.85 volts. (From A. Melissinos, *Experiments in Modern Physics*, Academic Press, New York (1966).)

spectroscopic evidence that mercury's first excited state is 4.86 eV above the lowest state.

You may now wonder why the first drop in energy is at 6.6 eV rather than at 4.9 eV. The answer lies in the work function. Apparently 1.7 eV is required to free an electron from the cathode, so that the accelerating voltage must be 6.6 volts to give an electron a kinetic energy of 4.9 volts after it is free.

For further evidence, Hertz observed the spectrum of light emitted by mercury vapor when the electrons were accelerated. He saw that certain spectral lines appeared only after V_{acc} reached a value at which a drop in current occurred. At this point the excited state emitted photons of energy $eV_{acc} - 1.7$. (If eV_{acc} was large enough to excite a state with $n > 2$, then other photons whose total energy was equal to $eV_{acc} - 1.7$ eV could also be emitted.)

1.5 THE CORRESPONDENCE PRINCIPLE

If Bohr's theory correctly gives the photon frequencies for transitions involving small values of n, should it not give the right answer when, for example, $n = 10^6$ and $n' = 10^6 - 1$? It should and it does; this is verified with gases of very low density (e.g., in interstellar space), where atoms with large radii (large n) avoid collisions long enough to radiate photons.

On the other hand, for orbits of macroscopic dimensions, we know that classical theory gives the correct frequency for the radiation: the frequency is equal to the frequency of revolution of the electron in its orbit. In other words, in this domain, *both theories are correct*. The general principle, called the *correspondence principle*, applies not only to Bohr's theory but to all subsequent quantum theories. It can be stated as follows:

The result of any quantum theory, in the limit as the quantum number approaches infinity, agrees with classical theory.

Thus Bohr's frequencies approach the classical frequencies as n increases.

1.5.1 Comparison of Frequencies

The Bohr frequency for a transition in ^{1}H between energy level n and level $n - 1$ is

$$\nu_{n-1, n} = \frac{m_r c^2 \alpha^2}{2h} \left[\frac{1}{(n-1)^2} - \frac{1}{n^2} \right] \tag{1.23}$$

or

$$\nu_{n-1,n} = \frac{m_r c^2 \alpha^2}{2h} \left[\frac{2n-1}{n^4 - 2n^3 + n^2} \right] \qquad (1.24)$$

As n increases, the expression in brackets approaches $2n/n^4$, or $2/n^3$, and

$$\nu_{n-1,n} \rightarrow \frac{m_r c^2 \alpha^2}{n^3 h} \quad \text{as } n \rightarrow \infty \qquad (1.25)$$

For comparison, if a classical particle of mass m_r has an angular momentum of $nh/2\pi$ in a circular orbit, its frequency of revolution is

$$\nu = \frac{m_r c^2 \alpha^2}{n^3 h} \qquad (1.26)$$

as you can verify. Thus for large values of n, both theories agree. More precisely, the frequency given by Bohr's theory for a transition from the nth orbit to the $(n-1)$th orbit lies between the frequencies of revolution of a particle of mass m_r in each of these orbits. This follows from the inequality

$$\frac{m_r c^2 \alpha^2}{n^3 h} < \frac{m_r c^2 \alpha^2}{2h} \left(\frac{2n-1}{n^4 - 2n^3 + n^2} \right) < \frac{m_r c^2 \alpha^2}{(n-1)^3 h}$$

which can be transformed to the more obvious

$$n^3 > \frac{2(n^4 - 2n^3 + n^2)}{2n-1} > (n-1)^3$$

The classical particle spirals in from the nth Bohr orbit to the $(n-1)$th orbit, continuously emitting radiation whose frequency increases from $m_r c^2 \alpha^2/n^3 h$ to $m_r c^2 \alpha^2/(n-1)^3 h$; the Bohr atom emits the intermediate frequency given by Eq. (1.24). The classical frequencies differ by about 3 parts in n parts, for large n, and the Bohr frequency is in the middle. Thus the Bohr frequency differs from each classical frequency by about 1 part in 9000 for an orbit of about 1 cm radius (for which $n = 13,747$).

1.5.2 Rate of Radiation

The correspondence principle also requires that the rate at which energy is radiated by an accelerated charge should be correctly given by the classical expression as well as by any valid quantum theory. Classically,

$$\text{Rate of radiation} = \frac{e^2/c^3}{6\pi\varepsilon_0}\mathbf{a}^2 \tag{1.27}$$

where $\mathbf{a}$ is the acceleration of the charge. In the quantum theory the actual time of emission of a photon is random, but the *average* time that the atom remains in the higher-energy state should be equal to the time required, at the classical rate, to radiate an energy equal to that of the emitted photon. We can compute that average time by requiring that it agree with Eq. (1.27), and the result agrees with observations.

1.5.3 Elliptical Orbits

The correspondence between Bóhr's theory and classical theory breaks down if there are quantum transitions in which the quantum number changes by $\delta n > 1$. By generalizing Eq. (1.23) to

$$\nu_{n-\delta n, n} = \frac{m_r c^2 \alpha^2}{2h}\left[\frac{1}{(n-\delta n)^2} - \frac{1}{n^2}\right] \tag{1.28}$$

we find that

$$\nu_{n-\delta n, n} \rightarrow \frac{m_r c^2 \alpha^2 \delta n}{n^3 h} \quad \text{as } n \rightarrow \infty \tag{1.29}$$

For large n, because of the factor δn, the emitted frequencies are integral multiples (harmonics) of the fundamental frequency $\nu_{n-1, n}$. Classically, no harmonics are radiated by a charged particle in a *circular* orbit, but they are emitted when an electron is in an *elliptical* (eccentric) orbit. For this reason, the correspondence principle *requires* that a quantum theory provide for noncircular orbits. A modification of the Bohr model, the Bohr–Sommerfeld model, introduced elliptical orbits, using a second quantum number k. The number n still gave the energy, but the number k, which could be any integer less than or equal to n, gave the angular momentum.

[This theory, although not correct in the values of angular momentum, accounted for the observation of so-called *fine structure* in atomic spectra —the occurrence of closely spaced multiple lines where Bohr's model only predicts a single line. These multiple lines indicated that some levels were really a set of levels with a small energy difference between them. These small differences could be caused by differences in the eccentricity of the orbits, related to the values of k, as shown in Figure 1.7. We defer further

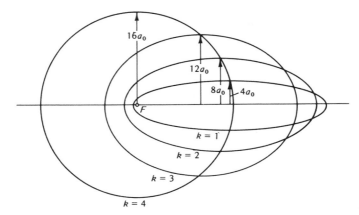

FIGURE 1.7 Allowed Bohr–Sommerfeld orbits for $Z = 1$, $n = 4$. Notice that all have the same energy and the same major axis, equal to $n^2 a_0$.

discussion of angular momentum until we explore the Schrödinger equation (in which the angular momentum quantum number is denoted by the letter l).]

The introduction of k does not completely solve the problem posed by Eq. (1.29), involving emission of harmonic frequencies, because we have not eliminated the possibility that δn could be 2 (or more) in a transition between two circular Bohr orbits. In such a transition the emitted frequency would be twice the frequency of revolution, in contradiction to classical behavior and thus in violation of the correspondence principle.

The contradiction is resolved by a *selection rule* that forbids a transition in which k changes by more than 1. (This rule is consistent with the fact that a photon has an intrinsic angular momentum of $h/2\pi$, equal to the change in angular momentum of the atom when $\delta k = 1$. In Chapter 16 we shall see further reasons for this rule.) In a circular Bohr orbit, $k = n$; if $\delta k = 1$, then δn cannot be greater than 1 (because n can never be smaller than k), and thus there are no harmonics. However, if the orbit is an eccentric ellipse, $k < n$, and harmonics are possible because δn can be greater than 1. We shall explore other selection rules later and derive them theoretically.

1.6 X RAYS: PRODUCTION AND DETECTION

X rays are high-energy photons that result from acceleration of electrons. (Gamma rays usually come from the nucleus; at any given wavelength, X rays and gamma rays are indistinguishable.) Because of its short wavelength, each X-ray photon has more energy than a photon of visible light. These "rays" were discovered by Röntgen in 1895 during his work with cathode rays (electron beams). Soon afterwards he took the first X-ray photographs.

1.6.1 Bragg Diffraction

The wave properties of X rays are not readily apparent. They were finally established when von Laue, Friedrich, and Knipping obtained a diffraction pattern from X rays passing through a ZnS crystal. Bragg then provided a relatively uncomplicated trick for analyzing the pattern. The simplest way to visualize Bragg's result is to consider a crystal as a collection of parallel planes, oriented in many different directions. An X ray of wavelength λ, striking two planes separated by a distance d, is reflected as in Figure 1.8 if the angle θ between the X ray and the plane satisfies the *Bragg equation*:

$$2d \sin \theta = n\lambda \qquad (1.30)$$

The process is diffraction rather than reflection, so Figure 1.8 is not part of a proof but rather an aid for remembering the Bragg equation. [For a derivation that does not assume mirror reflection, see L. R. B. Elton and D. F. Jackson, *Am. J. Phys.* **34**, 1036 (1966).] X-ray analyses based on Eq. (1.30) are now used routinely to determine crystal structures. By diffracting X rays from a ruled grating, one measures λ; one then uses the known λ to determine the value of d for a given crystal. In that way Avogadro's

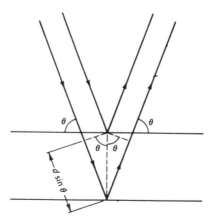

FIGURE 1.8 Bragg diffraction, somewhat simplified. The ray "reflected" from the lower plane has a longer path than the ray reflected from the upper plane. The path difference, $2d \sin \theta$, must equal $n\lambda$ for constructive interference.

number was determined in 1931; the value found then differs from the currently accepted value (6.02214×10^{23}) by only 0.05%, which was the estimated uncertainty at the time.[10]

1.6.2 X-Ray Production

Figure 1.9 shows the general features of the wavelength distribution of X rays produced when cathode rays (electrons) strike a target. Notice the sharp cutoff of the spectrum at the minimum wavelength λ_0, and the sharp peaks labeled K_α and K_β. The continuous spectrum, called *Bremsstrahlung* (braking radiation), is caused by the acceleration of the electrons as they lose speed in the target. As the electron stops in the target, its kinetic energy is converted into photons and/or heat. The cutoff wavelength λ_0 is determined by the maximum photon energy E_{max}, using the Einstein relation

$$E_{max} = h\nu_{max} = hc/\lambda_0 \qquad (1.31)$$

No single photon's energy can be greater than the amount of energy lost by an electron; therefore E_{max} is approximately equal to the kinetic energy of an electron in the beam, and this is verified by experiment.

The wavelength (e.g., λ_α) of each sharp peak is independent of the electrons' kinetic energy; of course, there could be no α peak if λ_α were less than λ_0.

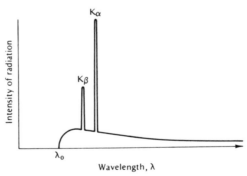

FIGURE 1.9 Spectrum of X rays produced when an electron beam strikes a target consisting of a single element.

---

[10] J. A. Bearden, *Phys. Rev.* **37**, 1210 (1931); E. R. Cohen and B. N. Taylor, *Physics Today* **44**, 9 (August 1991). The latest value has an estimated uncertainty of about 0.00006%.

Example Problem 1.3   Find the value of λ_0 for the X rays emitted by a television set, if the accelerating voltage in the tube is 25,000 volts.

Solution. $E_{max} = 25,000$ eV, so $\lambda_0 = hc/E_{max} = 1240/25,000$ nm = 0.05 nm.

1.7 MOSELEY'S LAW

The peaks in the X-ray spectrum were studied in 1913 by Moseley, who deduced that the peaks were produced by the same process that leads to emission of the optical spectra of the elements. He analyzed the results for targets consisting of single elements, using the Bohr model of the atom.[11] Surprisingly, although the Bohr model applies only to one-electron systems, the dependence of wavelength on Z can be related to that model.

The great significance of Moseley's work was in his *determination of atomic numbers*. Before this work, the periodic table of the elements was rather mysterious. For example, potassium should follow argon, just as sodium follows neon; but the atomic *mass* of potassium is smaller than the atomic *mass* of argon. Moseley resolved this paradox when he developed a formula for deriving the atomic *number* Z from the K_α frequencies ν_α. His values of Z gave a perfect correlation with the order determined by chemical properties.

Table 1.1 shows Moseley's results for various target elements. Notice that Z consistently has a value very close to $\sqrt{\nu_\alpha/(3/4)\nu_0} + 1$. This expression can be derived from the transition energy from the $n = 2$ level

TABLE 1.1   Wavelengths of K_α and K_β X rays, as reported by Moseley

Element	λ_α (nm)	λ_β (nm)	$\sqrt{\nu_\alpha/(\frac{3}{4}\nu_0)}$	Z
Ca	0.3357	0.3085	19.00	20
Ti	0.2766	0.2528	20.99	22
V	0.2521	0.2302	21.96	23
Cr	0.2295	0.2088	22.98	24
Mn	0.2117	0.1923	23.99	25
Fe	0.1945	0.1765	24.99	26
Co	0.1796	0.1635	26.00	27
Ni	0.1664	0.1504	27.04	28
Cu	0.1548	0.1403	28.01	29
Zn	0.1446	—	29.01	30

[11] H. G. J. Moseley, *Phil. Mag.* **26**, 1024 (1913); **27**, 703 (1914).

to the $n = 1$ level in a one-electron ion:

$$h\nu = h\nu_0 Z^2\left(\frac{1}{1^2} - \frac{1}{2^2}\right) \tag{1.32}$$

where $h\nu_0$ is the ionization energy of the hydrogen atom (13.60 eV), ν_0 being the largest frequency emitted by this atom [see Eq. (1.21)].

We can apply Eq. (1.32) to a multi-electron atom if we account for the mutual repulsion of the electrons. We do this empirically by assuming a nuclear charge of $e(Z - s)$ rather than eZ, the number s to be determined by fitting the data. Replacing Z by $Z - s$ in Eq. (1.32), dividing by h, and assuming that the frequency ν_α results from the transition $n = 2$ to $n = 1$, we have

$$\nu_\alpha = \nu_0(Z - s)^2\left(\frac{1}{1^2} - \frac{1}{2^2}\right) = \frac{3}{4}\nu_0(Z - s)^2 \tag{1.33}$$

Solving for Z gives

$$Z = \sqrt{\nu_\alpha/(3/4)\nu_0} + s \tag{1.34}$$

in agreement with Table 1.1 if $s = 1$.

A similar formula applies to X-ray lines produced by transitions from $n = 3$ states to $n = 2$ states. These lines, the L_α lines, have much longer wavelengths; they are not shown in Figure 1.9. We leave as an exercise for the reader to deduce the origin of the K_β lines and deduce their wavelengths.

1.8 COMPTON SCATTERING

In 1920 J. A. Gray reported, "When an ordinary beam of X or γ rays is scattered ... the scattered rays become less penetrating than the primary rays, the greater the angle of scattering."[12] The significant part of this report is the dependence of "penetrating" power (i.e., wavelength) on the angle of scattering.

A. H. Compton then deduced that this type of scattering involved conservation of momentum as well as energy. Compton's scattering equation, relating the change in wavelength to the angle of scattering, can be derived by applying the conservation laws for energy and momentum,

[12] J. A. Gray, *J. Franklin Inst.* **190**, 633 (1920).

assuming that *a photon of energy E has momentum p = E/c.* This assumption is consistent with the fact that a classical plane electromagnetic wave has a momentum density equal to its energy density divided by its speed.

In deriving Compton's equation, we take the view that the photon behaves exactly like a particle as it transfers momentum and energy to an electron. Although this view leads to the equation that is verified experimentally, we must not take it too literally. In the process, a photon (or wave) of one frequency disappears, to be replaced by another photon of a different frequency traveling in a different direction.

Because momentum is a vector, three conservation equations are involved. We assume that the original photon is traveling along the *x* axis with energy $h\nu$ and momentum $h\nu/c$ when it interacts with *an electron that is at rest.* The electron recoils with momentum *p* in a direction that makes an angle ϕ with the *x* axis, while the final photon, with energy $h\nu'$, has momentum $h\nu'/c$ in a direction that makes an angle θ with that axis. (See Figure 1.10.) The three equations thus are:

Conservation of p_x:
$$\frac{h\nu}{c} - \frac{h\nu'}{c}\cos\theta = p\cos\phi \qquad (1.35)$$

Conservation of p_y:
$$\frac{h\nu'}{c}\sin\theta = p\sin\phi \qquad (1.36)$$

Conservation of energy:
$$h\nu - h\nu' + mc^2 = \sqrt{m^2c^4 + p^2c^2} \qquad (1.37)$$

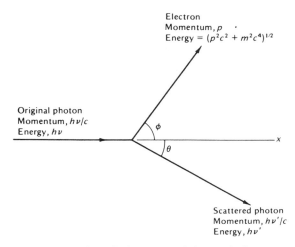

FIGURE 1.10   Angles of recoil of photon and electron in Compton scattering.

using the relativistic expression for the total energy of an electron with rest mass m and momentum p. (See Appendix C, Relativistic Dynamics.)

By squaring both sides of each equation, then adding corresponding sides of the momentum equations to eliminate ϕ, we obtain the two equations

$$h^2\nu^2 + h^2\nu'^2 - 2h^2\nu\nu' \cos \theta = p^2c^2 \qquad (1.38)$$

$$h^2(\nu - \nu')^2 + 2h(\nu - \nu')mc^2 = p^2c^2 \qquad (1.39)$$

Equating the left-hand sides of (1.38) and (1.39), and combining terms, we have

$$(\nu - \nu')mc^2 = h\nu\nu'(1 - \cos \theta) \qquad (1.40)$$

Finally, division by $mc\nu\nu'$ gives

$$\lambda' - \lambda = \frac{h}{mc}(1 - \cos \theta) \qquad (1.41)$$

Notice that the *change* in wavelength, $\delta\lambda$, is never negative under the stated condition that the electron is initially at rest. Also notice that $\delta\lambda$ is independent of the original wavelength, and its maximum value, when $\theta = \pi$ radians, is $2h/mc$, or only 0.00485 nm. Thus the effect is not seen with visible light, but it is enormous for X rays and gamma rays. For example, gamma rays whose energy equals the electron's rest energy mc^2, and whose wavelength is thus $h/mc = 0.00243$ nm, are emitted when a positron–electron pair are annihilated (Section 14.2). (The quantity h/mc is called the *Compton wavelength of the electron*.) Scattering of such a photon at the angle π yields a photon wavelength of 0.0073 nm, *triple* the original wavelength (and thus the photon energy is *one-third* of the original photon energy, since $E = hc/\lambda$.)

Two important Compton-type collisions require modification of this analysis.

1. *Object of different mass.* In this discussion the letter m has denoted the mass of the electron. When a photon scatters from a particle of different mass (e.g., a proton), m should represent the mass of that particle. The recoiling "particle" can be an entire atom, if the photon interacts with it without freeing an electron from the atom. This happens in a certain percentage of Compton events at any given angle, resulting in what is called an "unmodified line" because $\delta\lambda$ is so small ($\lambda' - \lambda \leq$ .000001 nm).

2. *Object that is not at rest.* The initial energy in Eq. (1.37) includes only the *rest* energy of the electron; therefore Eq. (1.41) does not apply to

interactions in which the electron is moving. In practice, the motion of the electrons has a significant effect. The *average* wavelength of the final photons at a given angle is approximated by Eq. (1.41); individual wavelengths vary because of the Doppler effect. Analysis of this so-called *Doppler broadening* gives information about motions of electrons in solids.

SUMMARY

This chapter has introduced the concept of the quantum and its experimental foundation in the photoelectric effect, optical and X-ray spectra of atoms, and X-ray scattering, which shows that photons carry momentum as well as energy. We have seen that quantum theories give the same results as classical theory for objects of macroscopic size and that electromagnetic energy is observed as waves or photons, depending on the circumstances. We also begin to see that a number of applications of these phenomena are possible.

ADDITIONAL READING

R. Eisberg and R. Resnick, *Quantum Physics of Atoms, Molecules, Solids, Nuclei, and Particles*, Chapter 2, Wiley, New York (1974).

K. Krane, *Modern Physics*, Wiley, New York (1983). Chapter 3 has many good diagrams illustrating apparatus from experiments discussed here.

J. D. McGervey, *Introduction to Modern Physics*, Academic Press, New York, (1983). Chapters 3 and 4 deal with early development of the quantum concept.

J. R. Taylor and C. D. Zafaritos, *Modern Physics for Scientists and Engineers*, Prentice-Hall, Englewood Cliffs, NJ (1991). Chapters 5 and 6 cover much of the material of this chapter.

EXERCISES

1. Electrons are ejected from a certain metal only when the wavelength of incident light is less than or equal to 400 nm. What wavelengths would eject electrons with a kinetic energy of 3 eV from this metal?

2. If the incident radiation has wavelength $\lambda = 310$ nm, what is the maximum kinetic energy of electrons ejected from a target of
    (a) Cs ($W = 1.9$ eV)   (b) Na ($W = 2.3$ eV)   (c) Al ($W = 4.1$ eV)

3. In a different universe where the laws of physics are the same, but the values of the physical constants are different from ours, a photoelec-

tric measurement gave the following results, using their equivalent units for wavelength λ (lerps) and maximum electron energy K (kroks)

λ	300	450	600	?
K	7.5	4.5	?	0

(a) Fill in the blanks; for λ, fill in the *shortest* λ that gives $K = 0$.

(b) Find the value of W for this target, in the appropriate unit.

4. A *positronium* (Ps) atom consists of an electron and a positron (a particle with the same charge as a proton but the same mass as an electron). They are bound together like the proton and electron in the hydrogen atom.

   (a) Find the radii of the Bohr orbits for Ps, in terms of the quantum number n. Compare with the orbit radii for the hydrogen atom.

   (b) Find the numerical values for the Ps energy levels, in terms of n.

5. The negative muon (μ^-) has the same charge as an electron, but its mass is about 207 electron masses. Compute the reduced mass, orbit radii, and energy levels for muonic ^{1}H and ^{2}H, in which a muon replaces the electron.

6. Compute the longest wavelength of the Balmer series for ^{1}H and for ^{2}H.

7. Compute the shortest wavelength of the Brackett series (Figure 1.4).

8. Use the correspondence principle and the classical rate of radiation [Eq. (1.27)] to estimate the average lifetime of the $n = 2$ state in ^{1}H.

9. Equation (1.13) shows that the orbit radius in a hydrogen-like atom or ion is proportional to the electron mass m, *not* the reduced mass m_r. Show that the "radius" found by replacing m by m_r in Eq. (1.13) is equal to the distance $r + R$ between the nucleus and the electron.

10. Mercury has another excited state at 6.7 eV, which can be reached by using the grid potentials of Figure 1.5a. Explain why this is so. In this case, at what voltages (besides those shown in Figure 1.6) should you expect to see a drop in current? What wavelengths would the tube then emit?

11. Calculate the maximum energy that an electron can lose in an *elastic* collision with a mercury atom.

12. A negative muon is captured by an atom of aluminum ($Z = 13$). Find the energy of the X ray emitted when the muon goes from the $n = 2$ orbit to the $n = 1$ orbit. Why should you *not* use the factor s of Eq. (1.33)?

13. Compute the wavelength of the K_α line of aluminum.

14. Find an expression for K_β lines analogous to the expression $\sqrt{\nu_\alpha/(3/4)\nu_0}$ used in analyzing K_α lines [Eq. (1.34)]. What transition produces K_β lines?

15. Compute the percentage change in wavelength and the change in the energy of a photon resulting from Compton scattering of (a) yellow light ($\lambda = 500$ nm) and (b) a 200-MeV X ray, through an angle of $\pi/2$.

16. A 50-keV photon moving in the $+x$ direction strikes an electron moving in the $-x$ direction with a kinetic energy of 800 keV.
    (a) Use the conservation laws directly to compute the final photon energy if the photon is scattered backward ($\theta = \pi$).
    (b) Verify your answer by using the Doppler shift formula $f'/f\sqrt{(c+v)/(c-v)}$ for photon approaching absorber as follows: Find the original photon's frequency in the electron's rest frame, then find the scattered photon's frequency in this frame by using the Compton formula, then use the Doppler formula again to find that photon's frequency in the original reference frame.

17. Some high-energy photons observed in cosmic rays may result from the *inverse Compton effect*: a low-energy (≈ 2 eV) photon *gains* energy by scattering from a high-energy *proton*. Compute the maximum final photon energy after such an interaction when the photon energy is 2 eV and the proton's momentum is 10^9 eV/c. *Hint*: Use the method of Exercise 16(b).

18. For Exercise 15(b), find the electron's recoil angle ϕ.

19. For a photon of energy E, scattered from an electron initially at rest, show that the final photon energy equals $E/[1 + (E/mc^2)(1 - \cos\theta)]$.

20. A photon whose energy is $0.25\ mc^2$ (127.8 keV) interacts via the Compton effect with an electron (mass m). The electron is initially at rest.
    (a) Find the smallest energy that the resulting photon could have.
    (b) For $\theta = -\pi/2$, find the emerging photon's energy, the kinetic energy of the electron, and the components of the electron's momentum.

21. A photon whose energy is $4mc^2$ (2.044 MeV) is scattered at an angle $\theta = \pi/2$ from an electron that was at rest. *Without using the Compton formula*, find
    (a) the scattered photon's energy,
    (b) the electron's final kinetic energy as a multiple of mc^2, and
    (c) the electron's final momentum components as multiples of mc.

CHAPTER 2

Waves and Particles

Having seen the dual nature of light as both wave and particle (its so-called *wave–particle duality*), one might ask whether electrons and other so-called particles can have wave properties. This question was asked by Louis de Broglie, and it was answered in the famous experiment of Davisson and Germer, who showed that electrons can behave like waves.

2.1 THE DE BROGLIE WAVE HYPOTHESIS

Much of our task in this book will be to study the far-reaching implications of wave equations developed by Erwin Schrödinger. These equations are based on de Broglie's hypothesis that for every particle there is a wave whose wavelength depends on its state of motion.

This wavelength, according to de Broglie, should be inversely proportional to the momentum of the particle, obeying the same relation that holds for photons:

$$\boxed{\lambda = h/p} \tag{2.1}$$

an equation that ranks with Einstein's $E = mc^2$ in significance.

[Accepting the possibility that particles have wavelengths, we still may wonder why the wavelength should depend on the momentum in this way. A photon has energy as well as momentum, and the *photon*'s wavelength can be expressed equally well in terms of energy E by $\lambda = hc/pc = hc/E$, since $E = pc$ for photons. However, the choice $\lambda = h/p$ is dictated by the requirement that the equation must be valid in all frames of reference. See Appendix C for further discussion.]

2.2 ELECTRON DIFFRACTION: THE DAVISSON–GERMER EXPERIMENT

Experimental evidence for the wave nature of electrons was found by Davisson and Germer, who observed the scattering of a beam of electrons from a nickel crystal. They found that at certain electron energies the beam was reflected strongly at one specific angle. For example, at a kinetic energy of 54 eV, electrons were reflected (scattered) at a 50° angle.

The effect is similar to that of Bragg scattering. Nickel crystals have the face-centered cubic (fcc) structure shown in Figure 2.1. The electron beam direction was perpendicular to the surface, a [111] plane in which there are rows of atoms spaced a distance $s = 0.2158$ nm apart. (See Figure 2.2.)[1]

If the scattering is actually Bragg diffraction, where are the scattering planes? In nickel, the [311] planes are a likely candidate, making an angle of 22° with the [111] planes, or an angle $\theta' = 68°$ with the incident beam.[2] The distance d between these planes is $0.2158 \sin 22°$, or 0.08084 nm.

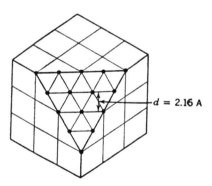

$d = 2.16$ A

FIGURE 2.1 Structure of the face-centered cubic (fcc) nickel lattice, showing a [111] plane at the surface. (Adapted from A. P. French, *Principles of Modern Physics*, Wiley, New York, 1958, p. 180.)

---

[1]Some books say that this scattering is strictly a surface effect, because the electron penetrates only a short distance into the crystal. But pure surface diffraction yields a peak in the angular distribution at all incident energies. Davisson–Germer peaks are seen only at certain energies.

[2]In terms of unit vectors **i**, **j**, and **k**, the normals to the [311] and [111] planes are $\mathbf{A} = \mathbf{i} + 3\mathbf{j} + 3\mathbf{k}$ and $\mathbf{B} = \mathbf{i} + \mathbf{j} + \mathbf{k}$, respectively. The angle between these vectors is $\cos^{-1} 7/\sqrt{57} = 22.00°$.

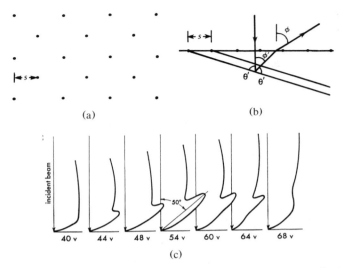

FIGURE 2.2 Electron diffraction by a nickel lattice whose top surface is a [111] plane. (a) Top view. Spacing $s = 0.2518$ nm. (b) Side view. Electron wave enters from top, then is "reflected" by crystal planes that make an angle θ' with the top surface. The reflected wave is refracted and emerges at an angle of ϕ from the vertical. (c) Intensity versus angle ϕ, for various energies of the incident electrons. (From H. Semat, *Introduction to Atomic and Nuclear Physics*, 4th edition, Holt, Rinehart and Winston, New York, 1992.)

Equation (1.30) then gives the required wavelength for Bragg scattering as

$$\lambda' = 2d \sin \theta' = 0.1617 \sin 68° = 0.150 \text{ nm} \qquad (2.2)$$

where λ' is the wavelength of the electron *when it is inside the nickel*. The angle between the incident beam and the scattered beam *within the crystal* is $\phi' = 44°$. But the electron wave is refracted as it leaves the crystal; thus the *observed angle* ϕ between the scattered and incident electrons is given by Snell's law:

$$\sin \phi = \lambda/\lambda' \sin \phi' \qquad (2.3)$$

where λ is the electron wavelength outside the nickel. The potential energy of an electron in nickel is known to be about 12 eV less than that of a free electron; thus the kinetic energy of these electrons increases from 54 eV to 66 eV when they enter the nickel. The wavelength λ can therefore be computed. We find that $\lambda/\lambda' = 1.1055$, and substitution into Eq. (2.3) gives $\phi = 50.2$. Thus the scattering angle derived from the Bragg condition and the de Broglie relation (2.1) agrees with the observed angle for 54-eV electrons on nickel.

We have found the wavelength λ' from the Bragg condition for scattering from [311] planes, rather than by computing it from the known kinetic energy of the electrons. Let us do this computation, to see if the result is consistent with Eq. (2.1). We have $\lambda' = hc/p'c = hc/\sqrt{2K'mc^2}$ (approximation valid to 1 part in 30,000; exact expression is $\sqrt{K'^2 + 2K'mc^2}$) = 1240/8213 nm, or $\lambda' = 0.151$ nm (using $K' = 66$ eV and $mc^2 = 511$ keV), in satisfactory agreement with the value of 0.150 nm found in Eq. (2.2) (considering the uncertainty of the measurement).

It might be objected that electrons of such low energy do not penetrate deeply enough into nickel to see the crystal structure. But it is not necessary for the electron wave to penetrate very far in order to feel this effect.

Example Problem 2.1   The smallest object that you can see with the unaided eye is a speck of dust about 0.1 mm in diameter, about 1 microgram in mass. What is the wavelength when it is moving at a speed of about 0.1 mm/s?

Solution. The momentum is $p = mv = 10^{-9}$ kg $\times 10^{-4}$ m/s $= 10^{-13}$ kg-m/s. The wavelength is therefore

$$\lambda = \frac{h}{p} \cong \frac{6.6 \times 10^{-34} \text{ J-s}}{10^{-13} \text{ kg-m/s}} = 6.6 \times 10^{-21} \text{ m}.$$

2.3 FOURIER SYNTHESIS AND WAVE PACKETS

2.3.1 Connection between the Wave and the Particle

In an experiment with de Broglie waves, as with other waves, the observed intensity of the wave at any given point should be proportional to the wave amplitude at that point. But the intensity of the wave is detected only by the transfer of energy to a detector, and this transfer is made in *discrete* amounts. For example, photons may be detected by the photodielectric effect, either in a photocell or in your eye, just as electrons can deposit a charge on an electroscope. But a fraction of a photon is never observed, and if the frequency is fixed, each photon has the same energy, just as each electron has the same charge. In such situations, the intensity is a measure of the *number of particles* detected.

If the intensity of the wave is low enough, the particles can be counted individually. They are not detected at a uniform rate or in any regular pattern. They appear at random, but in such a way that the average rate is what would be expected from the wave intensity, with allowance for normal statistical fluctuations. We must conclude that the *wave* determines the *probability* of detecting the *particle*.

This particle–wave cannot be localized at one point, because then there would be no wavelength to measure. To measure a wavelength we need a way to send the wave along two different paths, to produce some kind of interference. It is hard to conceive of a particle that travels on two different paths simultaneously, but nature is not only queerer than we imagine, it is queerer than we *can* imagine. The electron is both particle and wave, in some sense.

Unfortunately, if we imagine electrons to be classical particles (something like marbles, only smaller) we conjure up an image that exists only in our own imaginations. Nevertheless, certain features of electron behavior agree with this image. For example, an electron's speed can be determined by observing it at two points and measuring the time required for it to go from one point to the other.[3] Clearly, an electron does not behave like an infinitely long wave.

Let us then consider the electron (or other particle) to be a sort of localized wave. In that case it cannot have a definite wavelength (and thus it cannot have a definite momentum). Unless a wave is infinite in extent, it cannot have a precise wavelength. But it is possible to have a localized wave, called a *wave packet*, that has some particle properties. To understand how such waves behave, we need to examine *Fourier analysis*, beginning with the analysis of waves that are infinite in extent.

2.3.2 Fourier Series

A Fourier series is a sum of sine and cosine functions that approximates a periodic function. Given a periodic function $\psi(x)$ whose period is $2a$, we have, from the definition of a periodic function,

$$\psi(x + 2a) = \psi(x) \quad \text{for all values of } x \tag{2.4}$$

We *assume* that $\psi(x)$ can be written as the sum of a series of sine and cosine functions, such that this series, for all values of x, approaches the value of $\psi(x)$ as the number of terms in the series goes to infinity. To satisfy this requirement, each term in the series must be periodic with period $2a$. We therefore choose sine and cosine functions that have periods of $2a, 2a/2, 2a/3, \ldots$, so that they all have period $2a$ as well, by

[3]See, for example, the Physical Science Study Committee film, *The Ultimate Speed*, produced by W. Bertozzi, for a measurement of electron speed.

definition (2.4). Thus

$$\psi(x) = A_0 + \sum_1^\infty \left(A_n \cos \frac{n\pi x}{a} + B_n \sin \frac{n\pi x}{a} \right) \qquad (2.5)$$

Subsequent development is simplified by rewriting Eq. (2.5) in exponential form, using the identities

$$\cos \frac{n\pi x}{a} = \frac{e^{in\pi x/a} + e^{-in\pi x/a}}{2} \qquad (2.6a)$$

$$\sin \frac{n\pi x}{a} = \frac{e^{in\pi x/a} - e^{-in\pi x/a}}{2i} \qquad (2.6b)$$

Substitution into Eq. (2.5) gives

$$\psi(x) = A_0 + \sum_1^\infty A_n \frac{e^{in\pi x/a} + e^{-in\pi x/a}}{2} + B_n \frac{e^{in\pi x/a} - e^{-in\pi x/a}}{2i}$$

$$= A_0 + \sum_1^\infty \left(\frac{A_n}{2} + \frac{B_n}{2i} \right) e^{in\pi x/a} + \left(\frac{A_n}{2} - \frac{B_n}{2i} \right) e^{-in\pi x/a}$$

$$= \sum_{-\infty}^\infty C_n e^{in\pi x/a} \qquad (2.7)$$

where $C_0 = A_0$ and

$$C_n = \frac{A_n}{2} + \frac{B_n}{2i}, \qquad C_{-n} = \frac{A_n}{2} - \frac{B_n}{2i}, \qquad \text{for } n > 0 \qquad (2.8)$$

Notice that the values of C_n are defined for negative as well as positive values of n. We can find the coefficient C_m for any *specific* m as follows:
     We multiply both sides of Eq. (2.7) by $e^{-im\pi x/a}$, and then we integrate on x from $x = -a$ to $x = +a$:

$$\int_{-a}^a \psi(x) e^{-im\pi x/a}\, dx = \int_{-a}^a \sum_{n=-\infty}^\infty C_n e^{in\pi x/a} e^{-im\pi x/a}\, dx \qquad (2.9)$$

We assume that it is legitimate to integrate the series term by term (i.e., to interchange the summation and integral signs). The result is a sum of integrals of the form $\int_{-a}^a C_n e^{i(n-m)\pi x/a}\, dx$. In this sum, n takes on all integer values, while m is one specific integer.

Each integral is zero except the integral for which n is equal to m. Setting $n = m$ reduces the integral to the value $2aC_m$, and Eq. (2.9) eventually becomes

$$C_m = \frac{1}{2a} \int_{-a}^{a} \psi(x) e^{-im\pi x/a} \, dx \qquad (2.10)$$

If the right-hand side is integrable, as it will be for any $\psi(x)$ in which we might be interested, we have found the value of the mth coefficient. To write the Fourier series in real form, we solve Eqs. (2.8) for A_n and B_n, obtaining

$$A_0 = C_0; \qquad A_n = C_n + C_{-n} \quad \text{and} \quad B_n = i(C_n - C_{-n}) \quad \text{for } n > 0 \qquad (2.11)$$

Example Problem 2.2   Compute the coefficients A_n and B_n for the function

$$\psi(x) = 1 \qquad (-b < x < +b)$$
$$\psi(x) = 0 \qquad (b < |x| < a)$$
$$\psi(x + 2a) = \psi(x) \qquad \text{for all values of } x$$

Solution. Substitute $\psi(x) = 1$ and limits $-b$ and $+b$ into Eq. (2.10) to obtain

$$C_m = \frac{1}{2a} \int_{-b}^{b} e^{-im\pi x/a} \, dx = \frac{1}{2im\pi}(e^{im\pi b/a} - e^{-im\pi b/a})$$

$$= \frac{1}{m\pi} \sin \frac{m\pi b}{a} \qquad (m \neq 0)$$

Substitution into Eq. (2.11) then yields

$$A_n = \frac{2}{n\pi} \sin \frac{n\pi b}{a} \quad \text{(for } n \neq 0), \qquad B_n = 0, \qquad A_0 = C_0 = \frac{1}{2a} \int_{-b}^{b} dx = \frac{b}{a}$$

Explicitly, if we let $b/a = 1/3$, substitution into Eq. (2.5) gives the series

$$\psi(x) = \frac{1}{3} + \frac{2}{\pi} \sin \frac{\pi}{3} \cos \frac{\pi x}{a} + \frac{2}{2\pi} \sin \frac{2\pi}{3} \cos \frac{2\pi x}{a}$$

$$+ \frac{2}{3\pi} \sin \frac{3\pi}{3} \cos \frac{3\pi x}{a} + \cdots$$

$$= \frac{1}{3} + \frac{\sqrt{3}}{\pi} \cos \frac{\pi x}{a} + \frac{\sqrt{3}}{2\pi} \cos \frac{2\pi x}{a} + 0$$
$$- \frac{\sqrt{3}}{4\pi} \cos \frac{4\pi x}{a} - \frac{\sqrt{3}}{5\pi} \cos \frac{5\pi x}{a} + \cdots$$

The computer-generated series for $b/a = 3$ is shown in Fig. 2.3.

2.3.3 The Fourier Integral

The Fourier series is useful for analyzing any *periodic* wave of any shape. But the wave packet (Figure 2.4) representing the motion of a particle does not fit the definition of a periodic function [(Eq. 2.4)] for any finite value of a; the particle is in a single region of space.

However, if the period a is sufficiently large, we can still adjust the wave to fit our conception of a particle. We saw in the previous example problem that we can construct a Fourier series for a function that is zero over a large fraction of its period, and that fraction can be made as large as we wish by making the fraction b/a small. The smaller the value of b/a, the more terms we must use to make the series fit the function to a given degree of accuracy, but we can make the series go to zero over any specified range. Thus we can construct what looks like a nonperiodic function to an observer whose vision is limited, and we can remove that limit by letting a approach infinity.

We approach a nonperiodic function by defining the quantity $k_n \equiv n\pi/a$ (n being an integer), and the difference between successive values of

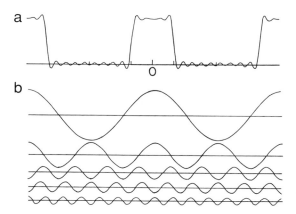

FIGURE 2.3 The function $\psi(x)$ of Example 2.2 for $b/a = 1/3$, showing (a) the sum of 10 terms in the Fourier series and (b) some individual terms in the series.

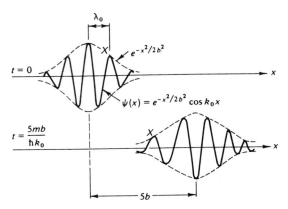

FIGURE 2.4 Real part of the wave packet represented by Eq. (2.18), when $t = 0$ and x is approximately zero. Later, when the packet has moved to $x = 5b$, the packet is broader and asymmetric, because its component waves have different speeds. Other properties of this wave are explored in Exercises 15 and 16.

k_n as $\delta k \equiv \pi/a$. In these terms, we can rewrite Eq. (2.10), with m replaced by n, as

$$C_n = \frac{\delta k}{2\pi} \int_{-a}^{a} \psi(x')e^{-ik_n x'} \, dx' \tag{2.12}$$

where x' has been substituted for x as the "dummy variable" in the integral, to avoid confusion when this expression is inserted into Eq. (2.7). We now rewrite Eq. (2.7) as

$$\psi(x) = \sum_{n=-\infty}^{\infty} C_n e^{ik_n x} \tag{2.13}$$

which becomes, when we substitute from Eq. (2.12),

$$\psi(x) = \frac{1}{2\pi} \sum_{n=-\infty}^{\infty} e^{ik_n x} \, \delta k \int_{-a}^{a} \psi(x')e^{-ik_n x'} \, dx' \tag{2.14}$$

As $a \to \infty$, the difference $\delta k = \pi/a \to 0$, and the sum becomes an integral. The result is the *Fourier integral theorem*:

$$\psi(x) = \frac{1}{2\pi} \int_{-\infty}^{\infty} e^{ikx} \, dk \int_{-\infty}^{\infty} \psi(x')e^{-ikx'} \, dx' \tag{2.15}$$

If we define a new function

$$\phi(k) = \frac{1}{\sqrt{2\pi}} \int_{-\infty}^{\infty} \psi(x)e^{-ikx} \, dx \qquad (2.16)$$

then

$$\psi(x) = \frac{1}{\sqrt{2\pi}} \int_{-\infty}^{\infty} \phi(k)e^{ikx} \, dk \qquad (2.17)$$

The function $\phi(k)$ is the *Fourier transform* of $\psi(x)$. This function is a continuous amplitude function that takes the place of the discrete amplitudes C_n. The continuous function $\phi(k)$ is needed because a nonperiodic function must be built up from waves with a continuous range of wavelengths.

2.3.4 Probability Density and the Gaussian Wave Packet

An instructive example of a wave packet is

$$\psi(x) = e^{-x^2/4\sigma^2} e^{ik_0x} \qquad (2.18)$$

whose real part is shown as the $t = 0$ part of Fig. 2.4 (with σ equal to $b/\sqrt{2}$).

The fact that $\psi(x)$ is complex does not disqualify it as a function that describes a particle's motion, because we do not observe $\psi(x)$ directly. We observe the particle when it interacts with a particle detector, and the probability of this interaction is determined by the *probability density*, analogous to the energy density in a classical wave.

For a classical wave the energy density is proportional to the square of the amplitude of the wave. By analogy, for a wave packet representing a particle, we take the probability density to be proportional to $|\psi(x)|^2$, the *absolute* square of the *probability amplitude* $\psi(x)$. By definition,

$$|\psi(x)|^2 \equiv \psi^*(x)\psi(x) \qquad (2.19)$$

where $\psi^*(x)$, the complex conjugate of $\psi(x)$, is the result of replacing i by $-i$ everywhere in $\psi(x)$. The right-hand side of Eq. (2.19) must therefore be a positive real number (or zero). In this case, using Eq. (2.18), we have

$$|\psi(x)|^2 = e^{-x^2/2\sigma^2} \qquad (2.20)$$

Notice that as $\sigma \to \infty$, $\psi(x) \to e^{ik_0x}$, a function that has a single wavelength (equal to $2\pi/k_0$) and a constant probability density. The Fourier transform of $\psi(x)$ then goes to zero everywhere except at $k = k_0$. But when σ is finite, the Gaussian factor $e^{-x^2/4\sigma^2}$ introduces other values of k, whose distribution is given by substituting the expression for $\psi(x)$ into the Fourier transform equation (2.16), as follows:

$$\phi(k) = \frac{1}{\sqrt{2\pi}} \int_{-\infty}^{\infty} e^{-x^2/4\sigma^2} e^{-i(k_0-k)x} \, dx \qquad (2.21)$$

The integral is evaluated by *completing the square* in the exponent to obtain

$$\phi(k) = \frac{1}{\sqrt{2\pi}} \int_{-\infty}^{\infty} e^{-[x/2\sigma - i(k_0-k)\sigma]^2} e^{-i(k_0-k)^2\sigma^2} \, dx \qquad (2.22)$$

The second exponential is independent of x; removing it from the integral and changing variables yields

$$\phi(k) = \frac{2\sigma}{\sqrt{2\pi}} e^{-(k_0-k)^2\sigma^2} \int_{-\infty}^{\infty} e^{-u^2} \, du \qquad (2.23)$$

where $u = x/2\sigma - i(k_0 - k)\sigma$ and $du = dx/2\sigma$.

The integral in Eq. (2.23) is equal to $\sqrt{\pi}$; the final result is

$$\phi(k) = \sigma\sqrt{2} \, e^{-(k_0-k)^2\sigma^2} \qquad (2.24)$$

Thus $\phi(k)$ is also of Gaussian form, peaked at $k = k_0$. Figure 2.5 shows $\phi(k)$ in terms of the same parameter $b = \sigma\sqrt{2}$ used in Fig. (2.4). Notice

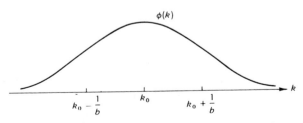

FIGURE 2.5 The function $\phi(k)$, the Fourier transform of the function $\psi(x)$ shown in Figure 2.4.

that

If σ increases, $\phi(k)$ becomes narrower and $\psi(x)$ becomes broader.

The Delta Function   If $\sigma \to \infty$ in Eq. (2.18), then $\psi(x)$ is infinitely broad, and its Fourier transform, $\phi(k)$, is zero for all values of k except $k = k_0$. Conversely, if $\sigma \to 0$, then $\psi(x)$ is zero for all values of x except $x = 0$.

It is useful to have a special symbol for a function that is zero everywhere except at one point. Such a function is the *Dirac delta function*, $\delta(x)$, whose properties are, *by definition*:

$$\delta(x) = 0 \quad \text{if } x \neq 0; \qquad \delta(x) = \infty \quad \text{if } x = 0 \qquad (2.25)$$

$$\int_{x_1}^{x_2} \delta(x)\, dx = 1 \quad \text{and} \quad \int_{x_1}^{x_2} f(x)\delta(x)\, dx = f(0) \quad \text{if } x_1 < 0 < x_2 \quad (2.26)$$

Multiplying the Gaussian function of Eq. (2.20) by the factor $1/\sigma$, then taking the limit as $\sigma \to \infty$, will produce a delta function. You should verify that it satisfies Eqs. (2.25) and (2.26). See Appendix A for further discussion of the Gaussian function.

To describe a function that is localized at $x = x_0 \neq 0$, we simply use $\delta(x - x_0)$. Similarly, $\delta(k - k_0)$ is nonzero only when $k = k_0$; its Fourier transform is an infinitely long sine wave of wavelength $2\pi/k_0$.

The symbol $\delta(x)$ must not be confused with δx; δx denotes the uncertainty in the value of x (to be discussed next), and $\delta(x)$ is a *function* of x.

2.4 UNCERTAINTY RELATIONS

The probability density $|\psi(x)|^2$ may be integrated on x between x_1 and x_2 to find the total probability that a given observation will find the particle to be in that range. The function $|\psi(x)|^2 = e^{-x^2/2\sigma^2}$ [Eq. (2.20)] is particularly significant, because it is of the same form as the Gaussian (normal) probability distribution given in Appendix A, where σ is the standard deviation. Measurement of the position of the particle described by this probability density will yield a result between $x = -\sigma$ and $x = +\sigma$ about two-thirds of the time. It is thus customary to identify σ as the uncertainty in the value of x; that is, $\sigma = \delta x$ for this probability density, and

$$\int_{x_1}^{x_2} |\psi(x)|^2\, dx \text{ is proportional to the probability that } x_1 < x < x_2 \quad (2.27)$$

Just as $|\psi(x)|^2$ gives the probabilities for measuring the particle's position, $|\phi(k)|^2$ gives the probabilities for measuring the particle's momentum $p_x = \hbar k$. Thus we have, analogous to statement (2.27),

$$\int_{k_1}^{k_2} |\phi(k)|^2 \, dk \text{ is proportional to the probability that } k_1 < k < k_2 \quad (2.28)$$

We use the wording "is proportional to" because we have not yet taken account of a *normalizing factor* to make the probability equal to 1 when the limits are $x_1 = -\infty$ and $x_2 = +\infty$. (The particle must be somewhere; see Chapter 3.)

Let us now find the uncertainty δk, using the same criterion that we adopted for δx. From Eq. (2.24) we find that

$$|\phi(k)|^2 = 2\sigma^2 e^{-2(k_0 - k)^2 \sigma^2} \quad (2.29)$$

Writing this in the form of the normal probability distribution, in which the exponent is equal to the square of the deviation (in this case $k_0 - k$) divided by twice the square of the standard deviation (in this case δk) would require that

$$|\phi(k)|^2 = 2\sigma^2 e^{-(k_0 - k)^2 / 2(\delta k)^2} \quad (2.30)$$

Comparison of Eqs. (2.30) and (2.29) shows that $2\sigma^2 = 1/2(\delta k)^2$, or $\delta k = 1/2\sigma$.

We see that the uncertainty δk is inversely proportional to the uncertainty δx, which makes the product of these uncertainties a constant, equal to $1/2$, independent of the value of σ. Since $p = \hbar k$, the uncertainty in momentum is $\delta p = \hbar \delta x$, and we may write

$$(\delta p)(\delta x) = \hbar(\delta k)(\delta x) = \hbar/2 \quad (2.31)$$

Although Eq. (2.31) applies only to this special case, this case illustrates general features of the relationship between $\psi(x)$ and $\phi(k)$, as follows:

1. To have wavelike properties, a particle must somehow "occupy" a range of positions, rather than a single definite point.
2. To have particle properties, i.e., to be localized, while retaining wavelike properties, a particle must possess a range of values of momentum, rather than a single definite momentum.
3. The range of position possibilities is inversely proportional to the range of momentum possibilities.

None of these features prevents a particle from being temporarily localized at a specific point. In that case the particle's wave packet is described by a delta function, and its momentum uncertainty is infinite.

2.4.1 The Heisenberg Uncertainty Principle

In 1927, Heisenberg proposed a generalization of Eq. (2.31). The Heisenberg uncertainty principle states that, in any simultaneous measurement of the position and momentum of a single particle,

$$(\delta p_x)(\delta x) \geq \hbar/2 \qquad (2.32)$$

where δx is the uncertainty in the measurement of a particular coordinate x, and δp_x is the uncertainty in a simultaneous measurement of the momentum along the x axis.

The Heisenberg microscope shows how relation (2.32) might be tested experimentally. This apparatus, the simplest possible form of microscope, examines an attempt to perform the most precise measurements that one can imagine, and thus it identifies the practical problem. (See Figure 2.6.) A single lens is used to focus light onto a detector that can register a single

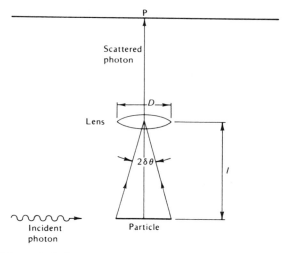

FIGURE 2.6 Using a single lens to detect the position of a particle. Light from any part of the extended area marked "particle" may reach point P as a result of diffraction through an angle between $\pi/2 - \delta\theta$ and $\pi/2 + \delta\theta$.

photon scattered from a single electron. We can, in principle, measure the electron's position as accurately as we wish, because (with perfect apparatus):

- The uncertainty δx is, in principle, determined only by diffraction of the light as it passes through the lens; if the photon's direction is uncertain by the amount $\delta\theta$ (the diffraction angle) and l is the distance from the electron to the lens, then $\delta x = l\,\delta\theta$.
- If λ is the wavelength of the light and D is the lens diameter, then the diffraction angle $\partial\theta$ is given by $\delta\theta \cong \lambda/D$.
- Therefore

$$\delta x \cong l\lambda/D \qquad (2.33)$$

We see that, in principle, we can make δx as small as we wish, by increasing D or by decreasing l or λ. But how does this affect the momentum uncertainty δp_x? Let us try to measure p_x the old-fashioned way, by measuring the electron's position twice to find how far it travels in a known time interval.

If we can measure both positions as accurately as we wish, where does the uncertainty in p_x arise? The measured momentum in this case can be perfectly *precise*, but it is not *accurate*.[4] We have not measured the momentum that we wanted to find; we have measured the momentum that the electron had *after it was struck by the photon*. To find the *relevant* momentum—the momentum that the electron *originally* had—we must correct for the momentum that the photon gave to the electron.

This correction can be made, but not precisely. The most accurate correction is made by assuming that the photon is scattered at the angle of $\pi/2$, giving all of its x component of momentum to the electron, a momentum equal to $h\nu/c$.

The uncertainty arises because the photon could have been "focused" at P as long as it passed through *any part of the lens*, so the scattering angle could have differed from $\pi/2$ by as much as the half of the angle subtended by the lens, or approximately $D/2l$, assuming that this is a small angle. If we also assume that the photon's wavelength does not change, the scattered photon has an x component of momentum equal in magnitude to as much as $(h\nu/c)(D/2l)$. This means an uncertainty of

$$\delta p_x \cong h\nu D/2lc \qquad (2.34)$$

---

[4] Precision and accuracy differ. My paycheck is precise to the last penny, but it is not accurate; it is far less than what I am worth.

Combining (2.33) and (2.34) gives

$$(\delta x)(\delta p_x) \cong (l\lambda/D)(h\nu D/2lc) = h/2 \qquad (2.35)$$

independent of the size of the lens, the wavelength of the light, and the distance l. Equation (2.35) indicates a larger uncertainty product than the minimum required by inequality (2.32), in part because here we have defined the uncertainties as maximum values rather than standard deviations.

2.4.2 The Diffraction Grating

Instead of considering the electron to be a point particle whose position is uncertain, let us now approach the uncertainty from the wave-packet point of view. What do we see when a wave packet is analyzed by a diffraction grating?

Suppose that wave packet $\psi(x)$ extends over a distance of only N_w wavelengths, being given by

$$\psi(x) = \sin(2\pi x/\lambda) \quad \text{for } 0 < x < N_w \lambda; \; \psi(x) = 0 \text{ elsewhere} \quad (2.36)$$

It strikes a grating whose spacing is d (Figure 2.7). From the diffraction angle θ we then determine the value of the wavelength λ. Elementary

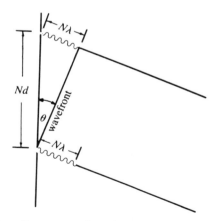

FIGURE 2.7 Diffraction of a wave packet of wavelength λ and total length $N_w \lambda$. The diffraction angle θ is a result of constructive interference of waves from only N slits. The diffracted wave leaving slit 1 travels a distance that is $N\lambda$ longer than that traveled by the wave leaving slit $N + 1$; thus these two waves can never interfere when they reach the detector.

books[5] show that the *resolving power* of a diffraction grating, i.e., the uncertainty $\delta\lambda$ in the measurement of λ, is related to the number of slits, N, by

$$\delta\lambda = \lambda/N \qquad (2.37)$$

But the diffraction pattern for a wave packet of total length $N_w\lambda$, Eq. (2.35), is produced by interference of waves from no more than N_w slits, regardless of the total number N of slits in the grating. The reason for this is that the path difference between a wave from slit 1 and a wave from slit $1 + N_w$ must equal $N_w\lambda$ for constructive interference (Figure 2.7). If the wave packet's total length is $N_w\lambda$, the contribution from slit 1 is gone when the contribution from slit $1 + N_w$ arrives at the detector, so there is no interference between them.

Therefore, the uncertainty in the measured λ for the wave of Eq. (2.35) is given by

$$\delta\lambda = \lambda/N_w \qquad (2.37a)$$

when the wave packet's total length is $N_w\lambda$.

We can now find a relation between the momentum uncertainty and the position uncertainty of a wave packet. Since the wave extends from $x = 0$ to $x = N_w\lambda$, we can say that the particle is located at $x = N_w\lambda/2 \pm N_w\lambda/2$, and therefore

$$\delta x \cong N_w\lambda/2 \qquad (2.38)$$

We estimate the magnitude of the momentum uncertainty by subtracting the average momentum from the maximum momentum, assuming that $\delta\lambda \ll \lambda$:

$$\delta p_x = h/(\lambda - \delta\lambda) - h/\lambda \cong h(\delta\lambda)/\lambda^2 = -h[(\lambda/N_w)/\lambda^2] = -h/N_w\lambda \qquad (2.39)$$

Multiplying these two uncertainties, we find that the magnitude of the uncertainty product is

$$(\delta x)(\delta p_x) \cong (N_w\lambda/2)(h/N_w\lambda) \cong h/2 \qquad (2.40)$$

a result consistent with Eq. (2.35). In both cases the estimated uncertainty product is larger than the minimum uncertainty product $\hbar/2$, partly

[5]See, for example, Hugh D. Young, *University Physics*, 8th edition, p. 1058, Addison Wesley, Reading, MA (1992).

because the uncertainties here are maximum uncertainties rather than standard deviations. It must also be remembered that expression (2.32) is an inequality. The equality holds only when $\psi(x)$ is a Gaussian wave packet and the uncertainties are the standard deviations for $\psi(x)$ and its Fourier transform $\phi(k)$.

2.5 TRAVELING WAVES

In discussing waves we have not yet explicitly written the time coordinate in our equations. A wave traveling along the x axis with velocity v can be written as a function of $x - vt$. This means that we can take account of time by simply replacing x by $x - vt$ in the preceding expressions for $\psi(x)$ [for example, in Eq. (2.13)]. This describes a wave that moves, without changing shape, through a distance $v\delta t$ in a time interval δt. (See Figure 2.8.)

2.5.1 Group Velocity

De Broglie waves must change shape as they move, because the wave packet of a particle is a superposition of components that have different momenta and thus move with different speeds. Such a wave has been shown in Figure 2.4. Thus to describe a de Broglie wave we cannot use a single velocity v in our equations. We introduce the symbol v_p for the velocity of one of these component waves; v_p, called the *phase velocity*, is a

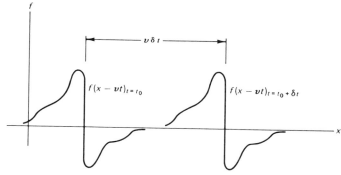

FIGURE 2.8 An arbitrary function $f(x - vt)$ shown at two times that differ by the amount δt. Increasing x by the amount $v\,\delta t$ while increasing t by δt leaves the quantity $x - vt$ unchanged, and thus the value of the function is unchanged. Therefore the curve showing the function at time t_0 is identical in shape to the curve of the function at time $t_0 + \delta t$.

function of the wavelength λ, because the momentum p depends on wavelength.

We cannot write the function ψ in Eq. (2.13) or (2.17) as simply $\psi(x - v_p t)$, because there is no unique value of v_p for such a wave packet. We take the velocity of each component wave into account by rewriting Eq. (2.17) as

$$\psi(x,t) = \frac{1}{\sqrt{2\pi}} \int_{-\infty}^{\infty} \phi(k) e^{ik(x - v_p t)} \, dk \tag{2.41}$$

where v_p, being dependent on λ, must be a function of k.

When $\phi(k)$ extends over a narrow range of values of k, the values of v_p can be nearly equal and the wave packet retains its shape well enough that one can observe the motion of the packet as a whole. But this motion, described by the *group velocity* v_g, does not equal the phase velocity. An example will show why this is so.

Let us introduce the quantity $\omega \equiv \nu/2\pi$, called the *angular frequency*. Since $\nu = v/\lambda$ for a wave of phase velocity v, and $k = 2\pi/\lambda$, we can write any single wave as a function of $kx - \omega t$ instead of $x - vt$. Then the sum of two waves of equal amplitude but with slightly different values of k can be written

$$\psi(x,t) = \sin(kx - \omega t) + \sin[(k + \delta k)x - (\omega + \delta\omega)t] \tag{2.42}$$

This can be rearranged, with the aid of some trigonometry, to

$$\psi(x,t) = 2\cos((\delta k/2)x - (\delta\omega/2)t)\sin([k + (\delta k/2)x] - [\omega + (\delta\omega/2)]t) \tag{2.43}$$

If both δk and $\delta\omega$ are very small ($\delta k \ll k$, $\delta\omega \ll \omega$), Eq. (2.43) reduces to

$$\psi(x,t) \approx 2\cos[(\delta k/2)x - (\delta\omega/2)t]\sin(kx - \omega t) \tag{2.44}$$

essentially a sine wave whose amplitude varies slowly with x and t (Figure 2.9). The factor $\sin(kx - \omega t)$ can be written $\sin k[x - (\omega/k)t]$, which is a function of $x - vt$ with v equal to ω/k. In this case v is identified with the velocity of the peaks in this sine wave, which by definition move with the *phase* velocity. Thus

$$v_p = \omega/k \tag{2.45}$$

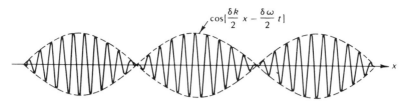

FIGURE 2.9 Sum of two waves of equal amplitude whose frequencies differ by $\delta\omega$. As required by Eq. (2.44), the amplitude of the crests oscillates between $2\cos[(\delta k/2)x - (\delta\omega/2)t]$ and $-2\cos[(\delta k/2)x - (\delta\omega/2)t]$, as shown by the broken lines. The envelope moves with speed $\delta\omega/\delta k$, the crests with speed ω/k.

Following the same line of reasoning, we see that the envelope, determined by the factor $2\cos[(\delta k/2)x - (\delta\omega/2)t]$, has velocity $\delta\omega/\delta t$. By definition, the velocity of this envelope is the *group* velocity. Thus

$$v_g = \delta\omega/\delta k \qquad (2.46)$$

For de Broglie waves, light waves in air, and many other types of wave, waves of different frequency move with different speeds. The speeds of the two waves that are added in Eq. (2.42) are, respectively, ω/k and $(\omega + \delta\omega)/(k + \delta k)$. If these speeds are different, then

$$\omega/k \neq (\omega + \delta\omega)/(k + \delta k) \quad \text{and therefore} \quad \delta\omega/\delta k \neq \omega/k \quad (2.47)$$

Consequently, the phase velocity is not equal to the group velocity, and the envelope of a wave packet moves with a speed that is different from the speed of the individual crests in the wave. (This happens with water waves; throw a stone into a lake and see for yourself.) This example suggests a general formula for the group velocity of a wave packet as

$$v_g = \left\{\frac{d\omega}{dk}\right\}_{k=k_0} \qquad (2.48)$$

where k_0 is the average value of k for the constituent waves in the packet. This formula gives the velocity of the envelope of a wave packet, as long as the packet consists mainly of waves for which k is close to k_0.

2.5.2 Group Velocity and Phase Velocity for a de Broglie Wave

The group velocity of a de Broglie wave can be shown to be equal to the velocity that we measure for the particle represented by that wave. To show this, we start with the definitions

$$\omega = 2\pi\nu \quad \text{and} \quad k = 2\pi/\lambda \qquad (2.49)$$

and we relate them to the energy and momentum of the particle, to extract the velocity of the particle. We know that the momentum p is equal to h/λ, or $\hbar k$. Let us assume that the energy E of the particle is equal to $h\nu$, or $\hbar\omega$, as it is for a photon. With $E = \hbar\omega$ and $p = \hbar k$, we find that

$$v_g = d\omega/dk = d(E/\hbar)d(p/\hbar) = dE/dp \qquad (2.50)$$

From relativistic dynamics (Appendix C) we have

$$E^2 = p^2c^2 + m^2c^4 \qquad (2.51)$$

Taking differentials of both sides gives

$$2E\,dE = 2pc^2\,dp \qquad (2.52)$$

and thus

$$v_g = dE/dp = pc^2/E \qquad (2.53)$$

which is the velocity of a particle of momentum p and energy E according to special relativity. Thus the group velocity is the velocity of the particle.

The phase velocity is different, being equal to ω/k. From Eqs. (2.49) we have

$$v_p = \omega/k = (E/\hbar)/(p/\hbar) = E/p \qquad (2.54)$$

From Eq. (2.54) we can deduce that *the phase velocity is greater than c*. (See Problem 2.15.)

2.6 COMPLEMENTARITY

If there is an ultimate limitation to our powers of observation, we must wonder whether what we wanted to observe exists at all. For example, although we can measure momentum and position simultaneously, we cannot measure both with any desired precision. After a precision position measurement, the resulting momentum uncertainty prevents us from knowing where the particle will go next. Any measurement always disturbs the particle's trajectory. Therefore:

> We can never measure the exact *trajectory* of a particle.

We might imagine that a particle still *possesses* a trajectory, even though we cannot determine what it is. But this view leads to a serious dilemma, which reveals a fundamental flaw in Bohr's circular-orbit postulate.

Consider the diffraction of electrons by two parallel slits, which has been observed to give results analogous to two-slit diffraction of light. The diffraction pattern can be built up from observations of single electrons.[6] The interference pattern is seen only when both slits are open, but the electrons, being observed one at a time, cannot interfere with each other. We must conclude that *each* electron is influenced in some way by *both* slits. This is brought out clearly by closing one of the two slits, reducing the total number of electrons that are observed. At points where very few electrons had been observed (because there had been destructive interference between waves from the two slits), the detectors now count *more* electrons than they had before. This seems incomprehensible if the electron is a classical particle that must go through one slit or the other.

The conclusion is inescapable. A single electron, as a wave, somehow goes through both slits, yet it is always observed as an indivisible particle; we never detect a fraction of an electron in a detector. Surely, you think, there must be something wrong with this picture; if it is indivisible, the electron must go though one slit or the other, not both. But is there any experiment to disprove the existence of this so-called *wave–particle duality*?

Let us try a "thought experiment," in which we try to observe the particle as it passes through one slit or the other, while we also try to observe the interference pattern. Presumably we could do this by illuminating the slits. As an electron passes through a slit, a photon may be scattered from it, thereby telling us which slit the electron went through on its way to the detectors that register the interference pattern.

If we succeed in doing this, we will find that the interference pattern has disappeared. When a photon scatters from an electron, the electron recoils, and the amount of recoil is sufficiently uncertain to prevent our determining where the electron *would have gone* if the photon had not struck it. The result is a distribution that shows no interference minima (Figure 2.10).

---

[6]An interference pattern can also be built up from observations of single photons. See S. Parker, Double-Slit Experiment with Single Photons, *Am. J. Physics* **40**, 1003 (1972).

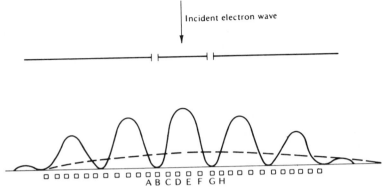

FIGURE 2.10 Possible intensity distribution from diffraction of electrons by narrow slits. Solid line shows the distribution when both slits are open. Dashed line shows the distribution when only the right-hand slit is open. Boxes represent electron detectors. Detectors B and G, near nodes of the interference pattern, count more electrons when one slit is closed than they do when both slits are open.

In practice, not all of the electrons would be struck by photons. In that case, we could in principle identify those electrons by correlating the observation of a scattered photon with the time of arrival of an electron at the detector. The electrons that have been struck will be distributed smoothly; the ones not struck will show interference minima!

Knowing that the momentum of a photon depends on its frequency, you might try to use kinder, gentler photons, so that they will not cause the electron to recoil so much. That will work; at a sufficiently small photon frequency, the interference pattern will appear. But as soon as the photon wavelength is long enough to preserve the electrons' interference pattern, you will find a broad diffraction pattern when you observe the *photons*, making it impossible to determine at which slit any given photon interacted with an electron.

Bohr described this situation by saying that the wave and particle aspects of matter are *complementary*. Each aspect can be considered as a potentiality; an experiment is required to convert the potentiality into something that we can observe. The act of observation brings out one aspect and suppresses the other aspect.

2.6.1 Hidden Variables

Many great physicists have found Bohr's description to be unsatisfactory. Some believe that the reliance on probabilities shows that something is missing from this model of nature, that the behavior of any given electron must be governed by some other "hidden" variable that remains to be identified.

For example, consider diffraction of an electron by a single slit. It is equally probable that the electron will be diffracted toward the right or toward the left, so the result is like flipping a coin. But the coin flip is not always random; some skillful people can flip a coin and make it come up heads or tails at will, by controlling the initial variables—position, velocity, angular velocity—that determine the coin's motion.[7] But there is no evidence of any variables that determine which way an electron will go. Any such variables are totally unknown (hidden) and quite likely nonexistent.

Short of actually finding a hidden variable, can we tell whether variables of this kind could be consistent with experiment? Surprisingly, there are ways of testing for the existence of such variables even though the variables may remain hidden. Some experimental tests are described in Chapter 12.

ADDITIONAL READING

D. Bohm, *Quantum Theory*, Prentice-Hall, New York, 1951. Chapters 1–3 contain thorough and excellent discussions of wave packets, wave properties of matter, particle properties of light, and the role of probability in quantum theory. Later books do not address the conceptual difficulties of quantum theory so extensively. See also the additional reading for Chapter 1.

EXERCISES

1. Compute the energy of a photon whose wavelength is the same as that of a 50-eV electron.

2. Compute the speed and de Broglie wavelength of
   (a) an electron whose kinetic energy is 5.0 eV
   (b) an electron whose kinetic energy is 5.0 MeV
   (c) a proton whose kinetic energy is 5.0 MeV
   (d) a grain of sand of 1 mg mass and 5.0 MeV kinetic energy.

3. If you do not try to restrict the motion, the wave and particle pictures give the same end results. If the potential energy of a particle suddenly increases as it crosses a boundary, it receives an impulse perpendicular to the surface; therefore, for example, an electron beam changes direction (is refracted) when it crosses an interface where its potential

---

[7]It has been reported that magician Persi Diaconis can do this.

energy changes. Show that Snell's law of refraction, which can be written as in Eq. (2.3) [$\sin \phi / \sin \phi' = \lambda / \lambda'$], is obeyed in this case. (*Hint:* Write λ / λ' in terms of the kinetic energy K and potential energy V on each side of the interface, write the angles ϕ and ϕ' in terms of the momentum components parallel and perpendicular to the interface (using $p^2 = 2mK$), and relate the two results.)

4. The relation between the wavelength λ and the angle ϕ in the Davisson–Germer experiment is $\lambda = d \sin \phi$. Show that this formula can be derived by assuming that all of the scattering occurs at the surface of the crystal. In that case, would you expect to see the results shown in Figure 2.2? Explain.

5. Find the kinetic energy, in eV, of neutrons that would be scattered from the [111] face of nickel at the same angle as the 54-eV electrons are scattered in the Davisson–Germer experiment.

6. Show that the wavelength λ of a particle of rest energy E_0 and kinetic energy K is given by $\lambda = hc\sqrt{K^2 + 2E_0 K}$.

7. Compute the coefficients in the Fourier series for

$$\psi(x) = 1 \quad \text{for } 0 < x < 2b$$

$$\psi(x) = 0 \quad \text{for } -a < x < 0 \text{ and } 2b < x < a$$

$$\psi(x + 2a) = \psi(x) \quad \text{for all } x$$

Comment on the difference between this series and the one of Example 2.2.

8. Compute the coefficients in the Fourier series for

$$\psi(x) = Ax \quad \text{for } 0 < x < 2b$$

$$\psi(x) = 0 \quad \text{for } -b < x < 0$$

$$\psi(x + 2b) = \psi(x) \quad \text{for all } x$$

and plot the sum of the first 10 nonzero terms in this series.

9. (a) Find the Fourier transform $\phi(k)$ of the function

$$\psi(x) = e^{ik_0 x} \quad \text{if } 0 < x < 2\pi \Re/k_0;$$

$$\psi(x) = 0 \quad \text{if } x < 0 \text{ or } x > 2\pi \Re/k_0$$

(b) Show that $\phi(k)$ is peaked at $k = k_0$, with the peak width inversely proportional to $\mathscr{R}$. What is the connection between this result and the resolving power of a diffraction grating (Section 2.4)?

10. (a) Compute the Fourier transform $\phi(k)$ of the function

$$\psi(x) = 1 \quad \text{for} \ -b < x < b; \qquad \psi(x) = 0 \quad \text{everywhere else}$$

(b) Show that the series found in Example 1 is obtained from your $\phi(k)$ by making each coefficient equal to

$$\frac{1}{\sqrt{2\pi}} [\phi(k) + \phi(-k)] \, \delta k$$

with $k = n\pi/a$ and $\delta k = \pi/a$.

11. If you drop a 2-g marble from a height of 1 m, trying to hit a crack in the floor, by how much does the uncertainty principle limit your accuracy?

12. In a two-slit diffraction experiment with photons, we place photon detectors (free electrons) by each slit, to tell us which slit the photon passed through on its way to the screen. To identify the slit, each electron must tell us the y coordinate of the photon with an uncertainty of less than $a/2$, where a is the distance between the slits. The angle θ between interference maxima is approximately λ/a, if the photon wavelength is λ. (See Figure 2.11.) Show that when the slit is identified by electrons that are localized to this extent, the resulting uncertainty in momentum destroys the interference pattern.

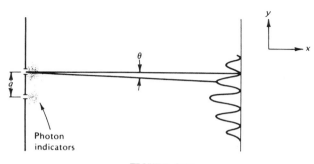

FIGURE 2.11

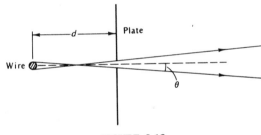

FIGURE 2.12

13. Electrons are boiled off a thin-wire filament and are attracted to a flat plate in which there is a narrow slit parallel to the wire. (The wire is perpendicular to the plane of Figure 2.12.) The beam of electrons thus formed diverges slightly because of the thickness of the wire and the width of the slit, but it seems that you could make the beam as narrow as you wish by making the distance d sufficiently large. Calculate
    (a) the geometrical divergence, and
    (b) the minimum divergence θ required by the uncertainty relation, given that the wire diameter and slit width are each 200 nm, $d = 3$ cm, and the accelerating potential between the wire and the slit is 5 volts.

14. At one time it was thought that atomic nuclei might contain electrons. Use the uncertainty principle to find how long an electron could stay inside a nucleus whose radius is 5 fm, before its momentum would take it away.

15. (a) Show from Eqs. (2.53) and (2.54) that, for a de Broglie wave, $v_g v_p = c^2$. Thus, since $v_g < c$, v_p must be greater than c.
    (b) The phase velocity v_p cannot be observed directly. Thus the fact that it is greater than c does not contradict the "speed limit" imposed by relativity; no information can be transmitted at such a speed. The phase velocity is in fact an arbitrary number, because the potential energy, and hence the total energy, are arbitrary numbers; only *changes* in energy are observable. We could obtain an entirely different result for the phase velocity by defining ω in terms of the kinetic energy, or

$$\hbar\omega = E - mc^2 \rightarrow p^2/2m \quad \text{for } v \ll c$$

Show that this definition leads to $v_g = p/m$ and $v_p = p/2m$ for $v \ll c$.

16. Show that the complete function describing the wave packet of Figure 2.4 is

$$\psi(x,t) = \text{constant} \times \exp\left[-(x - v_g t)^2(1 - iat)/(2b^2 + 2b^2a^2t^2)\right]$$
$$\times \exp\left[ik_0(x - v_p t)\right]$$

with $a = \hbar/mb^2$ and $b = \lambda_0 = 2\pi/k_0$.
Hint: $\psi(x,0)$ is given by $\psi(x)$ in Eq. (2.18) with $\sigma^2 = b^2/2$. Find the Fourier transform of $\psi(x,0)$ by substituting it for $\psi(x)$ in Eq. (2.18) to find $\phi(k)$ and show that

$$\psi(x,0) = \frac{1}{\sqrt{2\pi}} \int_{-\infty}^{\infty} \phi(k)e^{ikx}\,dk = \frac{1}{\sqrt{2\pi}} \int_{-\infty}^{\infty} e^{-(k-k_0)^2 b^2/2}e^{ikx}\,dk$$

Then obtain $\psi(x,t)$ by replacing x by $x - v_p t$ inside the integral, with v_p equal to $\hbar k/2m$. Use this result to show that Figure 2.4 is drawn correctly for time $t = 5mb/\hbar k_0$, in particular that the phase changes more rapidly with x as x increases, and that the standard deviation of the Gaussian envelope has increased to 1.28 mb at that time.

CHAPTER 3

The Schrödinger Equation in One Dimension

Knowing that a particle can be described by a wave, we now must find a way to handle the situation when the particle's momentum depends on its position, which happens whenever the potential energy varies with position. In this chapter we will see how to construct and apply the Schrödinger equation. From this equation we can determine the wave function $\psi(x)$, given any one-dimensional potential energy function $V(x)$. Later we will see how to extend the Schrödinger equation to two and three dimensions.

Schrödinger developed his wave equation on the basis of the de Broglie relation $\lambda = h/p$. His equation is not only consistent with that relation, it also correctly describes phenomena of a totally new character, which were then observed in numerous ways. For example, the Schrödinger equation predicts a non-zero probability of finding a particle in a region where its kinetic energy would be negative. This is not a flaw in the Schrödinger equation; rather, this possibility is necessary for the occurrence of phenomena that are now commonly observed and that form the basis for modern devices, as we shall see.

The Schrödinger equation has been used in the computation of energy levels of a wide variety of systems, with results that are in excellent agreement with experiment. There is no doubt that this equation correctly describes nonrelativistic systems on the atomic scale.

3.1 CONSTRUCTION OF THE WAVE EQUATION FOR A FREE PARTICLE

We do not *derive* the Schrödinger equation; we *construct* it. The proof of the equation is in the experimental results that it predicts. We start with the wave function for a particle of definite momentum $p = h/\lambda = \hbar k$, where $\hbar = h/2\pi$ by definition. This wave function may be written as

$$\psi(x, t) = A \cos(kx - \omega t) \tag{3.1}$$

where, as in Chapter 2, ω is the angular frequency ($\omega = 2\pi\nu$).

We wish to connect Eq. (3.1) with the potential energy V and kinetic energy K of the particle. Using the nonrelativistic expression $K = p^2/2m$, we can write

$$E = p^2/2m + V = \hbar^2 k^2/2m + V \tag{3.2}$$

Now we can construct a differential equation that leads to expression (3.1). We begin by taking derivatives with respect to x. With k assumed independent of x, we have

$$\frac{\partial \psi}{\partial x} = -Ak \sin(kx - \omega t)$$

and therefore

$$\frac{\partial^2 \psi}{\partial x^2} = -Ak^2 \cos(kx - \omega t) = -k^2\psi \tag{3.3}$$

But, from Eq. (3.2),

$$k^2 = \frac{2m(E - V)}{\hbar^2} \tag{3.4}$$

Therefore, combining (3.3) and (3.4), we have

$$\boxed{\frac{\partial^2 \psi}{\partial x^2} = -\frac{2m(E - V)\psi}{\hbar^2}} \tag{3.5}$$

Equation (3.5) is the *time-independent Schrödinger equation* in one dimension.

So far this is nothing new, but our construction is mathematically valid only when k is constant, which means V is constant. That is not very useful. The great step taken by Schrödinger was to *postulate* that Eq. (3.5) is valid even when V depends on x. Like many great theoretical advances, this step can be validated only by experimental tests, and a huge variety of tests have been made. They have shown no conflict with any prediction of the Schrödinger equation in the nonrelativistic domain.

3.1.1 Time Dependence of the Wave Function

The time dependence of the wave function in Eq. (3.1) occurs in the term ωt in the argument of the cosine. To incorporate that into the wave function, we recall that the energy of the de Broglie wave is equal to $\hbar \omega$. Thus, rearranging Eq. (3.5), we have

$$-\frac{\hbar^2}{2m}\frac{\partial^2 \psi}{\partial x^2} + V\psi = E\psi = \hbar \omega \psi \qquad (3.6)$$

We wish to write Eq. (3.6) in terms of derivatives with respect to time as well as space. This leads to a dilemma. If we let ψ equal $\cos(kx - \omega t)$ and take the *first* derivative, we obtain the factor ω that we need for Eq. (3.6), but we do not obtain the same function ψ; we get the sine function instead. If we take the *second* derivative, as we did for the x coordinate, we get back to the function ψ, but it is multiplied by ω^2 instead of ω.

Schrödinger solved the dilemma by using the complex function $e^{i(kx - \omega t)}$ instead of the cosine for the function $\psi(x, t)$. Then we find that Eq. (3.5) is still satisfied, and that

$$\frac{\partial \psi}{\partial x} = -i\omega\psi = -(iE/\hbar)\psi \qquad (3.7)$$

Combining Eqs. (3.6) and (3.7) gives

$$-\frac{\hbar^2}{2m}\frac{\partial^2 \psi}{\partial x^2} + V\psi = i\hbar\frac{\partial \psi}{\partial t} \qquad (3.8)$$

Equation (3.8), the *time-dependent Schrödinger equation* in one dimension, is a *linear* equation, for which the sum of two solutions is also a solution. This is essential to describe phenomena involving wave superposition and interference.

The imaginary part of ψ creates no difficulty, because only $|\psi|^2$, not ψ, is observable. The probability of observing a particle at a given location is proportional to $|\psi|^2$, or $\psi^*\psi$, which is always real. (See Section 2.5.)

3.2 BOUNDARY CONDITIONS, CONTINUITY CONDITIONS, AND THE SQUARE WELL

Many implications can be deduced from the *boundary conditions* that govern solutions of the Schrödinger equation. Any solution that does not satisfy the boundary conditions must be ruled out. These conditions are based on the reasonable requirement that ψ be finite. [We discussed an infinite ψ in connection with the delta function (Chapter 2), but that function is infinite at a single point, so it is not relevant to boundary conditions.] We also assume that in any physical problem both E and V will be finite.

In addition to boundary conditions, there are *continuity conditions* that must be obeyed. If E, V, and ψ are all finite everywhere, we see from Eq. (3.5) that $\partial^2\psi/\partial x^2$ must be finite. Consequently ψ and $\partial\psi/\partial x$ must be continuous everywhere, *even if V is discontinuous*, because a discontinuity in $\partial\psi/\partial x$ makes $\partial^2\psi/\partial x^2$ infinite.

3.2.1 Illustration: Square-Well Potential

To illustrate the use of the Schrödinger equation and its boundary conditions, we first apply it to the one-dimensional square-well potential, an example that does have important applications in spite of its somewhat artificial nature. This potential, shown in Figure 3.1, is given by

$$V = V_0 \quad (|x| > a); \qquad V = 0 \quad (|x| \le a) \qquad (3.9)$$

and we wish to find the wave functions and allowed energies for energy $E < V_0$.

Classically, E can have any non-negative value, and for energies less than V_0 it can never be found where $|x| > a$, because its kinetic energy would be negative at those points. In classical terms, a particle with $E < V_0$ bounces back and forth between the two sides of the well, always moving with the same speed as it remains in the "box" defined by the potential.

In quantum theory, we must determine the particle's behavior by finding the wave function, which must be a solution of the Schrödinger equation and must obey the continuity conditions at the boundaries of the

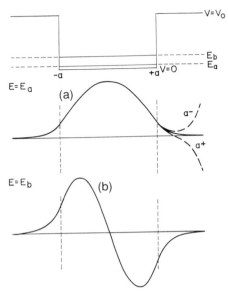

FIGURE 3.1 Solutions of the Schrödinger equation for various energies in a square well of dept V_0 and width $2a$. (a) Wave function for the lowest allowed energy E_a. (b) Function for the next-lowest allowed energy E_b. Notice that (b), with a larger energy than (a), penetrates farther into the forbidden region. Other curves show solutions which are unacceptable as wave functions.

well. A general expression for the wave function of a particle of energy E is

$$\psi(x, t) = u(x)e^{-i\omega t} = u(x)e^{-iEt/\hbar} \tag{3.10}$$

just the time-independent Schrödinger equation again, written in terms of $u(x)$, the space-dependent part of ψ.

To determine $u(x)$ for the given square well, we substitute into Eq. (3.6) the potential given in Eq. (3.9). With $\psi = u(x)e^{-i\omega t}$ we obtain

$$\frac{\partial^2 u}{\partial x^2} = -\frac{2m}{\hbar^2}(E - V_0)u \quad \text{for } |x| > a \tag{3.11}$$

$$\frac{\partial^2 u}{\partial x^2} = -\frac{2m}{\hbar^2}Eu \quad \text{for } |x| \le a \tag{3.12}$$

In Eq. (3.11) the coefficient of u is positive, because we are seeking solutions for the case $E < V_0$. Thus the equation can be written in the form

$$\frac{\partial^2 u}{\partial x^2} = \alpha^2 u \qquad (3.13)$$

where α is positive by definition. The general solution is then

$$u = Ae^{\alpha x} + Be^{-\alpha x} \text{ for } |x| > a \qquad (3.14)$$

where $\alpha = [2m(V_0 - E)/\hbar^2]^{1/2}$ [(in agreement with Eq. (3.11)], and A and B are two new constants to be determined by the boundary conditions. The condition that the wave function be finite tells us immediately that $A = 0$ for $x > a$, and $B = 0$ for $x < a$, so that the function will approach zero asymptotically as $x \to \infty$ and as $x \to -\infty$. Thus

$$u = Ae^{\alpha x} \quad \text{for } x < -a \quad \text{and} \quad u = Be^{-\alpha x} \text{ for } \quad x > a \qquad (3.15)$$

On the other hand, in Eq. (3.12) the coefficient of u is negative, because E must be positive. Thus we can use the form

$$\frac{\partial^2 u}{\partial x^2} = -k^2 u \qquad (3.16)$$

as in Eqs. (3.3) and (3.4), where k is real by definition. Thus the solution is, in general

$$u = C \cos kx + D \sin kx \quad \text{for } |x| \leq a \qquad (3.17)$$

where $k = (2mE/\hbar^2)^{1/2}$, and C and D are constants to be determined by boundary conditions and normalization.

Before doing the algebra of finding A, B, C, and D, let us look for the big picture, by trying to deduce from first principles the general shape of the wave function and why only discrete energies are allowed. What happens if we start to plot the wave function point by point, without an explicit solution? We can choose values for u and $\partial u/\partial x$ at any point, choose an energy E, and then use Eq. (3.11) to find the value of $\partial^2 u/\partial x^2$ at that point. The value of u at $x + \delta x$ is then given by $u(x) + (\partial u/\partial x)\,\delta x$ (using the original value of $\partial u/\partial x$), and the new value of $\partial u/\partial x$ is $\partial u/\partial x + (\partial^2 u/\partial x^2)\,\delta x$.

This procedure will give us a curve that satisfies Eq. (3.8), because we used that equation to find each value of $\partial^2 u/\partial x^2$. But such a curve will be unlikely to satisfy all of the boundary conditions, as a simple example will show. Let us start at a point where $x < -a$ and E is so small that the corresponding wavelength inside the well is greater than $4a$. As long as $x < -a$, the value of $\partial^2 u / \partial x^2$ is positive, because α^2 is positive [Eq. (3.15)].

At $x = -a$ we have a point of inflection; $\partial^2 u / \partial x^2$ changes sign, because it is now given by Eq. (3.12). At this boundary there is no problem, but there is a problem at the other boundary. When $x \geq a$, $\partial^2 u / \partial x^2$ is positive again, and large enough so that the slope of the curve eventually becomes positive. Beyond this point the slope is positive and *steadily increasing*, because $\partial^2 u / \partial x^2$ is also positive. Therefore this u goes to infinity as $x \to \infty$ (Figure 3.1, curve a$^-$), and the function is not acceptable.

If we plot u with larger values of E, we eventually reach the situation shown in Figure 3.1, curve a$^+$, where the function crosses the axis. In that case $u \to -\infty$ as $x \to \infty$. But at one specific energy, when $E = E_a$ (Figure 3.1a), the values of u, $\partial u / \partial x$, and $\partial^2 u / \partial x^2$ combine to produce a curve that goes asymptotically to zero as $x \to \infty$, and E_b is therefore the lowest allowed energy level.

Notice that the curve for the lowest level has no node; it never crosses the x axis. This is required for the lowest level in any well. If the curve does cross the axis (as in Fig. 3.1b, curve b), it is always possible to find an allowed level with lower energy. For example, the curve of Figure 3.1a does not cross the axis, because $\partial^2 u / \partial x^2$ is smaller in the region $|x| < a$.

The acceptable wave function reveals a possibility that is bizarre by classical standards; it extends into so-called *forbidden regions* where the kinetic energy, $E - V$, is negative. This means that it is possible to observe the particle in such a region. But observing a particle with negative kinetic energy seems clearly impossible.

The resolution of this apparent paradox is based on the uncertainty principle. One must measure the particle's position with considerable accuracy to determine that it is in the forbidden region. This requires a photon with such a short wavelength that the particle acquires a large amount of energy as a result of the measurement. Thus the particle acquires enough energy from the photon to make its kinetic energy positive after the measurement.

3.2.2 Calculation of Energy Levels

First consider the special case $V \to \infty$. In that case α becomes infinite, and the solution [Eq. (3.17)] is zero when $x < -a$ or $x > a$. Then, since u must be continuous, the solution for $-a < x < a$ [Eq. (3.18)] must give $u = 0$ at

$x = -a$ and at $x = a$. Therefore, setting x equal to $+a$, we have

$$u = C \cos ka + D \sin ka = 0 \qquad (3.18)$$

and when $x = -a$, since $\sin(-ka) = \sin ka$,

$$u = C \cos ka - D \sin ka = 0 \qquad (3.19)$$

Adding these two equations gives

$$2C \cos ka = 0 \qquad (3.20)$$

and subtracting (3.19) from (3.20) gives

$$2D \sin ka = 0 \qquad (3.21)$$

To satisfy *both* of these equations, we must have *either*

$$\cos ka = 0 \text{ and } D = 0 \quad \text{or} \quad \sin ka = 0 \text{ and } C = 0 \qquad (3.22)$$

If $D = 0$ then $\quad u = C \cos kx \quad k = n\pi/2a \quad$ and n is an *odd* integer

$$(3.23)$$

If $C = 0$ then $\quad u = D \sin kx \quad k = n\pi/2a \quad$ and n is an *even* integer

$$(3.24)$$

The energies are found from the definition $k = (2mE/\hbar^2)^{1/2}$, which gives

$$E = \hbar^2 k^2/2m = n^2 h^2/32ma^2 \qquad (3.25)$$

Knowing that the wave must go to zero at $x = \pm a$, we could have deduced this result in another way. The wavelength must then equal $4a/n$, and the result follows from the relations $p = \hbar/\lambda$ and $E = p^2/2m$. The resulting wave functions are shown in Figure 3.2 for the lowest three energy levels ($n \leq 3$).

Figure 3.2 also shows the quantity $|u(x)|^2$, identified as the probability density (Section 3.3). Notice that it goes to zero at certain "nodes" inside the well. If you wonder how a particle can go from side to side in a well without being found near a node, think of a standing wave on a vibrating string. Energy flows back and forth along the string, but at a node you cannot *extract* this energy. Similarly, in *interacting* with a particle in a box, you seldom find it near a node.

Of course you never find the particle *at* a node or at any other point, because a point has no dimensions. Classically or quantum mechanically,

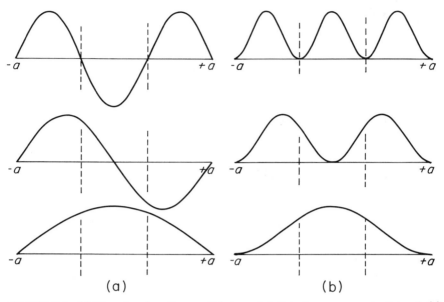

FIGURE 3.2 (a) Wave functions for a particle in an infinitely deep square potential well. (b) Probability densities derived from these wave functions.

the probability of finding a particle *at* a precise point is always zero, because there is an infinite number of possible points. The meaningful number is the probability of finding the particle in a given interval or volume. See Section 3.3.

When V is finite, the algebra is tedious, involving four equations, but it illustrates the use of boundary conditions. From the continuity of u we have

$$\text{at } x = -a: \qquad Ae^{-\alpha a} = C \cos ka - D \sin ka \qquad (3.26)$$

$$\text{at } x = +a: \qquad Be^{-\alpha a} = C \cos ka + D \sin ka \qquad (3.27)$$

The continuity of du/dx gives us

$$\text{at } x = -a: \qquad \alpha Ae^{-\alpha a} = Ck \sin ka - Dk \cos ka \qquad (3.28)$$

$$\text{at } x = +a: \qquad -\alpha Be^{-\alpha a} = -Ck \sin ka + Dk \cos ka \qquad (3.29)$$

Adding (3.26) and (3.27) yields

$$(A + B)e^{-\alpha a} = 2C \cos ka \qquad (3.30)$$

Subtracting (3.26) from (3.27) yields

$$(B - A)e^{-\alpha a} = 2D \sin ka \tag{3.31}$$

Adding (3.28) and (3.29) yields

$$\alpha(A - B)e^{-\alpha a} = 2Dk \cos ka \tag{3.32}$$

Subtracting (3.29) from (3.28) yields

$$\alpha(A + B)e^{-\alpha a} = 2Ck \sin ka \tag{3.33}$$

To eliminate the constants A, B, C, and D we divide corresponding sides of (3.33) by (3.30), and corresponding sides of (3.32) by (3.31), obtaining the two equations

$$\alpha = k \tan ka \quad \text{provided that } A \neq -B \text{ and } C \neq 0 \tag{3.34}$$

$$-\alpha = k \cot ka \quad \text{provided that } A \neq B \text{ and } D \neq 0 \tag{3.35}$$

It is clearly impossible for *both* (3.34) and (3.35) to be valid, because $\tan ka$ and $\cot ka$ must have the same algebraic sign. Thus either $C = 0$ and Eq. (3.34) is not valid (because it is a result of division by zero), or $D = 0$ and Eq. (3.35) is invalid for the same reason. Therefore we have two possibilities:

1. Equation (3.35) is *wrong*, $D = 0$, $A = B$, $\alpha = k \tan ka$, and the solution is

$$u = Ae^{\alpha x} \quad \text{for } x < -a \tag{3.36a}$$

$$u = Ae^{-\alpha x} \quad \text{for } x > +a \tag{3.36b}$$

$$u = C \cos kx \quad \text{for } |x| < a \tag{3.36c}$$

In this case $u(x)$ is an *even function*; $u(x) = u(-x)$ for all values of x.

2. Equation (3.34) is *wrong*, $C = 0$, $A = -B$, $-\alpha = k \cot ka$, and the solution is

$$u = Ae^{\alpha x} \quad \text{for } x < -a \tag{3.37a}$$

$$u = -Ae^{-\alpha x} \quad \text{for } x > +a \tag{3.37b}$$

$$u = D \sin kx \quad \text{for } |x| < a \tag{3.37c}$$

In this case $u(x)$ is an *odd function*; $u(x) = -u(-x)$ for all values of x. Notice that in Figure 3.1, curve (a) shows an even solution and curve (b) shows an odd solution. Solutions of the two types alternate as E increases.

In either case, we have found one equation [either (3.34) or (3.35)] relating the two unknown quantities k and α. A second equation comes from the definitions $k = (2mE/\hbar^2)^{1/2}$ and $\alpha = [2m(V_0 - E)/\hbar^2]^{1/2}$, which when squared and added give

$$k^2 + \alpha^2 = 2mV_0/\hbar^2 \equiv \beta^2 \qquad (3.38)$$

where we define the constant β for convenience. We could now use a computer to solve Eqs. (3.38) and (3.34) or (3.35) simultaneously and find the allowed values of k, in terms of V_0 and a, to any desired degree of accuracy. But to gain more insight, let us use a graphical method of solution. In doing this, it is convenient to substitute Eq. (3.38) into Eq. (3.34), obtaining

$$k^2 + k^2 \tan^2 ka = \beta^2$$

which is equivalent to

$$k^2 \sec^2 ka = \beta^2$$

or

$$\cos ka = \pm ka/\beta a \quad \text{(and } \tan ka > 0) \qquad \text{for the even solution} \quad (3.39a)$$

By a similar sequence of steps, we obtain

$$\sin ka = \pm ka/\beta a \quad \text{(and } \cot ka < 0) \qquad \text{for the odd solution} \quad (3.39b)$$

Figure 3.3 shows a graphical solution of Eq. (3.39a), with $\cos ka$ plotted as well as $+ka/\beta a$ and $-ka/\beta a$ against βa. Of the seven points of intersec-

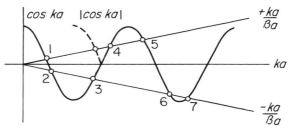

FIGURE 3.3 Graphs of $\cos ka$ and $\pm ka/\beta a$, showing possible values of ka for the "even solution" of the square-well potential. Dashed curve shows $|\cos ka|$ in the third quadrant. (See text for discussion.)

tion, only the odd-numbered ones are acceptable, because they are the only ones for which tan ka is greater than zero, as required by Eq. (3.39a).

We could draw a graph showing $\sin ka$, $+ka/a$, and $-ka/\beta a$, which would give us the solutions of Eq. (3.40), with allowed values of ka lying in the even quadrants. But the sine curve is not necessary, because all quadrants contain identical pieces of the sine or cosine curve. For example, in the third quadrant $\cos ka$ intersects the line $-ka/\beta a$; this is the same as the intersection of $|\cos ka|$ with $+ka/\beta a$ (dashed curve in Figure 3.3).

We can therefore find every solution by drawing *one* cosine curve in *one* quadrant, then drawing additional straight lines to represent the continuation of the graph of $+ka/\beta a$ (Figure 3.4b). We can read off each solution from such a graph if we simply remember that the lowest line segment shows ka between 0 and $\pi/2$, the second line segment gives ka between $\pi/2$ and π, etc.

Figure 3.4 shows seven allowed energy levels. The number of allowed levels depends on the value of βa; as βa increases, the slope of the line $ka/\beta a$ becomes smaller and there are more levels. For an infinitely deep well, βa is infinite, the line $ka/\beta a$ has zero slope, and there is an infinite number of solutions. These are at $ka = n\pi/2$, where n is a positive integer (as we have already seen for the case $V_0 \to \infty$).

Example Problem 3.1   Determine how many bound states there are for an electron in a square well of depth $V_0 = 20.0$ eV and width $2a = 3.00$ nm, and find the energies of the two lowest levels.

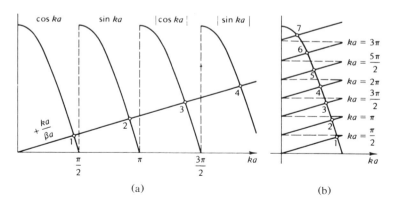

FIGURE 3.4   (a) Figure 3.3 redrawn to show odd solutions as well as even solutions. (b) Same figure, remapped into the first quadrant.

Solution. From Figure 3.4b we see that there is one bound state for $ka \leq \pi/2$, a second state for $\pi/2 < ka \leq \pi$, and a state for each increase of $\pi/2$ until the last state with $n\pi/2 < ka \leq \beta a$, making $n + 1$ states in all. Thus the number of states, N, is the next integer larger than $\beta a/(\pi/2)$. Using the data given, we compute βa from Eq. (3.38), with $mc^2 = 511$ keV and $hc = 1240$ eV-nm, to obtain

$$\beta a = (2mV_0)^{1/2}(a/\hbar) = (2mc^2V_0)^{1/2}(2\pi a/hc)$$

$$= (4.521 \times 10^3 \text{ eV}) \times (9.425 \text{ nm})/(1240 \text{ eV-nm}) = 34.36$$

and so $\beta a/(\pi/2) = 21.87$, making N equal to 22.

We could now construct a graph like that of Figure 3.4b and read the values of ka from it. But the two lowest values must lie very close to $\pi/2$ and π, respectively, because the slope of the line $ka/\beta a$ is so small in this case. In this region the cosine curve can be approximated as a straight line, as follows:

Because $\cos ka = \sin(\pi/2 - ka) \approx \pi/2 - ka$, Eq. (3.39) becomes

$$\pi/2 - (ka)_1 = (ka)_1/\beta a$$

so that

$$(ka)_1 = (\pi/2)(\beta a)/(1 + \beta a) = 1.571 \times 34.36/35.36 = 1.53 \quad (3.40)$$

The second value, $(ka)_2$, is still in the linear region of the cosine curve, so

$$(ka)_2 = \pi(\beta a)/(1 + \beta a) = 3.142 \times 34.36/35.36 = 3.06$$

The energy levels follow from the definition of k: $k = (2mE/\hbar^2)^{1/2}$.

We also know that β is defined by $\beta = (2mV_0/\hbar^2)^{1/2}$. Taking the ratio of these, we have

$$k/\beta = (E/V_0)^{1/2}$$

or

$$E = V_0(k/\beta)^2 \equiv V_0(ka/\beta a)^2$$

thus

$$E_1 = 20(1.53/34.36)^2 \text{ eV} = 0.040 \text{ eV}$$

and

$$E_2 = 20(3.06/34.36)^2 \text{ eV} = 0.16 \text{ eV}$$

to two significant figures.

Notice that these levels are just slightly lower than those for the infinitely deep well, for which $(ka)_1 = \pi/2$ (instead of 1.53), and $(ka)_2 = \pi$ (instead of 3.06). Inspection of the wave functions shows why this should be so. In the infinitely deep well the wave function does not penetrate into the forbidden region; thus the wavelength is shorter and the energy greater than that of the corresponding level in a well of finite depth.

3.2.3 Parity of Solutions

It is no mere accident that $u(x) = \pm u(-x)$ for all of the solutions of the square-well problem; it is a necessary consequence of the symmetry of the potential—the fact that in this case $V(x) = V(-x)$. The proof follows from the Schrödinger equation:

$$\frac{\partial^2}{\partial x^2} u(x) = -\frac{2m}{\hbar^2} [E - V(x)]u(x) \tag{3.41}$$

which must hold for all values of x. Thus we can replace x by $-x$ and obtain an equally valid equation:

$$\frac{\partial^2}{\partial x^2} u(-x) = -\frac{2m}{\hbar^2} [E - V(-x)]u(-x) \tag{3.42}$$

If $V(x) = V(-x)$, then we can replace $V(-x)$ by $V(x)$ in Eq. (3.42) to obtain

$$\frac{\partial^2}{\partial x^2} u(-x) = -\frac{2m}{\hbar^2} [E - V(x)]u(-x) \tag{3.43}$$

Equation (3.43) is identical to Eq. (3.41); the only difference is that the wave function is $u(x)$ instead of $u(-x)$. But in one dimension, there is only one wave function for a given value of E. This means that $u(-x)$ can differ from $u(x)$ only by multiplicative constant C:

$$u(x) = Cu(-x) \quad \text{for all values of } x \tag{3.44}$$

Since Eq. (3.44) holds for all values of x, it must be valid if we replace x by

$-x$ to obtain

$$u(-x) = Cu(x) \quad \text{for all values of } x \tag{3.45}$$

with the same constant C. Thus $u(x)u(-x) = C^2 u(-x)u(x)$ and $C = \pm 1$.

If $C = 1$ then $u(x) = u(-x)$ and we say that $u(x)$ has *even parity*. If $C = -1$, then $u(x) = -u(-x)$ and $u(x)$ has *odd parity*. This knowledge can be helpful in many situations, for example, the double well to be discussed in Section 4.1.

3.3 PROBABILITY, EXPECTATION VALUES, AND OPERATORS

Let us now deal quantitatively with the concept of the probability density, given by $|\psi(x,t)|^2$. (For a wave with a constant energy E represented by the factor $e^{-iEt/\omega}$, the probability density is independent of time, and $|\psi|^2 = |u|^2$.) If the wave function is to describe a single particle then the integral of the probability density over all values of x must be

$$\int_{-\infty}^{\infty} |\psi_N|^2 \, dx = 1 \tag{3.46}$$

because there is unit probability that the particle will be found somewhere. A wave function ψ_N satisfying Eq. (3.46) is called a *normalized* function. Given an unnormalized wave function ψ_u, you can normalize it by multiplying it by a normalizing factor N. Thus $\psi_N = N\psi_u$, and substitution into (3.46) gives

$$|N|^2 \int_{-\infty}^{\infty} |\psi_u|^2 \, dx = 1 \tag{3.47}$$

and

$$|N|^2 = \left\{ \int_{-\infty}^{\infty} |\psi_u|^2 \, dx \right\}^{-1} \tag{3.48}$$

The statement in section 3.2, that it is "reasonable" to require that the wave function be finite, can now be understood in the context of Eq. (3.48). In order to have a normalized probability density, we must require that the integral in Eq. (3.48) be finite. This would permit a wave function that is a delta function, because the *integral* of the delta function is finite, even though the function $\delta(x)$ itself is infinite at $x = 0$.

3.3.1 Expectation Values

Although we cannot, in general, predict the exact result of a measurement of a variable such as the x coordinate of a particle, the probability density does allow us to predict the *average* result of a large number of measurements of a variable such as x on systems whose wave functions are identical. The arithmetic mean of all the measured values of x, called the *expectation value* of x, is, by definition

$$\langle x \rangle = \int_{-\infty}^{\infty} x |\psi|^2 \, dx \tag{3.49}$$

which is often written

$$\langle x \rangle = \int_{-\infty}^{\infty} \psi^* x \psi \, dx \tag{3.50}$$

the x being placed between ψ^* and ψ for consistency with later developments. The same relation holds for the expectation value of any power of x:

$$\langle x^n \rangle = \int_{-\infty}^{\infty} \psi^* x^n \psi \, dx \tag{3.51}$$

and therefore

$$\langle f(x) \rangle = \int_{-\infty}^{\infty} \psi^* f(x) \psi \, dx \tag{3.52}$$

where $f(x)$ is any function of x that can be expanded in a power series.

The wave function gives the same sort of information about momentum. We recall that the Fourier transform of $\psi(x)$, the function $\phi(k)$, is the probability amplitude for k. From the identification of $|\phi(k)|^2$ as the probability *density* for the variable k, we can deduce that the expectation value of k is obtained by an expression analogous to Eq. (3.50):

$$\langle k \rangle = \int_{-\infty}^{\infty} \phi(k)^* k \phi(k) \, dk \tag{3.53}$$

Multiplying both sides of (3.53) by the constant $\hbar$ gives us the expectation value of $\hbar k$—the momentum component p_x:

$$\langle p_x \rangle = \langle \hbar k \rangle = \int_{-\infty}^{\infty} \phi(k)^* \hbar k \, \phi(k) \, dk \qquad (3.54)$$

This clearly will give the right answer, but it is not necessary to compute $\phi(k)$ in order to find the value of $\langle p_x \rangle$. We can find a formula for $\langle p_x \rangle$ that involves the function ψ directly, as follows. We start with the formula for $\phi(k)$ in terms of $\psi(x)$ [Eq. (2.16)]:

$$\Phi(k) = \frac{1}{\sqrt{2\pi}} \int_{-\infty}^{\infty} \psi(x) e^{-ikx} \, dx$$

and the complex conjugate of this equation:

$$\phi^*(k) = \frac{1}{\sqrt{2\pi}} \int_{-\infty}^{\infty} \psi^*(x') e^{ikx'} \, dx'$$

where we use the "dummy variable" x' to avoid confusion in the subsequent equations. Substitution into Eq. (3.54) gives

$$\langle p_x \rangle = (\hbar/2\pi) \int_{-\infty}^{\infty} dk \int_{-\infty}^{\infty} \psi^*(x') e^{ikx'} \, dx' \int_{-\infty}^{\infty} k\psi(x) e^{-ikx} \, dx \quad (3.55)$$

But $ke^{-ikx} = i(\partial e^{-ikx}/\partial x)$, so we may write

$$\langle p_x \rangle = (\hbar i/2\pi) \int_{-\infty}^{\infty} dk \int_{-\infty}^{\infty} \psi^*(x') e^{ikx'} \, dx' \int_{-\infty}^{\infty} \psi(x) \left(\frac{\partial}{\partial x} e^{-ikx} \right) dx \qquad (3.56)$$

Integration by parts on x yields (since $\psi(\infty) = 0$)

$$\langle p_x \rangle = -(\hbar i/2\pi) \int_{-\infty}^{\infty} dk \int_{-\infty}^{\infty} \psi^*(x') e^{ikx'} \, dx' \int_{-\infty}^{\infty} e^{-ikx} \frac{\partial \psi(x)}{\partial x} \, dx \quad (3.57)$$

We can now arrange terms by simply (1) placing factors that are independent of x inside the integral on x, and (2) integrating on k before integrating on x. This yields

$$\langle p_x \rangle = \frac{\hbar}{i} \int_{-\infty}^{\infty} \left\{ \frac{1}{2\pi} \int_{-\infty}^{\infty} e^{-ikx} \, dk \int_{-\infty}^{\infty} \psi^*(x') e^{ikx'} \, dx' \right\} \frac{\partial \psi(x)}{\partial x} \, dx \quad (3.58)$$

The quantity in brackets can now be identified as $\psi^*(x)$, according to the Fourier integral theorem (Eq. 2.15). Thus the final result is

$$\langle p_x \rangle = \int_{-\infty}^{\infty} \psi^*(x) \frac{\hbar}{i} \frac{\partial \psi(x)}{\partial x} \, dx \qquad (3.59)$$

Notice the parallel between Eq. (3.50) and Eq. (3.59). The integrals are identical except for the replacement of the *variable* x by the *operator* $(\hbar/i)(\partial/\partial x)$, which is called the *momentum operator*, because it yields the expectation value of p_x when used in an equation like (3.59). To understand the presence of this operator, consider the result for an infinitely long wave with a single wavelength, for which the momentum is p_0, equal to $\hbar k_0$. For such a wave, $\psi(x) = e^{ik_0x}$, and

$$\frac{\hbar}{i} \frac{\partial \psi(x)}{\partial x} = \frac{\hbar}{i} \left[ik_0 e^{ik_0x} \right] = \hbar k_0 \psi(x) = p_0 \psi(x) \qquad (3.60)$$

Thus there is a connection between the differential operator $(\hbar/i)(\partial/\partial x)$ and the variable p_x. Application of this operator to a wave function has generated the value of p_x in the wave represented by this wave function. In other words, the rate at which the wave function changes with x is proportional to the x component of the momentum. This is a logical extension of the connection between wavelength and momentum, $\lambda = h/p$, given by de Broglie.

Many other dynamical variables are represented by operators in quantum mechanics. In general, the expectation value of variable o represented by the operator O is given, by analogy to Eq. (3.59), as

$$\langle o \rangle = \int_{-\infty}^{\infty} \psi^* O \psi \, dx \qquad (3.61)$$

For example, applying p_x twice *as an operator* gives the operator for p_x^2:

$$p_x^2 = p_x p_x = \frac{\hbar}{i} \frac{\partial}{\partial x} \frac{\hbar}{i} \frac{\partial}{\partial x} = -\hbar^2 \frac{\partial^2}{\partial x^2} \qquad (3.62)$$

In general,

$$p_x^n = \frac{\hbar}{i} \frac{\partial^n}{\partial x^n} \qquad (3.63)$$

You have already seen the p^2 operator in the Schrödinger equation, in which the total energy of a particle in one dimension is written as $p_x^2/2m + V$. (This expression is called the *Hamiltonian*, a name carried over from classical mechanics.) Writing this as an operator [making use of Eq. (3.62)], and applying it to ψ, gives us the time-independent Schrödinger equation:

$$\left\{ -\frac{\hbar^2}{2m} \frac{\partial^2}{\partial x^2} + V \right\} \psi = E\psi \qquad (3.64)$$

3.3.2 Eigenfunctions and Eigenvalues

The time-independent Schrödinger equation is an example of an *eigenvalue* equation, which can be defined as equation of the form

$$\text{Operator} \times \text{function} = \text{eigenvalue} \times \text{function}$$

with the characteristic features that the function is the same on both sides of the equation [e.g., the function ψ in Eq. (3.64)], and the eigenvalue is a constant [e.g., the energy E in Eq. (3.64)]. Equation (3.60) is another example, with e^{ikx} as an eigenfunction of the momentum operator. Notice that $\sin kx$ is *not* an eigenfunction of this operator. (See Problem 15.)

Eigenvalue equations abound in quantum mechanics. We justify them because they work, and they work in such a wide variety of systems that we have elevated the rules involving these equations to a pair of postulates.

POSTULATE 1 Any dynamical variable describing the motion of a particle can be represented by an operator.

POSTULATE 2 The only possible result of a measurement of a dynamical variable is one of the eigenvalues of the corresponding operator.

Let us briefly consider an implication of Postulate 2. After you measure a quantity, you should know what its value is; this is what we mean by making a measurement. This value can change as time passes, but at the moment of measurement, the value becomes known. Furthermore, quantum mechanics places no limit on the precision of a measurement of a single variable. This means that a second measurement of the same variable, made immediately after the first measurement, must yield the same value. But we can be certain of the result of a measurement on a system only if the wave function of the system is an eigenfunction of the measured variable. We are led to the following conclusion:

Immediately after measurement of a dynamical variable, the wave function must be an eigenfunction of the operator corresponding to this variable, and the corresponding eigenvalue must equal the measured value of the variable.

What about the wave function of the system before any measurement has been made? A third postulate describes our knowledge of such a system, in general:

POSTULATE 3 Any acceptable wave function ψ can be expanded in a series of eigenfunctions of an operator corresponding to any dynamical variable.

In this case, logic dictates that measurement of the variable will yield an eigenvalue corresponding to one or more of the eigenfunctions in the series. [As we show later, the probability of finding that particular eigenvalue will depend upon the coefficient(s) of the term(s) that is(are) an eigenfunction(s) of the operator corresponding to that variable.]

This logic leads to the conclusion that the original wave function is replaced by a new wave function, thus making it 100% certain that a second measurement will give the same result as the first measurement.[1] This replacement of the wave function is often referred to as the "collapse" of the wave function; the mechanism of that collapse has long been a subject of conjecture and debate.

One example of this situation, to be discussed in Section 5.1, is the case in which a particle moving along the x axis has a known kinetic energy but an unknown momentum; i.e., p^2 is known, but it is not known whether p is positive or negative. Thus the wave function has two terms, which "collapse" into a single term when p is measured. Other very important examples involve angular momentum measurements, to be treated in Section 7.2.

---

[1]P. A. M. Dirac, *Quantum Mechanics*, 4th Edition (Oxford, 1958), states (p. 35), "If the dynamical system is in an eigenstate of a real dynamical variable ξ, belonging to the eigenvalue ξ', then a measurement of ξ will certainly give the result ξ'. Conversely, if the system is in a state such that a measurement of a real dynamical variable ξ is certain to give the result ξ', then the state is an eigenstate of ξ and (ξ') is the eigenvalue of ξ to which this eigenstate belongs."

3.3.3 Expectation Values and Uncertainty

The quantities δx and δp that appear in the Heisenberg uncertainty relation (2.31) are defined experimentally to be the root-mean-square (rms) deviation from the mean value of a series of measurements on identically prepared systems. We now have the mathematical tools to calculate these standard deviations for any given wave function.

First we define Δx as the deviation in a single measurement of x. If the measured value is x, then by definition

$$\Delta x \equiv x - \langle x \rangle$$

Again by definition, the rms deviation is the square root of the mean of the squares of the individual deviations, or

$$\delta x \equiv \langle (\Delta x)^2 \rangle^{1/2} = \langle (x - \langle x \rangle)^2 \rangle^{1/2}$$

where the expectation values $\langle x \rangle$ and $\langle (\Delta x)^2 \rangle$ are computed as prescribed by Eq. (3.52). Thus

$$\langle (\Delta x)^2 \rangle = \int_{-\infty}^{\infty} \psi^*(\Delta x)^2 \psi \, dx \qquad (3.65)$$

For example, if ψ is a normalized Gaussian wave function, we can write

$$|\psi|^2 = \frac{1}{\sigma\sqrt{2\pi}} e^{-x^2/2\sigma^2} \qquad (3.66)$$

where σ is the standard deviation. (See Appendix A for details on statistical standard deviations.) We see from the symmetry of this function that $\langle x \rangle = 0$. Therefore $\Delta x = x$, and substitution into Eq. (3.65) gives

$$\langle (\Delta x)^2 \rangle = \frac{1}{\sigma\sqrt{2\pi}} \int_{-\infty}^{\infty} x^2 e^{-x^2/2\sigma^2} \qquad (3.67)$$

Introducing the variable $w = x^2/\sigma^2$, we have

$$\langle (\Delta x)^2 \rangle = \frac{2\sigma^2}{\sqrt{\pi}} \int_0^{\infty} w^{1/2} e^{-w} \, dw \qquad (3.68)$$

where the integral is in the form of the gamma function, defined as

$$\Gamma(n) \equiv \int_0^\infty w^{n-1} e^{-w} \, dw$$

In this case, $n = 3/2$, and $\Gamma(3/2)$ equals $\sqrt{\pi}/2$. Inserting this value into Eq. (3.68), we obtain

$$\delta x \equiv \langle (\Delta x)^2 \rangle^{1/2} = \sigma \tag{3.69}$$

in agreement with our definition of δx as a standard deviation.

Let us apply the analogous formula for momentum,

$$\delta p_x \equiv \langle (\Delta p_x)^2 \rangle^{1/2} = \langle (p_x - \langle p_x \rangle)^2 \rangle^{1/2} \tag{3.70}$$

to the same wave function (3.66). We know from symmetry that $\langle p_x \rangle = 0$ for this function, so that $p_x - \langle p_x \rangle = p_x$, and Eq. (3.70) becomes

$$\delta p_x = \langle p_x^2 \rangle^{1/2} \tag{3.71}$$

We can use the p_x^2 operator to find $\langle p_x^2 \rangle$, with

$$|\psi|^2 = \frac{1}{\sigma\sqrt{2\pi}} e^{-x^2/2\sigma^2}$$

as follows:

$$\langle p_x^2 \rangle = \int_{-\infty}^\infty \psi^* \left(-\hbar^2 \frac{\partial}{\partial x^2} \right) \psi \, dx \tag{3.72}$$

$$= \frac{-\hbar^2}{\sigma\sqrt{2\pi}} \int_{-\infty}^\infty e^{-x^2/4\sigma^2} \frac{\partial^2}{\partial x^2} (e^{-x^2/4\sigma^2}) \, dx \tag{3.73}$$

$$= \frac{-\hbar^2}{2\sigma^3\sqrt{2\pi}} \int_0^\infty (x^2 - 2) e^{-x^2/2\sigma^2} \, dx \tag{3.74}$$

This integral can be reduced to the sum of two gamma functions, with the final result that $\langle p_x^2 \rangle = \hbar^2/4\sigma^2$, or from (3.69) and (3.71)

$$\delta p_x = \hbar/2\sigma = \hbar/(2\delta x) \tag{3.75}$$

in agreement with the result obtained by using the Fourier transform [Eq.

(2.31)]. Now we see precisely what is meant by δx and δp_x experimentally, and we have the mathematical tools to evaluate δx and δp_x for other wave functions. (See Exercises 10 and 11.) We will always find that the uncertainty product is greater than $\hbar/2$ unless the wave function is Gaussian.[2]

ADDITIONAL READING

D. Bohm, *Quantum Theory* (referred to in Chapter 2) has helpful discussions on all aspects of the topics covered here.

P. A. M. Dirac, *Quantum Mechanics*, 4th edition, Oxford Science Publications, New York (1958, reprinted in paperback, 1991). Page 36 has a discussion of the measurement process and its relation to the postulates of quantum mechanics.

R. H. Dicke and J. P. Wittke, *Introduction to Quantum Mechanics*, Addison-Wesley, Reading, MA (1960). Pages 50–60 have many good figures showing various wells and wave functions for various levels in these wells.

EXERCISES

1. Convince yourself that Figure 3.1 is drawn correctly for the case $\beta a = \pi$, as follows. Estimate from the graph the values of E/V_0, k, and α for each of the two allowed levels, and calculate whether these values are consistent with the value of β and the equation $\sin ka = \pm k/\beta$ or $\cos ka = +k/\beta$.

2. Calculate the minimum depth V_0, in electron volts, required for a square well to contain two allowed energy levels, if the width of the well is $2a = 2$ fm and the particle's mass is $2.0 \text{ GeV}/c^2$.

3. Draw a graph for an *unbound* state in the square-well potential of Figure 3.1. Is there any restriction on the possible values of E for such a state?

4. An electron is trapped in a square potential well given by

$$V = V_0 \text{ for } |x| > 2 \text{ nm} \quad \text{and} \quad V = 0 \text{ for } |x| < 2 \text{ nm}$$

The value of k for this electron is $(23\pi/12)$ nm.
(a) What is the kinetic energy of this electron?
(b) How many *lower* energy levels are there for this electron in this well? (*Hint:* You can deduce this from the value of ka.)
(c) Use the equation $\sin ka = \pm k/\beta$ or $\cos ka = +k/\beta$, *whichever is appropriate*, to find the well depth V_0. Explain your choice of equation.

[2] L. I. Schiff, *Quantum Mechanics*, 2nd edition, 54–55, McGraw–Hill, New York, 1955.

5. An electron's ground-state kinetic energy in an 8-eV-deep square well is 2 eV. For an electron in this well,
   (a) How many bound levels are there?
   (b) Find the energy of the highest level and the width of the well.

6. Find the probability that a ground-state electron will be found within 0.0010 nm of the wall in an infinitely deep potential well of width 0.20 nm.

7. An electron is trapped in the well given by

$$V = \infty \qquad x < 0$$
$$V = 0 \qquad 0 \leq x \leq a$$
$$V = +V_0 \qquad x > a$$

Write a general expression for the wave function, and solve for the three lowest energy levels when $V_0 = 10$ eV and $a = 5.0$ nm. How many bound levels are there? For what range of values of a are there *no* bound levels?

8. For the ground state of Example Problem 1, find the probability that the electron will be found outside the well.

9. Show from Eq. (3.59) that the expectation value of p_x is zero for a particle in a square well, and explain the result.

10. For the wave function

$$\psi = A[\cos(\pi x/a) + 1] \qquad |x| \leq a$$
$$\psi = 0 \qquad |x| > a$$

(a) Show that A must equal $(3a)^{-1/2}$ to normalize ψ.
(b) Evaluate δx and δp_x.
(c) Find the boundaries of the classically allowed region (where the kinetic energy is positive).

11. A particle of mass m has a wave function whose unnormalized space part is

$$u(x) = 0 \quad \text{for } x < a; \qquad u(x) = (x - a)e^{-x/a} \quad \text{for } x \geq a$$

(a) Show that u and du/dx are both continuous at $x = a$.
(b) Draw the graph of this function, indicating the classically allowed region for this particle.
(c) Find the probability of finding the particle between $x = a$ and $x = 2a$.
(d) Evaluate δx and δp_x.

(e) Compute the expectation value of the kinetic energy of this particle.

12. A particle of mass m is in the potential well given by

$$V = A\,\delta(x) \quad \text{for } |x| < a; \qquad V = \infty \quad \text{for } |x| > a$$

where $\delta(x)$ is the Dirac delta function and A is a positive constant.
(a) What are the dimensions of the constant A?
(b) Show by a graph that the state with lowest energy must have a wavelength greater than $2a$. *Hint:* The slope of the wave function must be discontinuous at $x = 0$ where V is infinite. Does the slope increase or decrease at that point?
(c) Show that the function

$$u(x) = \sin(kx + ka) \quad \text{for } x \le a; \qquad u(x) = \sin(-kx + ka) \quad \text{for } x \ge a$$

satisfies the boundary conditions and the Schrödinger equation for this potential.
(d) Equate the change in slope of $u(x)$ at $x = 0$ to the integral of the second derivative of u in the interval between $-\varepsilon$ and $+\varepsilon$, and use this equation to derive the relation $ka = (-maA/\hbar^2)\tan ka$.
(e) Show that for the lowest energy level, $\pi/2 < ka < \pi$. Find the energy of this level in terms of the mass m and the values of a and A.

13. A particle of mass m is trapped by the potential $V(x) = -b\,\delta(x)$; b is a positive constant and $\delta(x)$ is the Dirac delta function. Follow the procedure outlined in Exercise 12(d) to find the ground-state energy. *For Exercises 14–17, use the computer program QMVGA.*

14. An electron is in the potential well

$$V = 0 \quad \text{for } x < -1 \text{ nm}; \qquad V = -2 \text{ eV} \quad \text{for } -1 \text{ nm} < x < 1 \text{ nm};$$
$$V = 0 \quad \text{for } x > 1 \text{ nm}$$

(a) Using the method of Example Problem 1, determine how many bound levels there should be in this well.
(b) Use Eqs. (3.38) and (3.39), and successive approximations, to find the energy of each bound level to four significant figures. Verify your answers by finding the wave functions that satisfy the boundary conditions, using the QMVGA program (with "V" equal to -2 eV, "WS" equal to 2 nm, and "W" equal to 6 nm). Use PRINT SCREEN to print out the wave functions that you have found. (Printing the screen may not work unless you have first executed the MS-DOS "graphics" command.)

(c) Determine which, if any, of these levels would also be allowed in a well with $V = \infty$ for $x < 0$; $V = -2$ eV for $0 < x < 1$ nm; $V = 0$ for $x > 1$ nm

15. Select a single square well of depth 4 eV, step width WS = 1 nm, and overall screen width W = 3 nm. Determine how many bound energy levels $(E < 0)$ there are, and find the energy of each level to four significant figures. Compare with the analytical solution for these levels. Does it matter that the computed wave function goes to zero at $x = 0$ and at $x = 3$ instead of at infinity?

16. Determine the three lowest *positive* energy levels for the 3-nm-wide well. (*Hint:* Find the longest possible wavelength for each part of the well.)

17. Draw wave functions of two *unbound* electron states for the potential of Figure 3.1. Choose W = 3 nm and WS = 1 nm with $V < 0$ and $E > 0$. Use PRINT SCREEN to print out these functions. Then
(a) Compute the kinetic energy of an electron whose wavelength is exactly 1 nm.
(b) Guided by your answer to part (a), find values of E and V such that the electron's wavelength in the potential step region will be $1/2$ or $1/3$ of that in the regions outside the step, and print out both graphs.
(c) Explain the difference between the wave amplitude in the step region and the amplitude outside the step region in terms of the correspondence principle.
(d) Find values of E and V such that the wavelength in the potential step region will be 2 or 3 times that in the regions outside the step. Can you do this with a negative value of V?

18. Show that the function $\sin kx$ is not an eigenfunction of the operator p_x, even though this function has a definite wavelength equal to $2\pi/k$. What are the possible results of a measurement of p_x for a particle whose wave function is $\sin kx$? How is this related to the answer to Exercise 9?

19. A particle of mass m has a one-dimensional wave function whose space part is $u(x) = N \sin(bx)\cos(bx)$, where $b = 2$ nm^{-1} and $mc^2 = 4.0 \times 10^4$ eV.
(a) By use of the appropriate operators determine whether $u(x)$ is an eigenfunction of either the momentum p_x or the energy E.
(b) This particle is confined between two walls at which $V = \infty$. Find the smallest possible separation a between these walls that will satisfy the boundary conditions with the function $u(x)$.
(c) For the value of a found in part (b), find the value of N.

CHAPTER 4

Further Analysis of One-Dimensional Bound Systems

In this chapter we consolidate our knowledge by applying the one-dimensional Schrödinger equation to a variety of potential-energy functions, starting with multiple square wells. We will see that even our extremely simplified examples of time-independent wave functions in one dimension can help us to understand many important phenomena. Not all of these examples are so simple that we can write exact solutions for them, but we shall see that numerical solutions plotted by computer are highly accurate and revealing.

We shall also see that it is not always necessary to determine the wave function in order to find the energy levels. The simple harmonic oscillator potential illustrates the use of a powerful technique, based on quantum-mechanical operators, to determine all of the energy levels. This technique also provides a way to generate a whole set of wave functions from a single function, without explicitly solving the Schrödinger equation for each one.

4.1 MULTIPLE SQUARE WELLS

In many practical situations the potential energy has a number of minima. The general character of the solutions of the Schrödinger equation for such systems can be explored by solving the Schrödinger equation for an artificial potential made up of a series of square wells.

4.1.1 The Double Well

We begin with a double well, an infinitely deep square well with a "bump" in the middle (Figure 4.1), whose potential energy V is given by

$$
\begin{array}{ll}
V = \infty & |x| > b \\
V = 0 & a \le |x| \le b \\
V = +V_0 & |x| < a
\end{array}
$$

Let us find the allowed energies E and the corresponding wave functions for $E < V_0$. From the results for the single square well (Section 3.2), we know that the solution $u(x)$ in the classically allowed regions $(a \le |x| \le b)$ can be written as a sinusoidal function. The general form of such a function can be written as follows:

$$
u(x) = -A \sin(kx + \phi) \quad \text{for } a \le x \le b \tag{4.1}
$$

where $k = \sqrt{2mE}/\hbar$ as defined in Section 3.2, the phase angle ϕ is determined by the boundary conditions, and the minus sign is chosen to make the function positive as x approaches $+b$.

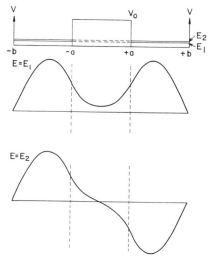

FIGURE 4.1 A symmetric double well and the wave functions for the two lowest energies in this well.

We also found for the single square well that the solution in the classically forbidden regions (in this case the central region where $|x| < a$ and $V = V_0$) is an exponential: either $e^{-\alpha x}$ or $e^{\alpha x}$ or a linear combination of these, where $\alpha = \sqrt{2m(V_0 - E)}/\hbar$, as in Section 3.2. Because of the symmetry of the well, the linear combination must be either even or odd. Two combinations are therefore possible:

$$u(x) = B \cosh \alpha x \quad \text{or} \quad u(x) = B \sinh \alpha x \quad \text{for } |x| \le a \qquad (4.2)$$

These are the even and odd combinations of $e^{\alpha x}$ and $e^{-\alpha x}$, respectively.

The boundary conditions at $x = a$, plus the condition that $u = 0$ at $x = b$, are now sufficient to determine the wave function. Two equations suffice to eliminate two of the three constants A, B, and ϕ, and the third equation determines the allowed values of k. The third of the three constants is determined by normalization of the wave function.

The two lowest-energy solutions are shown in Figure 4.1. The solution of higher energy has one node and negative parity, as in a single well.

Example Problem 4.1   Determine the two lowest allowed energies of an electron in the potential of Figure 4.1, if $a = 0.50$ nm, $b = 1.50$ nm, and $V_0 = 1.60$ eV.

Solution. We know that $u(x) = 0$ at $x = b$. Thus Eq. (4.1) gives

$$A \sin(kb + \phi) = 0 \quad \text{or} \quad \phi = -kb$$

From the boundary conditions at $x = a$ we also have, from the fact that $b = 3a$,

continuity of $u(x)$: $B \cosh \alpha a = -A \sin(ka + \phi) = -A \sin(-2ka)$

$$\qquad (4.3)$$

continuity of du/dx: $B\alpha \sinh \alpha a = -Ak \cos(ka + \phi)$

$$= -Ak \cos(-2ka) \qquad (4.4)$$

Dividing (4.4) by (4.3) yields

$$\alpha \tanh \alpha a = -k \cot(2ka) \qquad (4.5)$$

$$k^2 + \alpha^2 = 2mV_0/\hbar^2 = 42.0 \text{ nm}^{-2} \qquad (4.6)$$

From the definitions of k and α, from Figure 4.1, with $b - a = 2a$, we deduce[1] that $\pi/2 < 2ka < \pi$ for these two levels. Trial and error with Eqs. (4.5) and (4.6) and a calculator then shows that $2k_1a \cong 2.708$, and

[1] We see from Figure 4.1 that $\lambda/4 < (b - a) < \lambda/2$. The wavelength is $\lambda = 2\pi/k$, making $\lambda/4 = \pi/2k$. Thus, since $b - a = 2a$, we have $\pi/2k < 2a < \pi/k$.

$E_1 = \hbar^2 k_1^2 a^2 / 2ma^2 = 2.708^2 \hbar^2 / 8ma^2 = 0.2794$ eV. You can use the computer program QMVGA to verify this result.

For the second energy level, the wave function has odd parity. Therefore we use the odd function sinh αx in Eq. (4.2). Then, instead of Eq. (4.5), we have

$$\alpha \coth \alpha a = -k \cot(2ka) \qquad (4.7)$$

By trial and error we find that $2k_2 a = 2.712$ and $E_2 = 0.2802$ eV.

For large values of αa, both coth αa and sinh αa approach 1. Thus the two lowest levels merge into a single level. This is to be expected on physical grounds; as $\alpha a \to \infty$, the wave function in the forbidden region approaches zero. The even and odd wave functions then differ only in parity, which does not affect the wavelength in the allowed region.

4.1.2 Application to Condensed-Matter Physics

For the double well we have seen that there can be two energy levels that have very nearly the same energy. You can show (see Exercise 2) that the third level lies considerably higher than the first two. In general, for a series of n wells (as in Figure 4.2), the lowest n levels are closely spaced, relative to the difference between the nth level and the $n + 1$ level.

Using QMVGA, you can generate the wave function for any of the lowest 14 levels in the seven-well case. Three are shown in Figure 4.2. Notice that the wave function for $n = 8$ differs radically from the wave function for $n = 7$. Each node for $n = 7$ appears in a forbidden region, where the potential energy is high; each node for $n = 8$ is in a classically allowed region, where the potential energy is small. Thus the average potential energy is much higher for $n = 8$ than for $n = 7$, and an *energy gap* appears between these two levels.

This sort of energy gap occurs naturally in a regular array of potential wells. In a pure solid it accounts for the difference between an electrical conductor and an insulator. In a system containing many electrons, no two electrons can occupy the same quantum state, according to the *Pauli exclusion principle* (to be discussed in detail in Chapter 11). If the number of electrons is less than the number of states with energy below the gap, it is easy for the electrons to move to states of higher energy when a potential difference is applied, and the material is a good conductor. But if these numbers are equal, or nearly equal, the material is a poor conductor, because a large amount of energy is required to move an electron across the energy gap to a higher energy level.[2]

---

[2] The *Kronig–Penney* model approximates a crystal as a series of square potential wells, from which one can deduce various properties of metals.

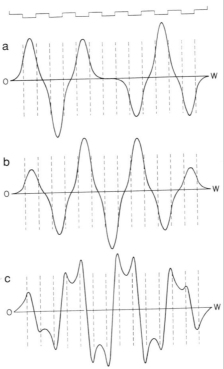

FIGURE 4.2 Wave functions for (a) quantum number $n = 6$, (b) $n = 7$, (c) $n = 8$, for an electron in a set of seven equally spaced potential wells of depth 4.0 eV. The respective energies are: $E_a = -3.0172$ eV, $E_b = -3.0119$ eV, and $E_c = -0.7816$ eV. Notice the asymptotic decay of each function in the forbidden regions.

4.2 THE SIMPLE HARMONIC OSCILLATOR AND STEPPING OPERATORS

The energy levels of the one-dimensional harmonic oscillator are found from the time-independent Schrödinger equation. In this equation the potential energy is given by the harmonic oscillator potential $Kx^2/2$, where K is the spring constant. We also know that the angular frequency ω is given classically by

$$\omega = \sqrt{K/m} \tag{4.8}$$

The Schrödinger equation is thus

$$\left\{ -\frac{\hbar^2}{2m}\frac{d^2}{dx^2} + \frac{1}{2}Kx^2 \right\} u = Eu \tag{4.9}$$

Computer-drawn solutions to this equation are shown in Figure 4.3. Notice the characteristic features: point of inflection where $E = V$, oscillation where $E > V$, asymptotic decay to zero where $E < V$, the lowest energy level lying above the minimum in the potential well.

The computer program QMVGA, used to draw Figure 4.3, can also be used to find any allowed energy level and determine how the energy of a level depends on the mass m and spring constant K.

4.2.1 Finding a Formula for the Energy Levels

A formal technique for solving an equation like (4.9) begins with the assumption that the wave function can be represented by a power series in x with undetermined coefficients, possibly multiplied by an exponential function. The coefficients are found by substituting this series into the Schrödinger equation, and this substitution also yields the allowed energy levels. But here we use another powerful method that provides a good example of the use of operators in quantum mechanics.

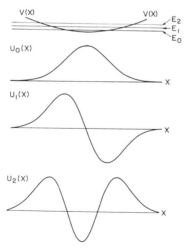

FIGURE 4.3   Harmonic oscillator potential and the wave functions for three allowed energy levels: E_0, E_1, and E_2.

We begin with substitutions that simplify the appearance of Eq. (4.9). Let

$$b \equiv 2mE/\hbar^2 \quad \text{and} \quad a^2 \equiv mK/\hbar^2 \tag{4.10}$$

The resulting equation is

$$\left\{ \frac{d^2}{dx^2} - a^2x^2 \right\} u = -bu \tag{4.11}$$

The form of this operator suggests that we try "factoring" it. Let us compute the result of the operation

$$(d/dx - ax)(d/dx + ax)u \tag{4.12}$$

Expansion of this expression gives us four terms, as follows:

$$\frac{d^2u}{dx^2} - ax\frac{du}{dx} + \frac{d}{dx}(axu) - a^2x^2u$$

and expanding the third term gives

$$\frac{d}{dx}(axu) = ax\frac{du}{dx} + au$$

with the result that the operator (4.12) becomes

$$\left(\frac{d}{dx} - ax \right)\left(\frac{d}{dx} + ax \right)u = \frac{d^2u}{dx^2} - ax\frac{du}{dx} + \frac{d}{dx}(axu) - a^2x^2u$$

$$= \frac{d^2u}{dx^2} - ax\frac{du}{dx} + ax\frac{du}{dx} + au - a^2x^2u$$

$$= \frac{d^2u}{dx^2} + au - a^2x^2u \tag{4.13}$$

or

$$(d/dx - ax)(d/dx + ax)u = (d^2/dx^2 - a^2x^2)u + au$$

and finally

$$(d/dx - ax)(d/dx + ax)u = -bu + au \tag{4.14}$$

when we use Eq. (4.11) to substitute for the first expression on the right.

The usefulness of Eq. (4.14) becomes clear when we operate on both sides with the operator $(d/dx + ax)$ and continue as follows:

$$(d/dx + ax)(d/dx - ax)(d/dx + ax)u = (-b + a)(d/dx + ax)u \quad (4.15)$$

This can be simplified by defining the function $u' \equiv (d/dx + ax)u$ to obtain

$$(d/dx + ax)(d/dx - ax)u' = (-b + a)u' \quad (4.16)$$

Expanding the left-hand side of (4.16) as we did with Eq. (4.11) gives us

$$\frac{d^2u'}{dx^2} + ax\frac{du'}{dx} - \frac{d}{dx}(axu') - a^2x^2u' = (-b + a)u' \quad (4.17)$$

or the equivalent equation (compare with Eq. (4.11)),

$$\left\{\frac{d^2}{dx^2} - a^2x^2\right\}u' = -(b - 2a)u' \quad (4.18)$$

We are back where we started from, with the harmonic oscillator equation, but now the eigenvalue is $-(b - 2a)$ instead of $-b$, and the eigenfunction is u' instead of u. By definition, $b = 2mE/\hbar^2$, and from Eq. (4.8) we also know that $a = m\omega/\hbar$. Thus we have

$$b - 2a = 2mE/\hbar^2 - 2m\omega/\hbar \quad (4.19)$$

Since $b - 2a$ has replaced b in the harmonic oscillator equation, we should set $b - 2a$ equal to $2mE'/\hbar^2$, where E' is the energy for this harmonic oscillator when the wave function is u'. Therefore, $2mE'/\hbar^2 = 2mE/\hbar^2 - 2m\omega/\hbar$, and multiplying both sides by $\hbar^2/2m$ yields

$$\boxed{E' = E - \hbar\omega} \quad (4.20)$$

The conclusion is: If $u(x)$ is an eigenfunction of the energy operator for the simple harmonic oscillator, then the function $u' = (d/dx + ax)u$ is also an eigenfunction of the same operator, with an energy eigenvalue that is smaller by $\hbar\omega$.

The operator $d/dx + ax$ is called a *lowering operator*. If we know one eigenfunction we can, by repeated application of the lowering operator, generate a series of eigenfunctions with eigenvalues equally spaced in energy. This process cannot go on indefinitely, because this series *must have a lowest level*, whose energy must be smaller than $\hbar\omega$. (For any energy larger than $\hbar\omega$, we can use the lowering operator to generate a lower level.) Furthermore, no level has $E < 0$, because $V \geq 0$ everywhere; you cannot observe a particle for which the kinetic energy $(E - V)$ is negative *everywhere*.

4.2.2 Finding the Lowest Energy Level

Let the eigenfunction for the lowest level be u_0 and the energy of this level be E_0. What happens if we apply the lowering operator to u_0? Since there is no lower level, the lowering operator must produce an eigenfunction that does not exist. Therefore it can only produce zero, and we write

$$(d/dx + ax)u_0 = 0 \qquad (4.21)$$

It is easy to separate the variables, obtaining $du_0/u_0 = -ax\,dx$; the solution is

$$\ln u_0 = -ax^2/2 + \text{constant} \qquad (4.22)$$

or

$$u_0 = Ae^{-ax^2/2} \qquad (4.23)$$

We may now find the lowest energy level by inserting u_0 into Eq. (4.11). Carrying out the differentiation gives us

$$-aAe^{-ax^2/2} + a^2x^2Ae^{-ax^2/2} - a^2x^2Ae^{-ax^2/2} = -bAe^{-ax^2/2} \qquad (4.24)$$

which reduces to $a = b$, or

$$m\omega/\hbar = 2mE_0/\hbar^2 \text{ with the final result that} \qquad (4.25)$$

$$\boxed{E_0 = \hbar\omega/2} \qquad (4.26)$$

We have found the lowest energy level, and we have also verified that Eq. (4.21) gives the eigenfunction u_0 for this level, because we have found an eigenvalue by substituting u_0 into Eq. (4.11).

Knowing u_0, we can find the other eigenfunctions by applying the *raising operator* $(d/dx - ax)$ to both sides of Eq. (4.16). The result reduces to

$$(d^2/dx^2 - a^2x^2)u'' = -bu'' \tag{4.27}$$

where u'', defined as $(d/dx - ax)u'$, is an eigenfunction of the original equation (4.11), with the *original eigenvalue* $-b$. Thus we have raised the energy eigenvalue up again by an amount $\hbar\omega$, back to its original value. We can use this raising operator repeatedly, starting with the function u_0, to work our way up through the whole series of eigenfunctions and generate the eigenfunction for any energy level that is possible.

You may wonder if there are additional levels that are missed by this procedure. There are not, as we show by the following reasoning:

- If one such level existed, we could use raising and lowering operators to generate a set of them, including one whose eigenvalue is less than $\hbar\omega$.
- The eigenfunction with the lowest energy must be a solution to Eq. (4.21). But Eq. (4.21) has only one solution, and that solution is u_0 [Eq. (4.23)].
- Therefore every eigenfunction in that set can be generated by the same operators that generated the original set, making that set identical to the set we have already found.

In summary, the energy levels of a harmonic oscillator in one dimension are

$$E_n = n\hbar\omega + \hbar\omega/2 \qquad (n = 0, 1, 2, \ldots) \tag{4.28}$$

The corresponding eigenfunctions $u_n(x)$ are given by the raising operator as

$$u_n = A_n(d/dx - ax)^n e^{-ax^2/2} \tag{4.29}$$

where A_n is a normalizing factor.

The normalized eigenfunctions for the three lowest levels are

$$u_0(x) = (a/\pi)^{1/4} e^{-ax^2/2} \tag{4.30a}$$

$$u_1(x) = (4a^3/\pi)^{1/4} x e^{-ax^2/2} \tag{4.30b}$$

$$u_2(x) = (a/4\pi)^{1/4} (2ax^2 - 1) e^{-ax^2/2} \tag{4.30c}$$

In each case the complete wave function $\psi_n(x, t)$ must include the time-dependent factor $e^{-iE_nt/\hbar}$. In general, the function $u_n(x)$ is the product of a polynomial of the form $a_n x^n + a_{n-2} x^{n-2} + \cdots$, multiplied by the factor $e^{-ax^2/2}$. These polynomials are called *Hermite polynomials*.[3]

4.2.3 General Considerations

The Hermite polynomials illustrate three general concepts that have been discussed earlier:

- *Parity*. The wave function in a symmetric potential must have either even or odd parity. Thus each polynomial consists exclusively of even powers of x for even parity, or odd powers of x for odd parity.
- *The Number of nodes*. In a one-dimensional wave function this equals the number of wave functions of lower energy. Thus in the functions of Eqs. (4.30), u_0 has no node, u_1 has a node at the origin, and u_2 has two nodes. The number of nodes equals the highest power of x in the polynomial.
- *The Correspondence principle*. This principle (Section 1.5) is clearly evident in the harmonic oscillator wave functions. If you a draw graphs of the functions (as you can do with the QMVGA program), you see that the functions for $n > 0$ are larger in amplitude where the kinetic energy is small. This feature is required by the correspondence principle (Section 1.5), because a classical particle must have a greater probability of being observed where it is moving more slowly and thus spends more time.

The correspondence principle plays another, less obvious role in the harmonic oscillator problem. Without doing any calculations, we might have deduced from this principle that the spacing between the energy levels should be $h\nu$ (or $\hbar\omega$) where ν is the classical oscillator frequency. Classically, a charged particle oscillating sinusoidally with frequency ν can only radiate electromagnetic waves of the same frequency, so it must only emit *photons* of that frequency if quantum theory is to be consistent with classical theory. Thus the energy difference between the allowed levels must be $h\nu$ (or $\hbar\omega$), just the value we have found.

4.2.4 Selection Rules

As pointed out in Section 1.5, in the hydrogen atom the quantum number n can change by more than one unit when the classical orbits are not

---

[3]E. Merzbacher [*Quantum Mechanics*, Wiley, New York (1961), Chapter 5] gives a detailed discussion of the harmonic oscillator, obtaining general expressions for these polynomials.

circular. If that happened for the harmonic oscillator, the correspondence principle would require that a classical harmonic oscillator could emit radiation of a higher frequency than its frequency of oscillation. That would be possible only if there were higher frequencies in the motion itself, i.e., if there were harmonics of the fundamental frequency.

In the hydrogen atom that is indeed the case. When the orbit is elliptical, the motion is not sinusoidal; it contains higher harmonics of the fundamental frequency of the motion. In the case of the hydrogen atom, only certain transitions are allowed; as you might guess, these are the ones that do not conflict with the correspondence principle. The *selection rules* that determine which transitions are allowed can of course be worked out from the principles of quantum mechanics, as we shall see in Chapter 14.

The observation that the photons emitted from a harmonic oscillator all have the same frequency tells us that a selection rule must be at work in this case as well. In Chapter 14 we shall see how this rule follows from the time dependence of the wave functions and the general rules governing transitions from one quantum state to another.

4.3 THE QUANTUM BOUNCER

The quantum bouncer is a particle in a one-dimensional linear potential energy field given by

$$V = kz \quad \text{for } z > 0; \qquad V = \infty \quad \text{for } z \leq 0 \qquad (4.31)$$

An example is a ball bouncing up and down under the influence of gravity. Its potential energy V near the earth's surface is proportional to its altitude z, and $k = mg$ in this case. If the energy of the ball is conserved, it continues to bounce repeatedly between $z = 0$ (ground level) and some maximum z, just as an electron in a well bounces between the sides of the well.

Figure 4.4 shows a possible wave function for an electron moving under the influence of this potential, for the case $k = 3.000$ eV/nm. For a

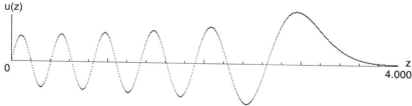

FIGURE 4.4 Wave function $u(x)$ for an electron in the potential $V = kz$, with $k = 3.000$ eV/nm, for $z > 0$.

given total energy E, the classically allowed region extends from $z = 0$ to $z = z_{max}$, the point of inflection, where $V = E$. Thus $z_{max} = E/k$. Notice these features:

- In the classically allowed region, the wave function oscillates with a wavelength that increases with increasing z, as kinetic energy decreases.
- As wavelength increases, amplitude also increases, because the probability density must increase as the particle's kinetic energy decreases.
- The energy eigenvalue can be deduced from the point of inflection, because $V = kz$, and at the point of inflection $E = V$. In Figure 4.4, $k = 3.0$ eV/nm and $E \cong 9.58$ eV; we can see that $z = E/k \cong 3.19$ nm is the inflection point.
- As always, the number of nodes equals the number of *lower* energy levels. The level shown in Figure 4.4, with 10 nodes, is the eleventh lowest ($n = 11$).

Example Problem 4.2   Find the approximate energy of the $n = 10$ level for the electron in Figure 4.4, simply by observing where the first node is for $n = 11$.

Solution. The first node for $n = 11$ is at $z \cong 0.20$ nm, where $V = 0.60$ eV. From the node at $z = 0.20$ to the inflection point at $z \cong 3.19$ nm is a distance of about 2.99 nm. Let us now shift the origin, letting

$$V = k(z - 0.20) \quad \text{for } z > 0.20; \qquad V = \infty \quad \text{for } z \leq 0.20 \quad (4.32)$$

This shift does not change the slope of V, but it makes V equal to infinity at $z = 0.20$ nm. Consequently, the wave function is cut off at that point, but its point of inflection, where $V = E$, is unchanged, remaining at $z = 3.19$ nm. From Eq. (4.32) we find that $V = 8.97$ eV when $z = 3.19$. We conclude that $E \cong 8.97$ for this function, which now has only nine nodes and thus has $n = 10$.

If you run QMVGA with $k = 3$ eV/nm to find the level for $n = 10$, you will find that the energy level is 8.98 eV, comfortably within the margin of error in estimating the location of the first node.

Figure 4.4 was produced by numerical integration of the Schrödinger equation, as were our previous graphs of wave functions. The program QMVGA does this by starting with $u = 0$ and an arbitrary value of du/dx (or du/dz in this case) at the origin, using the Schrödinger equation to find the value of d^2u/dx^2, then proceeding as explained in Appendix F.

4.3.1 Formula for the Energy Levels

The energy depends on the particle mass m and the constant k. If we know a single energy level, we can use a formula to find all the other levels for given values of k and m. The derivation of this formula on the basis of the correspondence principle is instructive.[4]

In the classical limit the levels are closely spaced, and we can treat the number n as a continuous variable. The difference in energy δE between one level and another may then be written

$$\delta E = (dE/dn)\, \delta n \qquad (4.33)$$

a relation that is accurate when δn is sufficiently small so that dE/dn is constant between the two energy levels. In the classical limit, when n is extremely large, $\delta n = \pm 1$ is indeed sufficiently small. In that case $-\delta E$ is the energy of the photon emitted when a transition occurs between one level and the next lower level. The frequency ν of this photon must equal the fundamental frequency of the motion, because no lower frequency can be emitted. Thus

$$-\delta E = dE/dn = E_{\text{photon}} = h\nu \quad \text{if } \delta n = -1 \qquad (4.34)$$

We can find the bouncer's frequency from freshman mechanics. Because the acceleration is constant, the average speed is half the maximum speed $v_{\max}$ (the speed at $z = 0$). The period of oscillation, $1/\nu$, is twice the time required to fall from the maximum height $z = z_m$ to $z = 0$, at an average speed of $v_{\max}/2$. Thus

$$1/\nu = 2 z_{\max}/(v_{\max}/2) = 4 z_{\max}/v_{\max} \quad \text{or} \quad \nu = v_{\max}/4 z_{\max} \quad (4.35)$$

Since the potential energy $V = 0$ when $z = 0$ and $v = v_{\max}$, the total energy E equals the kinetic energy at that point:

$$E = m v_{\max}^2/2 \qquad (4.36)$$

E also equals the potential energy at $z = z_{\max}$:

$$E = k z_{\max} \qquad (4.37)$$

---

[4] This derivation is based upon one given by Frank S. Crawford, Applications of Bohr's Correspondence Principle, *Am. J. Phys.* **57**, 621 (1989).

We can now eliminate v_{max} and z_{max} by taking the square root of both sides of (4.36) and dividing by (4.37) as follows:

$$E^{1/2} = v_{max}\sqrt{m/2} \quad \text{so that} \quad E^{1/2}/E = \frac{v_{max}\sqrt{m/2}}{kz_{max}} \tag{4.38}$$

or, using Eq. (4.35) for the frequency v,

$$E^{-1/2} = \frac{4\sqrt{m/2}}{k} v \tag{4.39}$$

and thus

$$hv = hkE^{-1/2}/\sqrt{8m} \tag{4.40}$$

Now we use Eq. (4.34):

$$hv = dE/dn = hkE^{-1/2}/\sqrt{8m} \tag{4.41}$$

and separate the variables to obtain

$$E^{1/2}\,dE = (hk/\sqrt{8m})\,dn \tag{4.42}$$

or, to simplify the dimensions,

$$E^{1/2}\,dE = \left(hck/\sqrt{8mc^2}\right) \tag{4.43}$$

Integrating both sides gives

$$2E^{3/2}/3 = \left(hck/\sqrt{8mc^2}\right)(n + n_0) \tag{4.44}$$

where n_0 is a constant of integration. Raising both sides to the 2/3 power gives (at last)

$$E = (9h^2c^2k^2/32mc^2)^{1/3}(n + n_0)^{2/3} \tag{4.45}$$

If we now determine the constant n_0, Eq. (4.45) can give us all of the energy levels. The solution of the Schrödinger equation for one energy will suffice for this. A formal mathematical solution is somewhat tedious, involving an infinite series known as an Airy function; however, we have already found solutions by computer, using the approximation that produced Figure 4.4.

Let us use a solution, for $k = 3$ eV/nm and $m = m_e$, to calculate n_0. From Eq. (4.44), with $n = 1$, $mc^2 = 0.511003$ MeV, and $E = 1.63611$ eV, we have

$$1 + n_0 = 2\sqrt{8mc^2} \, (1.63611 \text{ eV})^{3/2}/3hck \qquad (4.46)$$

$$= 2\sqrt{4.0880 \text{ eV}} \, (1.63611 \text{ eV})^{3/2}/(3 \times 1239.85 \text{ eV-nm} \times 3 \text{ eV/nm}) \qquad (4.47)$$

The units all cancel and we have the dimensionless number $n_0 = -0.2416$.

Example Problem 4.3 Test this result by computing the value of E_2 for $k = 3$.

Solution. From Eq. (4.45) we can easily find the ratio E_2/E_1; the factor $(9h^2c^2k^2/32mc^2)^{1/3}$ cancels out and we have

$$E_2/E_1 = [(2 + n_0)/(1 + n_0)]^{2/3} = (1.7584/0.7584)^{2/3} = 1.7518$$

so that

$$E_2 = 1.7518 \times 1.6361 \text{ eV} = 2.866 \text{ eV}$$

As a check we use QMVGA and find that $E_2 = 2.861$ eV. The slight difference is within the expected uncertainty for the approximation of QMVGA, which plots each curve as a series of straight segments. The curvatures are different for $n = 2$ and $n = 1$; thus the results are inconsistent with each other.

SUMMARY

We see that many techniques can be used to find wave functions and energy levels for quantum systems without directly solving the Schrödinger equation for each case. The operator method and numerical analysis used here are of broad applicability. We will use these methods to solve three-dimensional problems; they are particularly powerful in dealing with angular momentum. The operator method, essential for treating spin angular momentum and for dealing with elementary-particle interactions, will be developed further in Chapter 7.

ADDITIONAL READING

John D. McGervey, *Introduction to Modern Physics*, 2nd edition, Academic Press, New York (1983). Chapter 12 has details on the Kronig–Penney potential and its connection to the theory of metals and semiconductors.

E. Merzbacher, *Quantum Mechanics*, Wiley, New York (1961), gives many instructive details of the solutions for the potentials discussed here.

EXERCISES

1. Using the method of Example Problem 4.1, compute the lowest energy level for the double well defined by $a = 2.0$ nm, $b = 4.0$ nm, and $V_0 = 3\hbar^2/2ma^2$, and show that there is only one solution for which $E < V_0$.

2. For the double well given in Example Problem 1,
   (a) Verify that the values of k_1, k_2, E_1, and E_2 are computed correctly, by substitution into Eq. (4.5), using Eq. (4.6) to find α_1 and α_2.
   (b) For the third level, the wave function must have two nodes. From this fact, show that its wavelength must be shorter than $2a$ and that therefore $\pi < k_3 a < 3\pi/2$. Compute k_3 and E_3, and compare the three energies. Notice the much larger "gap" between E_3 and E_2, compared to the gap between E_2 and E_1.
   (c) Use the program QMVGA to verify that your values give acceptable wave functions. Compare with Figure 4.1.

3. (a) Generate wave functions for the two lowest energy levels in a double well like the one shown in Figure 4.1, using QMVGA with $N = 1$, $W = 3$ nm, $WS = 1$ nm, and $V = +1$ eV. Print out the result, including the energy level for each function. *Hint:* Each energy should be close to that of the lowest level in a single well of depth 1 eV and width 1 nm. That energy can be found by using the method of Sec. (3.2.2). Can you explain?
   (b) Determine the next two energy levels by generating the wave functions as in part (a). *Hint:* Begin with a trial energy level based on the approximation that the kinetic energy in each of these levels should be about four times the kinetic energy of each of the lowest two levels.
   (c) Repeat (a) and (b) with $V = +2$ eV. Do you see any change in the "splitting" between the first and second levels and between the third and fourth levels?

4. The potential

$$V(x) = \frac{\hbar^2}{2m}\left\{ \frac{4\sinh^2 x}{225} - \frac{2\cosh x}{5} \right\}$$

is a double well, having two minima.
(a) Draw a graph of $V(x)$ and determine the value of x at each minimum of V.
(b) Show that one solution of the Schrödinger equation for this potential is

$$u(x) = (1 + 4\cosh x)\exp\left\{ -\frac{2\cosh x}{15} \right\}$$

and show this level on your graph of $V(x)$.
(c) Draw a graph of $u(x)$. Show that it has the proper behavior regarding classical turning points and classically forbidden regions.

5. For the lowest energy level of a harmonic oscillator with spring constant K, find the expectation values of the potential energy $Kx^2/2$ and the kinetic energy $p_x^2/2m$. Compare with the total energy of this state.

6. (a) Verify that applying the raising operator to the harmonic oscillator eigenfunction $u_0(x)$ produces the function $u_1(x)$, and that applying this raising operator to $u_1(x)$ produces the function $u_2(x)$ [Eqs. (4.30)].
(b) Verify that both $u_1(x)$ and $u_2(x)$ satisfy Eq. (4.9), with the correct eigenvalue for each.

7. (a) Show that the functions of Eqs. (4.30) are correctly normalized.
(b) Apply the raising operator to the function u_2 of Eq. (4.30) to generate the next function, $u_3(x)$. Determine the normalizing factor. Is this function even or odd? Does it have the correct number of nodes?
(c) Show that $u_3(x)$ satisfies Eq. (4.09), and find the energy eigenvalue.

8. Use QMVGA to generate $u(x)$ for a simple harmonic oscillator with $n = 5$ and
(a) $K = 51.1$ eV/nm^2 and $m = m_e$ (default values)
(b) $K = 102.2$ eV/nm^2, $m = m_e$
(c) $K = 51.1$ eV/nm^2, $m = 2m_e$
Show that the energy has the right dependence on K and m. Print graphs of $u(x)$ and $P(x)$ for (a), and indicate the classically allowed

region. Is the dependence of $P(x)$ on x consistent with the correspondence principle?

9. Use QMVGA to generate $u(x)$ for a simple harmonic oscillator with
the default values of K and m and various values of n. Then verify
that the energy is proportional to $n + 1/2$ and the kinetic energy,
$E - V$, is zero at the boundaries of the classically allowed region.

10. Determine the energy levels for $n = 1$ and $n = 9$ for the potential of
Figure 4.4, by the method used in the text for the $n = 10$ level.

11. Apply QMVGA to the potential of Eq. (4.31) with $k = 1$. [See Frank S.
Crawford, Applications of Bohr's Correspondence Principle, *Am. J.
Phys.* **57**, 621 (1989).]
   (a) Find an allowed energy E between 5.0 and 5.4 eV. Determine the
   quantum number n for this level, and find the next higher energy
   level. Print out both wave functions and indicate significant features of the plots, with regard to points of inflection, allowed
   region(s), and wave amplitudes.
   (b) Repeat for $k = 2$. Verify, by comparison with part (a), that the
   energy levels are proportional to $k^{2/3}$ [as required by Eq. (4.44)] to
   give significant figures. (Set z_m equal to 5.4 nm.)
   (c) Repeat for $k = 1$ and $m = 2m_e$. Again by comparison with part
   (a), verify that the energy for a given value of the quantum number
   n is proportional to $m_e^{-1/3}$, again in agreement with a calculation
   based on the correspondence principle.
   (d) Use the results of (a) to verify that the energy E_n is proportional to
   $(n + n_0)^{2/3}$.

12. For the linear potential of the previous problem, generate the wave
function for $n = 13$. Estimate the wavelength in the region near $x = 0$,
and determine whether that wavelength is consistent with the values of
the potential V and energy E for this wavefunction.

13. For the potential $V = kx$ with $k = 2$, the computer is programmed
only for the energies of levels with $n \leq 13$. Find the energy for $n = 14$
by extrapolation from the $n = 13$ level; then plot the graph and adjust
the energy until the wave function satisfies the boundary condition for
large values of z.

14. Find the lowest five energy levels for the potential $V = k|x|$. *Hint:*
Either the function u or du/dx must be zero at the origin. Using
QMVGA, you can select those wave functions by observing only the
region $x > 0$. The lowest allowed energy in this potential is not an
allowed energy in the potential $V = kx$ (with $V = \infty$ for $x \leq 0$) because
it is not zero at the origin. But you can deduce its energy from the

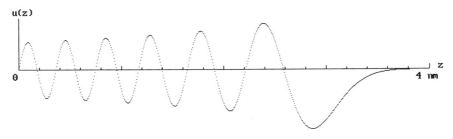

FIGURE 4.5

change in x between the last maximum and the point of inflection for any of the eigenfunctions.

15. (a) Solve Exercise 11 given that $V = \infty$ for $z \geq 6.0$. What is the change in the energy of the 13th level and the next higher level, compared with the results of Exercise 11?

   (b) Repeat (a) for the case $V = \infty$ for $z > 5.0$. Does the wave function differ qualitatively from the wave function of (a) for the same n?

16. The one-dimensional wave function for an electron in the potential energy

$$V = \infty \quad \text{for } z < 0; \qquad V = kz \quad \text{for } z > 0, \quad \text{where } k = 4.000 \text{ eV/nm}$$

   is shown above.

   (a) Find the quantum number n for this level, and justify your answer.

   (b) Mark on the graph the boundaries of the classically allowed region for the electron in this level. Give your reason for your choice.

   (c) Find the electron's energy E in eV, to two significant figures. [*Hint:* Use your answer to part (b).]

   (d) If a muon were in this potential with the same value of n, would the allowed region be narrower, wider, or the same size? Explain.

   (e) What would be the lower and upper limits on the possible results of a measurement of the magnitude of the momentum p of the electron in this state (in units of eV/c)? Show how you arrive at your answer. (*Hint:* There are two ways to arrive at the answer. One way is graphical.)

CHAPTER 5

The Free Particle as a Traveling Wave

The Schrödinger equation's impressive ability to determine energy levels of bound states is accompanied by an even more impressive record involving phenomena of traveling waves. In this category we have predictions of entirely new phenomena, such as "tunneling" through a potential barrier that would be impenetrable to a classical particle, "reflection" at barriers through which classical particles are 100% transmitted, and a multitude of interference phenomena.

5.1 THE FREE PARTICLE AND ITS BEHAVIOR AT A POTENTIAL STEP

Consider a beam of electrons traveling from left to right along the x axis under the influence of the potential energy V, where

$$V = 0 \quad \text{for } x < 0; \qquad V = -V_0 \quad \text{for } x > 0$$

Such a potential could be set up in the manner shown in Figure 5.1.

If one of these electrons has total energy E, its kinetic energy increases by the amount V_0 when it crosses the potential step at $x = 0$. Classically, all of the electrons would cross the step, gaining kinetic energy. But the wave function of the electrons does not permit all of them to continue across this step; some must be reflected, just as light is reflected when a light wave passes from air into water. The amplitude of the reflected wave is found by solving the time-dependent Schrödinger equation for all values of x.

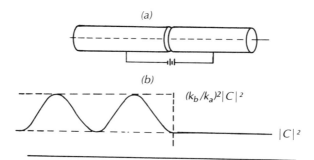

FIGURE 5.1 (a) An electron traveling along the axis of the cylinder feels a constant potential until it reaches the second cylinder, where there is a sudden drop in potential energy. (b) Behavior of the probability density $|\psi|^2$. Interference between the incident and reflected waves produces a standing wave for $x < 0$; with no interference, $|\psi|^2$ is constant for $x > 0$.

In doing this, we consider the region of negative x and the region of positive x separately, matching the functions in the two regions by means of the boundary conditions, just as we did in finding wave functions for the finite square well.

As we saw in Chapter 3, the time-dependent wave function for a particle with total energy $E = \hbar\omega$ and positive momentum $\hbar k$ is given by

$$\psi(x,t) = Ae^{ikx}e^{-iEt/\hbar} = Ae^{ikx}e^{-i\omega t} \qquad (5.1)$$

where A is the amplitude of the wave and k is defined to be positive.

For all values of x the time-dependent factor $e^{-i\omega t}$ is the same, because the total energy E is the same for all x. But the value of k depends on the momentum p, which changes as the electron crosses the potential step.

For negative x, $\qquad p = \pm\hbar k_a$, where $k_a = \sqrt{2mE}/\hbar \qquad (5.2)$

because the original wave has positive momentum and the reflected wave has negative momentum. The wave function for negative x is therefore the sum of these two waves, or

$$\psi(x,t) = Ae^{ik_a x}e^{-i\omega t} + Be^{-ik_a x}e^{-i\omega t} \quad \text{for } x < 0 \qquad (5.3)$$

where the amplitude B equals 0 when no reflection occurs.

For positive x, $p = +\hbar k_b$,   where $k_b = \sqrt{2m(E + V_0)}/\hbar$ (5.4)

because there is no reflected wave in this region, and the kinetic energy has increased by the amount of the step potential V_0. The wave function for positive x is therefore

$$\psi(x,t) = Ce^{ik_b x}e^{-i\omega t} \quad \text{for } x > 0 \tag{5.5}$$

Notice that k changes at the potential step, because the momentum changes, but the *angular frequency* ω *does not change*, because the total energy E does not change. This behavior is like that of a classical wave; when the wave speed changes at a boundary, the frequency is unchanged and the wavelength changes.

We can now use the continuity conditions at $x = 0$ to find the ratio B/A. This ratio determines the probability that an electron will be reflected at the potential step. The continuity of ψ at $x = 0$ requires that

$$A + B = C \tag{5.6}$$

and from the continuity of $\partial\psi/\partial x$ at $x = 0$ we have

$$ik_a A - ik_a B = ik_b C \tag{5.7}$$

Eliminating C, we find that

$$\frac{B}{A} = \frac{k_a - k_b}{k_a + k_b} \tag{5.8}$$

The probability R that an electron will be reflected at the potential step must equal $|B|^2/|A|^2$, because $|B|^2$ is the probability density in the reflected wave and $|A|^2$ is the probability density in the incident wave. Thus

$$R = \frac{|B|^2}{|A|^2} = \frac{|k_a - k_b|^2}{|k_a + k_b|^2} \tag{5.9}$$

Let us see if this result is plausible. If $k_a = k_b$, then there is no step, and $R = 0$, as it should. If the step is positive and greater than E, then R should equal 1, because no particle can be transmitted. This also is consistent with Eq. (5.9), because in that case k_b is imaginary; it can be written as $i\alpha$ [as in the square well; see Eq. (3.15)]. The right-hand side of

(5.9) then becomes

$$(k_a - i\alpha)(k_a - i\alpha)^* / (k_a + i\alpha)(k_a + i\alpha)^* = 1 \qquad (5.9a)$$

When k_a and k_b are real, we can write the probability T of transmission as

$$T = 1 - R = 1 - \frac{|k_a - k_b|^2}{|k_a + k_b|^2} = \frac{|4k_a k_b|}{|k_a + k_b|^2} \qquad (5.10)$$

which goes to zero as $k_b \to 0$ ($V \to \infty$ for $x > 0$), a result to be expected classically. But T also goes to zero as $k_b \to \infty$ ($V \to -\infty$ for $x > 0$), a result that has no classical explanation.

You might think that we could calculate the transmission probability T from the ratio $|C|^2/|A|^2$, because $|C|^2$ is the probability density in the transmitted wave. In fact, T is *proportional* to $|C|^2/|A|^2$, not *equal* to $|C|^2/|A|^2$. From the continuity conditions (Eqs. 5.6 and 5.7) we find that

$$\frac{|C|^2}{|A|^2} = \frac{4k_a^2}{(k_a + k_b)^2}$$

and comparison with Eq.(5.10) shows that

$$T = \frac{k_b}{k_a} \frac{|C|^2}{|A|^2} \qquad (5.11)$$

The transmission coefficient gives us the number of particles *per unit area per unit time* for the region $x > 0$, but Eq. (5.11) says that this number is the *product* of two factors: the particle density and the particle speed. The density factor is found from $|C|^2$; the speed from the momentum, or k_b. To visualize these factors, consider a stream of classical particles crossing a step (Figure 5.2). None are reflected, but the particle density changes because the speed changes; the density is inversely proportional to the speed.

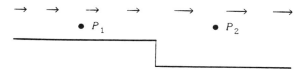

FIGURE 5.2 Classical behavior of a stream of particles crossing a potential step. After crossing the step, the particles have greater speed, and the distance between them has increased in direct proportion to their speed.

5.2 PROBABILITY CURRENT

In the preceding discussion we were led to a distinction between the probability *density*, given by the absolute square of the amplitude of the wave function, and the *intensity* of the wave, which is the product of the density and the wave velocity for any kind of wave. Let us generalize this concept and eventually extend it to three dimensions, starting from the time-dependent Schrödinger equation:

$$i\hbar \frac{\partial \psi}{\partial t} = -\frac{\hbar^2}{2m} \frac{\partial^2 \psi}{\partial x^2} + V\psi \tag{5.12}$$

and its complex conjugate:

$$-i\hbar \frac{\partial \psi^*}{\partial t} = -\frac{\hbar^2}{2m} \frac{\partial^2 \psi^*}{\partial x^2} + V\psi^* \tag{5.13}$$

We now derive a relation involving the probability density $\psi^*\psi$. We multiply both sides of Eq. (5.12) by ψ^* and both sides of Eq. (5.13) by ψ to obtain

$$i\hbar \psi^* \frac{\partial \psi}{\partial t} = -\frac{\hbar^2}{2m} \psi^* \frac{\partial^2 \psi}{\partial x^2} + V\psi\psi^* \tag{5.14}$$

$$-i\hbar \psi \frac{\partial \psi^*}{\partial t} = -\frac{\hbar^2}{2m} \psi \frac{\partial^2 \psi^*}{\partial x^2} + V\psi^*\psi \tag{5.15}$$

Subtracting (5.15) from (5.14) eliminates V and gives

$$i\hbar \left\{ \psi^* \frac{\partial \psi}{\partial t} + \psi \frac{\partial \psi^*}{\partial t} \right\} = -\frac{\hbar^2}{2m} \left\{ \psi^* \frac{\partial^2 \psi}{\partial x^2} - \psi \frac{\partial^2 \psi^*}{\partial x^2} \right\} \tag{5.16}$$

or

$$\frac{\partial}{\partial t} (\psi^*\psi) = i \left(\frac{\hbar}{2m} \right) \frac{\partial}{\partial x} \left\{ \psi^* \frac{\partial \psi}{\partial x} - \psi \frac{\partial \psi^*}{\partial x} \right\} \tag{5.17}$$

We now define the quantity

$$S_x = -i \left(\frac{\hbar}{2m} \right) \left\{ \psi^* \frac{\partial \psi}{\partial x} - \psi \frac{\partial \psi^*}{\partial x} \right\} \tag{5.18}$$

so that Eq. (5.17) becomes

$$\frac{\partial}{\partial t}(\psi^*\psi) = -\frac{\partial}{\partial x}S_x \tag{5.19}$$

or, in three dimensions,

$$\frac{\partial}{\partial t}(\psi^*\psi) = -\left\{\frac{\partial}{\partial x}S_x + \frac{\partial}{\partial y}S_y + \frac{\partial}{\partial z}S_z\right\} \equiv -\text{div }\mathbf{S} \tag{5.20}$$

where **S**, the *probability current density*, may be written

$$\mathbf{S} = -i\left(\frac{\hbar}{2m}\right)\{\psi^* \, \nabla\psi - \psi \, \nabla\psi^*\} \tag{5.21}$$

The left-hand side of Eq. (5.20) is the rate of change of probability density at a given point; the right-hand side is the net flow of probability density into that point. This equation is mathematically analogous to the continuity equation in electrodynamics or hydrodynamics:

$$\frac{\partial\rho}{\partial t} = -\text{div }\mathbf{J} \tag{5.22}$$

where ρ is the charge (or fluid) density and **J** is the current density. Equation (5.22) expresses the fact that a net outward flow of current results in a decrease of charge (or fluid); Eq. (5.20) expresses the fact that a net outward flow of probability current results in a decrease of probability.

5.2.1 Probability Current in a Plane Wave

For example, we now compute the probability current density S_x for a plane wave [Eq. (5.1)]:

$$\psi(x,t) = Ae^{ikx}e^{-i\omega t} \tag{5.1}$$

Differentiation yields $\dfrac{\partial\psi}{\partial x} = ik\psi$ and $\dfrac{\partial\psi^*}{\partial x} = -ik\psi^*$. Eq. (5.18) now gives

$$S_x = -(i\hbar/2m)(ik\psi^*\psi + ik\psi\psi^*) = (\hbar k/m)|\psi|^2 = v_x|\psi|^2 \tag{5.23}$$

Thus the probability *current* is equal to the product of the probability *density* and the velocity, a logical result.

Integrating both sides of Eq. (5.19) between the limits x_1 and x_2, we obtain

$$\frac{\partial}{\partial t} \int_{x_1}^{x_2} \psi^* \psi \, dx = - \int_{x_1}^{x_2} \frac{\partial S_x}{\partial x} \, dx = S_x(x_1) - S_x(x_2) \qquad (5.24)$$

The integral on the left is the total amount of "fluid" between $x = x_1$ and $x = x_2$. The fluid (probability) flows into the region between $x = x_1$ and $x = x_2$ at a rate equal to $S_x(x_1)$, and it flows out of this region at a rate equal to $S_x(x_2)$. The difference between these rates gives the rate at which probability builds up in this region.

Let us use these equations to gain a better understanding of Figure 5.1b. For $x > 0$ we square the wave function of Eq. (5.5) to obtain

$$|\psi|^2 = |Ce^{ik_b x} e^{-i\omega t}|^2 = |C|^2 \qquad (5.25)$$

The probability density is constant in both time and space, and the probability current is

$$S_x = v_b |C|^2 = (\hbar k_b / m) |C|^2 \qquad (5.26)$$

For $x < 0$ we have

$$|\psi|^2 = |Ae^{ik_a x} e^{-i\omega t} + Be^{-ik_a x} e^{-i\omega t}|^2 \qquad (5.27)$$

$$= |A|^2 + |B|^2 + AB^* e^{2ik_a x} + A^* B e^{-2ik_a x} \qquad (5.28)$$

Although A and B can in general be complex constants, we know that $|\psi|^2$ must be real. To show this explicitly we write

$$A = Ae^{i\phi} \quad \text{and} \quad B = -Be^{i\phi} \qquad (5.29)$$

where A, B, and ϕ are real, positive constants. The minus sign is required because B/A is negative in the present case. [See Eq. (5.8).]

With these definitions we have $AB^* = A^*B = -AB$, and Eq. (5.28) becomes

$$|\psi|^2 = A^2 + B^2 - AB(e^{2ik_a x} + e^{-2ik_a x}) \qquad (5.30)$$

$$= A^2 + B^2 - 2AB \cos 2k_a x \qquad (5.31)$$

The time dependence has disappeared, because $|e^{-i\omega t}|^2 = 1$. The x dependence is the standing wave shown in Figure 5.1b, with maxima and minima given by

$$|\psi|_{max}^2 = A^2 + B^2 + 2AB = (A + B)^2 \qquad (5.32a)$$

$$|\psi|_{min}^2 = A^2 + B^2 - 2AB = (A - B)^2 \qquad (5.32b)$$

producing the standing wave seen in Figure 5.1b. Notice these features:

- The wavelength of the standing wave, determined by the factor $\cos 2k_a x$, is π/k_1, just half the wavelength of ψ itself.
- At $x = 0$, the slope of $|\psi|^2$ is zero because this slope, like the slope of ψ itself, must be continuous. Thus $|\psi|^2$ must equal either $|\psi|_{max}^2$ or $|\psi|_{min}^2$ at that point.
- Because the value of $|\psi|^2$ must also be continuous at $x = 0$, $|C|^2$ must equal either $|\psi|_{max}^2$ or $|\psi|_{min}^2$.
- The probability current must be the same for all values of x, because there are no sources or sinks.

The last point can be used to derive a relation between the ratio $|\psi|_{max}/|\psi|_{min}$ and the ratio k_b/k_a. We begin by writing expressions for S_x for $x < 0$ and for $x > 0$. For $x < 0$, S_x is the sum of two currents, a positive current of $v_a A^2$ and a negative current of $v_b B^2$. The net current is

$$S_x = v_a(A^2 - B^2) = (\hbar k_a/m)(A + B)(A - B)$$

or, for $x < 0$

$$S_x = (\hbar k_a/m)|\psi|_{max}|\psi|_{min} \qquad (5.33)$$

while for $x > 0$

$$S_x = v_b|C|^2 = (\hbar k_b/m)|\psi|_{min}^2 \qquad (5.34)$$

Equating the two expressions gives

$$(\hbar k_a/m)|\psi|_{max}|\psi|_{min} = (\hbar k_b/m)|\psi|_{min}^2$$

which reduces to

$$k_a/k_b = |\psi|_{min}/|\psi|_{max} \qquad (5.35)$$

as shown in Figure 5.2(b).[1]

[1] Further discussion of these points is found in an excellent article by James E. Draper, *Am. J. Phys.* **47**, 525 (1979).

We shall develop these ideas further in the next section, where we will deal with the more practical case of a potential *barrier* rather than a single step.

Example Problem 5.1   Find the probability that an electron's momentum will be negative when it is observed in the region where its wave function is given by Eq. (5.3).

   Solution. The wave function is the sum of two parts: $\psi(x,t) = \psi_{\text{right}} + \psi_{\text{left}}$, where

$$\psi_{\text{right}} = Ae^{ik_a x}e^{-i\omega t} \quad \text{and} \quad \psi_{\text{left}} = Be^{-ik_a x}e^{-i\omega t}$$

Thus $|\psi_{left}|^2 = |B|^2$ and $|\psi|^2 = |A|^2 + |B|^2 + AB^* e^{2ik_a x} + A^* B e^{-2ik_a x}$. The probability that the electron's momentum is negative is proportional to $|\psi_{\text{left}}|^2$, or $|B|^2$; the total probability of observing the electron is proportional to $|\psi|^2$, which has an average value of $|A|^2 + |B|^2$ [See Eq. (5.31)]. Thus the probability that the electron *being observed* will have negative momentum is the ratio $|B|^2/(|A|^2 + |B|^2)$.

Example Problem 5.2

(a) Prove from Eq. (5.31) that $|C|^2 = |\psi|^2_{\text{min}}$ as indicated in Figure 5.1.

(b) Under what circumstances would $|C|^2$ equal $|\psi|^2_{\text{max}}$ at a potential step?

   Solution. (a) From Eq. (5.31) with $x = 0$ we have $|\psi|^2 = A^2 + B^2 - 2AB$, and this is equal to $|\psi|^2_{\text{min}}$ according to Eq. (5.32).

(b) If the potential step is *positive* rather than negative, then the speed is reduced at the step, and $k_b < k_a$. Then Eq. (5.8) tells us that B/A is positive, and B equals $+Be^{i\phi}$ rather than $-Be^{i\phi}$ [Eq. (5.24)]. Consequently, $AB^* = A^*B = AB$, and Eq. (5.31) becomes $|\psi|^2 = A^2 + B^2 + 2AB \cos 2k_a x$. When $x = 0$ we have $|\psi|^2 = |\psi|^2_{\text{max}}$, and thus $|C|^2 = |\psi|^2_{\text{max}}$.

5.3 BARRIER PENETRATION

We found in Chapter 4 that a particle can pass through a potential barrier even though its energy is less than the maximum potential energy inside the barrier region. Let us now consider a stream of particles incident from the left upon the potential barrier of Figure 5.3, and determine the probability that a particle will penetrate the barrier and emerge on the other side.

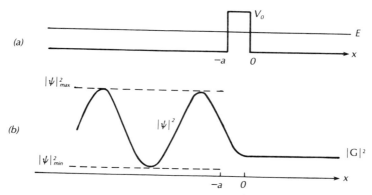

FIGURE 5.3 (a) Square potential barrier of width a and height $V_0 = \hbar^2/2ma^2$. (b) Probability density $|\psi|^2$ drawn to scale, for particles incident from the left with energy $E = V_0/2$. Notice that $|\psi|^2$ oscillates for $x < -a$, decays inside the barrier until it has zero slope at $x = 0$, and then becomes constant.

Classically, when the particle's energy E is less than V_0, this probability is zero, and when $E > V_0$ the probability is 1. But we have already seen that the classical probability is wrong when the barrier is infinitely wide and $E > V_0$. Now we shall see for the finite-width barrier of Figure 5.3 that the classical probability is also wrong when $E < 0$—that there is a nonzero probability of transmission through the barrier in that case and that this phenomenon is well verified experimentally.

We proceed as in Section 5.1, but now there are two complications:

- We must apply the continuity conditions at two boundaries instead of one.
- We have the classically forbidden region $-a < x < 0$, in which k is imaginary and is replaced by $\alpha \equiv ik$, as in the finite square well.

As usual, $\psi(x, t) = u(x)e^{-i\omega t}$, and we require that $u(x)$ and du/dx both be continuous. The expressions for $u(x)$ are

$$u = Ae^{ikx} + Be^{-ikx} \qquad (x < -a) \qquad (5.36a)$$

$$u = Ce^{\alpha x} + De^{-\alpha x} \qquad (-a < x < 0) \qquad (5.36b)$$

$$u = Ge^{ikx} \qquad (x > 0) \qquad (5.36c)$$

Notice two significant features of these expressions:

- In the forbidden region $u(x)$ must be the sum of two exponentials in order to satisfy the conditions at both boundaries.

- As before, the expression e^{ikx}, when multiplied by the time factor $e^{-i\omega t}$, represents a wave traveling in the positive x direction. In the same manner, e^{-ikx} represents a wave traveling in the negative x direction. Both waves are present for $x < -a$, because the initial wave is reflected at $x = -a$. The reflected wave is not present in the region where $x > 0$.

The continuity conditions at $x = 0$ and $x = -a$ yield four equations:

	Continuity at $x = -a$	Continuity at $x = 0$	
u	$Ae^{-ika} + Be^{ika} = Ce^{-\alpha a} + De^{\alpha a}$	$C + D = G$	(5.37)
du/dx	$ik(Ae^{-ika} - Be^{ika}) = \alpha(Ce^{-\alpha a} - De^{\alpha a})$	$\alpha(C - D) = ikG$	

We can eliminate C, D, and G to find the ratio

$$\frac{B}{A} = \left\{ \frac{(k^2 + \alpha^2)\sinh^2 \alpha a}{2ik\alpha \cosh \alpha a + (k^2 - \alpha^2)\sinh \alpha a} \right\} e^{-2ika} \qquad (5.38)$$

The reflection coefficient is therefore

$$R = \frac{BB^*}{AA^*} = \left\{ 1 + \frac{4k^2\alpha^2}{(k^2 + \alpha^2)\sinh^2 \alpha a} \right\}^{-1} \qquad (5.39)$$

To rewrite this in terms of E and V_0 we use the definitions $k^2 = 2mE/\hbar^2$ and $\alpha^2 = 2m(V_0 - E)/\hbar^2$, with the result

$$R = \left\{ 1 + \frac{4E(V_0 - E)}{V_0^2 \sinh^2 \alpha a} \right\}^{-1} \qquad (5.40)$$

The transmission coefficient is

$$T = 1 - R = \left\{ 1 + \frac{V_0^2 \sinh^2 \alpha a}{4E(V_0 - E)} \right\}^{-1} \qquad (5.41)$$

This result applies only to cases in which $E < V_0$, but it can easily be applied to particles of energy $E > V_0$. In such cases the equations are still valid, but the wave function oscillates in the no-longer-forbidden region $-a < x < 0$. This can be handled by treating α as an imaginary quantity and using the substitution $\sinh \alpha a = -i \sin i\alpha a$, where the argument of the sine function is the *real* quantity $i\alpha a = (a/\hbar)\sqrt{2m(E - V_0)}$. Then

$$R = \left\{1 + \frac{4E(E - V_0)}{V_0^2 \sin^2 i\alpha a}\right\}^{-1} \tag{5.42}$$

and

$$T = \left\{1 + \frac{V_0^2 \sin^2 i\alpha a}{4E(E - V_0)}\right\}^{-1} \tag{5.43}$$

An important special case occurs when $i\alpha a = n\pi$, n being any integer. In that case Eq. (5.43) shows that $T = 1$; there is perfect transmission.

The probability current equation (5.33) gives further insight. When waves are traveling in both directions, as in the region $x < -a$, the current is

$$S_x = (\hbar k/m)|\psi|_{\max}|\psi|_{\min} \qquad x < -a$$

For $x > 0$, $|\psi|$ is constant and

$$S_x = (\hbar k/m)|\psi|^2 = |G|^2 \qquad x < 0$$

But S_x is the same everywhere, so

$$|G|^2 = |\psi|_{\max}|\psi|_{\min} \tag{5.44}$$

Figure 5.4 shows T as a function of the ratio E/V_0 for one particular value of V_0. Notice the values of E/V_0 at which there is perfect transmission. Figure 5.5 shows a wave function for one case of perfect transmission. Notice that in this case, with no reflected wave, the probability density is constant to the left of the barrier, where $|\psi|_{\max} = |\psi|_{\min} = |G|$. Of course there is a reflected wave *inside* the barrier, causing the standing-wave pattern there.

When the reflection is present, we could deduce the value of R from an accurately drawn graph of $|\psi|^2$ as follows: From Eq. (5.9), $R = |B|^2/|A|^2$, where A and B again are complex quantities. But Eq. (5.8) no longer holds, and the phase angle ϕ between A and B is not necessarily π [as it is in Eq. (5.29)]. Thus we define

$$A = \mathrm{A}e^{i\phi} \quad \text{and} \quad B = \mathrm{B}e^{i(\phi + \theta)} \tag{5.45}$$

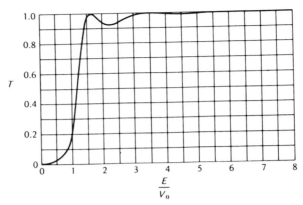

FIGURE 5.4 Transmission probability T as a function of energy E for a rectangular barrier of width a and height V_0 equal to $8\hbar^2/ma^2$, where m is the particle's mass. [Adapted from L. I. Schiff, *Quantum Mechanics*, 2nd edition. Copyright McGraw-Hill, New York (1955). Used with permission of McGraw-Hill Book Company.]

Equation (5.30) then becomes

$$|\psi|^2 = A^2 + B^2 - AB(e^{2ik_ax}e^{-i\theta} + e^{-2ik_ax}e^{i\theta}) \tag{5.46}$$

$$= A^2 + B^2 - 2AB\cos(2k_ax - \theta) \tag{5.47}$$

As before,

$$|\psi|^2_{max} = (A + B)^2 \quad \text{and} \quad |\psi|^2_{min} = (A - B)^2 \tag{5.48}$$

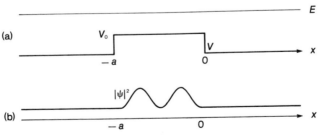

FIGURE 5.5 Perfect transmission through a potential barrier. (a) Total energy E and potential energy V. (b) Probability density $|\psi|^2$. Notice zero slope at $x = -a$ and at $x = 0$, to match the constant $|\psi|^2$ for $x < -a$ and $x > 0$. This requires that the wavelength of $|\psi|^2$ (half the wavelength of ψ) must be equal to a/n, where n is an integer. In this case, $n = 2$. Compare with Figure 5.1.

Therefore

$$\frac{|\psi|_{min}}{|\psi|_{max}} = \frac{A - B}{A + B} = \frac{1 - \sqrt{R}}{1 + \sqrt{R}} \qquad (5.49)$$

and we can solve for R, obtaining

$$R = \left\{ \frac{|\psi|_{max} - |\psi|_{min}}{|\psi|_{max} + |\psi|_{min}} \right\}^2 \qquad (5.50)$$

This relation can be applied to Figure 5.3, where $|\psi|_{min}/|\psi|_{max} \cong 1/4$, making $R \cong (3/5)^2 = 0.36$.

As previously mentioned, Figure 5.5 shows an example of perfect transmission, where $|\psi|_{min} = |\psi|_{max}$ and $R = 0$. As we see from Figure 5.4, *perfect* transmission cannot occur when $E < V_0$, but some transmission can occur. Quantum mechanical barrier penetration has been observed in many different situations: in oscillations of atoms in molecules, in tunneling of electrons through a potential barrier between two conductors or superconductors, and in the emission of alpha particles by atomic nuclei. Quantum mechanics not only explains these classically inexplicable processes, it enables us to calculate the rate at which they occur.

5.3.1 The SQUID

The superconducting quantum interference device, SQUID, is an extremely sensitive detector of magnetic fields, with important applications to medicine. The device is based on tunneling of electrons through a barrier formed by a thin layer of insulator between two superconducting surfaces.

ADDITIONAL READING

R. Eisberg and R. Resnick, *Quantum Physics of Atoms, Molecules, Solids, Nuclei, and Particles*, Wiley, New York (1974). Chapter 6, pages 215–226, gives an extensive treatment of barrier penetration and applications.

EXERCISES

1. An electron with kinetic energy of 4 eV reaches a potential step as defined in Section 5.1. Find the probability that it will be reflected

if $V_0 =$
(a) 5 eV (negative step)
(b) $-$ 5 eV (positive step)
(c) 96 eV

2. Draw a figure like Figure 5.1 of $|\psi|^2$ vs. x for a positive potential step.

3. An electron is in the region $x < 0$ (Figure 5.1), for which the space
   dependence of $\psi(x, t)$ [see Eq. (5.1)] is $u(x) = Ae^{ikx} + Be^{-ikx}$.
   (a) Write an expression in terms of A and B for the probability that
       the electron will have positive momentum when it is observed.
   (b) If $A = 4$ and $B = 2i - 3$, find the value of this probability.

4. Find the value of the probability current S_x in terms of the amplitude
   A for the electron of Exercise 1(a). Show that S_x has the same value
   for $x < 0$ and for $x > 0$.

5. An electron's potential energy is given by the graph below.

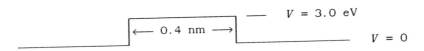

$\qquad\qquad$ $V = 3.0$ eV

$\leftarrow$ 0.4 nm $\longrightarrow$

$V = 0$

   (a) Find the smallest kinetic energy that an electron must have outside
       the barrier region in order to be perfectly transmitted through the
       3.0-eV, 0.4-nm-wide barrier.
   (b) For the kinetic energy of part (a), draw a graph of $|u(x)|^2$ and find
       the electron's total energy E (not including rest energy) in eV.

6. Extend the treatment of barrier penetration in Section 5.3 to the case
   of a negative "barrier." Make a graph of T versus E, similar to that of
   Figure 5.5, for the case $V_0 = -8\hbar^2/ma^2$.

7. Find the probability of transmission of an electron of initial kinetic
   energy $E = 2.0\ V_0$ through a barrier of width a, if $V_0 = 4\hbar^2/m_e a^2$.

8. Find the probability of transmission of a muon of initial kinetic energy
   $E = 2.0\ V_0$ through the barrier of Problem 7. Muon mass $= 207m_e$.

9. Solve Eqs. (5.37) for the ratio $|G|^2/|A|^2$. Is this ratio equal to the
   transmission coefficient T given in Eq. (5.41)? Why?

10. Compute the probability current S_x for the region $-a < x < 0$, using
    Eqs. (5.31). By finding expressions for C and D in terms of A, show
    that S_x is the same in this region as it is for $x < -a$.

11. Using the condition $i\alpha\pi = n\pi$, verify that perfect transmission occurs
    at the lowest two values of E/V_0 shown to have $T = 1$ in Figure 5.4.

CHAPTER 6

Three Dimensions and Angular Momentum

To deal with the real world we must extend our equations to three dimensions. The easiest way to do this is to use cartesian coordinates x, y, and z, but we cannot go far unless we switch to spherical coordinates, which are more appropriate for the Coulomb potential because of its spherical symmetry. We begin with the rectangular symmetry of a wave function for a particle confined inside a rectangular box. We shall then see how to deal with spherical coordinates and with the angular momentum operator that develops naturally from the use of these coordinates.

6.1 THE SCHRÖDINGER EQUATION IN THREE DIMENSIONS

The time-independent Schrödinger equation may always be written as the operator equation

$$(p^2/2m)\psi = (E - V)\psi \tag{6.1}$$

where $E - V$ is, as usual, the kinetic energy. If ψ is a function of cartesian coordinates x, y, and z, then the p^2 operator may be written

$$p^2 = p_x^2 + p_y^2 + p_z^2, = -\hbar^2(\partial^2/\partial x^2) - \hbar^2(\partial^2/\partial y^2) - \hbar^2(\partial^2/\partial z^2) \tag{6.2}$$

extending the prescription of Chapter 4 to all three coordinates, because these coordinates are all equivalent. This form of the operator is used when both ψ and V are written in cartesian coordinates. The complete

115

Schrödinger equation is then

$$-\hbar^2(\partial^2/\partial x^2 + \partial^2/\partial y^2 + \partial^2/\partial z^2)\psi(x,y,z,t) + V(x,y,z)\psi(x,y,z,t)$$
$$= E\psi(x,y,z,t) \tag{6.3}$$
$$-\hbar^2(\partial^2/\partial x^2 + \partial^2/\partial y^2 + \partial^2/\partial z^2)\psi(x,y,z,t) + V(x,y,z)\psi(x,y,z,t)$$
$$= i\hbar(\partial\psi/\partial t) \tag{6.4}$$

and, as before, we can write ψ as the product of a space-dependent part and a time-dependent part:

$$\psi(x,y,z,t) = u(x,y,z)e^{-iEt/\hbar} \tag{6.5}$$

6.1.1 Particle in a Box

The cartesian coordinate system is used when the potential energy is appropriate for it. For example, assume that the potential energy is zero inside a rectangular box and is infinite outside the box. (See Figure 6.1). The walls of the box are the planes $x = 0$, $x = a$, $y = 0$, $y = b$, $z = 0$, and $z = d$. Thus we have the boundary conditions

$$u = 0 \quad \text{when } x = 0, \text{ or } x = a, \text{ or } y = 0, \text{ or } y = b, \text{ or } z = 0, \text{ or } z = c \tag{6.6}$$

6.1.2 Separation of Variables

The standard technique for solving a partial differential equation like (6.3) is the *separation of variables*. We assume, without proof, that the function

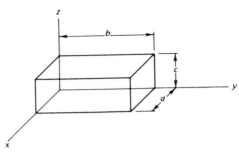

FIGURE 6.1 Impenetrable box containing a particle whose potential energy is zero inside the box and infinite outside the box.

$u(x, y, z)$ may be written as the product of three functions: $X(x)$ is a function of x only, $Y(y)$ is a function of y only, and $Z(z)$ is a function of z only:

$$u(x, y, z) = X(x)Y(y)Z(z) \tag{6.7}$$

Equation (6.3) now becomes, with $V = 0$ (inside the box)

$$-\frac{\hbar^2}{2m}(X''YZ + XY''Z + XYZ'') = EXYZ \tag{6.8}$$

where $X'' \equiv d^2X/dx^2$, $Y'' \equiv d^2Y/dy^2$, and $Z'' \equiv d^2Z/dz^2$. We can now easily separate the variables by dividing both sides of Eq. (6.8) by XYZ, obtaining

$$-\frac{\hbar^2}{2m}\left(\frac{X''}{X} + \frac{Y''}{Y} + \frac{Z''}{Z}\right) = E \tag{6.9}$$

We can now separate the equation into three independent equations, one in each of the three coordinate variables. The key to this operation is that *each term in this equation is constant, independent of the values of the variables.* For example, if x changes, only the factor X''/X could possibly change. But the value of X''/X cannot change if everything else in the equation is constant. Therefore X''/X must be constant; Y''/Y and Z''/Z must also be constant, for the same reason, and we can write the equations

$$X''/X = -k_x^2, \qquad Y''/Y = -k_y^2, \qquad Z''/Z = -k_z^2 \tag{6.10}$$

where the constants are written as $-k_x^2$, $-k_y^2$, and $-k_z^2$ to indicate that they are negative numbers (making the equations equivalent to the standard equation for simple harmonic motion). To satisfy the boundary conditions that $u = 0$ on the planes $x = 0$, $y = 0$, and $z = 0$, the respective solutions must be

$$X = A\sin(k_x x), \qquad Y = B\sin(k_y y), \qquad Z = C\sin(k_z z) \tag{6.11}$$

The constants k_x, k_y, and k_z are then determined by the boundary conditions $X(a) = 0$, $Y(b) = 0$, $Z(c) = 0$, with the result as in the one-dimensional square well

$$k_x a = n_x \pi, \qquad k_y b = n_y \pi, \qquad k_z c = n_z \pi \tag{6.12}$$

where n_x, n_y, and n_z are positive integers. The complete solution may be written

$$\psi(x, y, z, t) = \text{constant} \times \sin(n_x \pi x/a) \sin(n_y \pi y/b) \sin(n_z \pi z/c) e^{-iEt/\hbar}$$

$$(6.13)$$

where

$$E = \left(\frac{\hbar^2}{2m}\right)\left(k_x^2 + k_y^2 + k_z^2\right)$$

$$= \frac{\hbar^2}{2m}\left\{\left(\frac{n_x \pi}{a}\right)^2 + \left(\frac{n_y \pi}{b}\right)^2 + \left(\frac{n_z \pi}{c}\right)^2\right\} \qquad (6.14)$$

The lowest energy level, E_{111}, is found by setting $n_x = n_y = n_z = 1$, with the result $E_{111} = (h^2/8m)(a^{-2} + b^{-2} + c^{-2})$. It is important to notice that *no quantum number for a particle in a square well is ever equal to zero*, because the particle's wave function [Eq. (6.13)] would then be identically zero. Even if the resulting energy were not zero, the probability density would be zero; there would be no particle to possess that energy.

Equation (6.14) could have been deduced without using the Schrö-dinger equation, simply because $E = p^2/2m$, $p^2 = p_x^2 + p_y^2 + p_z^2$ and each component of **p** can be found from the de Broglie relation, using wave-lengths λ_x, λ_y, and λ_z determined by the dimensions of the box, as in a one-dimensional square well:

$$\lambda_x = 2a/n_x, \qquad \lambda_y = 2b/n_y, \qquad \lambda_z = 2c/n_z \qquad (6.15)$$

6.1.3 Degeneracy

For quantum numbers greater than 1, it often happens that more than one set of quantum numbers give the same energy eigenvalue. For a simple example, if $a = b = c$, then the next level above the ground level can be given by three sets of quantum numbers; these sets are 2,1,1, 1,2,1, and 1,1,2 for the numbers n_x, n_y, and n_z, respectively. When this happens, we say that the states are *degenerate*, and the *degeneracy* is equal to the number of states that have the same energy.

Example Problem 6.1   Find the three lowest energy levels and the degeneracy of each for an electron in a square well when $a = b = 5.00$ nm and $c = 4.00$ nm.

Solution. All of the energy eigenvalues contain the factor $h^2/8m$, or $h^2c^2/8mc^2$, which equals 0.376 eV-nm^2. Thus we have, from Eq. (6.14)

$$E_{111} = .376(a^{-2} + b^{-2} + c^{-2}) = 0.376[2/25 + 1/16]$$
$$= .062 \text{ eV} \qquad \text{Degeneracy 1}$$
$$E_{211} = E_{121} = 0.376[4/25 + 1/25 + 1/16] \text{ eV}$$
$$= 0.0987 \text{ eV} \qquad \text{Degeneracy 2}$$
$$E_{112} = 0.376[1/25 + 1/25 + 4.16] \text{ eV}$$
$$= 0.1241 \text{ eV} \qquad \text{Degeneracy 1}$$

6.1.4 The Three-Dimensional Harmonic Oscillator

The potential energy of a spherically symmetric three-dimensional harmonic oscillator may be written

$$V(x, y, z) = Kr^2/2 = (m\omega^2/2)(x^2 + y^2 + z^2) \qquad (6.16)$$

Therefore, writing the time-independent Schrödinger equation in terms of the Hamiltonian operator H, we have

$$Hu(x, y, z) = (H_x + H_y + H_z)u(x, y, z) = Eu(x, y, z) \qquad (6.17)$$

where the individual operators H_x, H_y, and H_z are given by

$$H_x = -\frac{\hbar^2}{2m}\frac{\partial^2}{\partial x^2} + \frac{m\omega^2 x^2}{2}, \qquad H_y = -\frac{\hbar^2}{2m}\frac{\partial^2}{\partial y^2} + \frac{m\omega^2 y^2}{2},$$

$$H_z = -\frac{\hbar^2}{2m}\frac{\partial^2}{\partial z^2} + \frac{m\omega^2 z^2}{2}$$

These are simply one-dimensional operators like the one in Eq. (4.9).

We can now easily separate the variables in Eq. (6.16) by writing the wave function as $u_{n_x n_y n_z}(x, y, z) = u_{n_x}(x)u_{n_y}(y)u_{n_z}(z)$, thereby obtaining three equations that are identical to Eq. (4.7) (except for the change of variable from x to y or z). Each solution, like the solutions of Eq. (4.9), may be written in the form of Eq. (4.29):

$$u_{n_x}(x) = (d/dx - ax)^{n_x} e^{-ax^2/2} \qquad (6.18)$$

where we apply the raising operator n_x times, and $n_x = 0, 1, 2, \ldots$.

Substitution into Eq. (6.17) gives us, after carrying out the operations,

$$(H_x + H_y + H_z)u_{n_x n_y n_z}(x, y, z)$$
$$= \left(n_x + \tfrac{1}{2} + n_y + \tfrac{1}{2} + n_z + \tfrac{1}{2}\right)\hbar \omega u_{n_x n_y n_z}(x, y, z) \qquad (6.19)$$

where each of the three quantum numbers n_x, n_y, and n_z can have any of the values $0, 1, 2, \ldots$. Thus the total energy is given by

$$E_n = [n + \tfrac{3}{2}]\hbar \omega \qquad (n = n_x + n_x + n_x) \qquad (6.20)$$

where n is any positive integer or zero. When $n = 0$ we have the ground-state energy $E_0 = 3\hbar \omega/2$, and the unnormalized eigenfunction is the product

$$u_0(x)u_0(y)u_0(z) = e^{-ax^2/2}e^{-ay^2/2}e^{-az^2/2} = e^{-ar^2/2} \qquad (6.21)$$

Notice the contrast between these quantum numbers and the quantum numbers for the three-dimensional square well. In the three-dimensional square well none of the values n_x, n_y, or n_z can be equal to zero; in the harmonic oscillator any or all of these can equal zero, as is clear from Eq. (6.21).

The eigenfunctions u_{100}, u_{010}, and u_{001} for $n = 1$ are, respectively,

$$u_{100} = u_1(x)u_0(y)u_0(z) = (\partial/\partial x - ax)e^{-ax^2/2}e^{-ay^2/2}e^{-az^2/2}$$
$$= -2axe^{-ar^2/2} \qquad (6.22a)$$
$$u_{010} = u_1(x)u_0(y)u_0(z) = (\partial/\partial y - ay)e^{-ax^2/2}e^{-ay^2/2}e^{-az^2/2}$$
$$= -2aye^{-ar^2/2} \qquad (6.22b)$$
$$u_{001} = u_1(x)u_0(y)u_0(z) = (\partial/\partial z - az)e^{-ax^2/2}e^{-ay^2/2}e^{-az^2/2}$$
$$= -2aze^{-ar^2/2} \qquad (6.22c)$$

where $r = x^2 + y^2 + z^2$. In Section 8.1 we shall see how the harmonic oscillator can be solved in the spherical coordinates r, θ, and ϕ.

6.2 SPHERICALLY SYMMETRIC POTENTIALS

6.2.1 Spherical Coordinates

The principles and techniques of the previous section are applicable to all coordinate systems. For the study of atoms, spherical coordinates r, θ, and

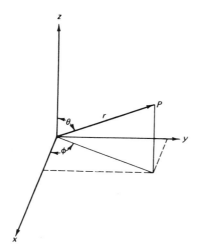

FIGURE 6.2 Spherical and rectangular coordinates of a point P.

ϕ are most useful. (See Figure 6.2.) These coordinates are related to the coordinates x, y, and z by

$$r^2 = x^2 + y^2 + z^2, \qquad \cos\theta = z/r, \qquad \tan\phi = y/x \qquad (6.23)$$

To write the Schrödinger equation in these coordinates, we need to replace derivatives with respect to x, y, and z by derivatives with respect to r, θ, and ϕ. We do this by using the chain rule to compute $\partial u/\partial x$, $\partial u/\partial y$, and $\partial u/\partial z$; for example,

$$\partial u/\partial x = (\partial u/\partial r)(\partial r/\partial x) + (\partial u/\partial\theta)(\partial\theta/\partial x) + (\partial u/\partial\phi)(\partial\phi/\partial x)$$

$$(6.24)$$

and corresponding equations for $\partial u/\partial y$ and $\partial u/\partial z$. The necessary partial derivatives can be shown, with the aid of Eqs. (6.23), to be

$$\frac{\partial r}{\partial x} = \sin\theta\cos\phi \qquad \frac{\partial r}{\partial y} = \sin\theta\sin\phi \qquad \frac{\partial r}{\partial z} = \cos\theta$$

$$\frac{\partial\theta}{\partial x} = \frac{1}{r}\cos\theta\cos\phi \qquad \frac{\partial\theta}{\partial y} = \frac{1}{r}\cos\theta\sin\phi \qquad \frac{\partial\theta}{\partial z} = -\frac{1}{r}\cos\theta$$

$$\frac{\partial\phi}{\partial x} = -\frac{1}{r}\frac{\sin\phi}{\sin\theta} \qquad \frac{\partial\phi}{\partial y} = \frac{1}{r}\frac{\cos\phi}{\sin\theta} \qquad \frac{\partial\phi}{\partial z} = 0$$

We find the second derivatives $\partial^2 u/\partial x^2$, $\partial^2 u/\partial y^2$, and $\partial^2 u/\partial z^2$ by using the chain rule again, replacing u by $\partial u/\partial x$, to obtain

$$\frac{\partial^2 u}{\partial x^2} = \left\{ \frac{\partial}{\partial r} \frac{\partial u}{\partial x} \right\} \frac{\partial r}{\partial x} + \left\{ \frac{\partial}{\partial r} \frac{\partial u}{\partial y} \right\} \frac{\partial \theta}{\partial x} + \left\{ \frac{\partial}{\partial r} \frac{\partial u}{\partial z} \right\} \frac{\partial \phi}{\partial x} \quad (6.25)$$

and corresponding equations for $\partial^2 u/\partial y^2$ and $\partial^2 u/\partial z^2$. It is understood that the first partial derivatives in Eq. (6.25) are already expressed in terms of spherical coordinates and derivatives with respect to those coordinates. For example,

$$\frac{\partial u}{\partial x} = \sin\theta\cos\phi\frac{\partial u}{\partial r} + \frac{1}{r}\cos\theta\cos\phi\frac{\partial u}{\partial \theta} - \frac{\sin\theta}{r\sin\phi}\frac{\partial u}{\partial \phi} \quad (6.26)$$

Completing the operations of Eq. (6.24) for all three coordinates, and substituting into Eq. (6.3), we finally have the time-independent Schrödinger equation in spherical coordinates:

$$-\frac{1}{r}\frac{\partial^2}{\partial r^2}(ru) - \frac{1}{r^2\sin\theta}\frac{\partial}{\partial\theta}\left\{ \sin\theta\frac{\partial u}{\partial\theta} \right\}$$

$$-\frac{1}{r^2\sin^2\theta}\frac{\partial^2 u}{\partial\phi^2} = \frac{2m}{\hbar^2}[E - V(r)]u \quad (6.27)$$

where u is now a function of r, θ, and ϕ. The reader should verify that each term in Eq. (6.27) has the same dimensions. Before attempting to generate solutions to this equation, let us discuss some general principles.

6.3 ANGULAR MOMENTUM OPERATORS AND EIGENVALUES

The law of *conservation of angular momentum* is obeyed in all systems in which the potential energy is independent of the angular coordinates, as we have assumed in Eq. (6.19) by writing the potential energy as $V(r)$. This is true in quantum mechanics as well as in classical mechanics. In both classical and quantum mechanics, the definition of angular momentum is the vector equation

$$\mathbf{L} = \mathbf{r} \times \mathbf{p} \quad (6.28)$$

However, in quantum mechanics the vector $\mathbf{p}$ is an operator. Thus the vector $\mathbf{L}$ must also be an operator, whose components can be deduced from the definition of the cross product. In cartesian coordinates these components are

$$L_x = yp_z - zp_y = -i\hbar\left\{y\,\frac{\partial}{\partial z} - z\,\frac{\partial}{\partial y}\right\} \qquad (6.29a)$$

$$L_y = zp_x - xp_z = -i\hbar\left\{z\,\frac{\partial}{\partial x} - x\,\frac{\partial}{\partial z}\right\} \qquad (6.29b)$$

$$L_z = xp_y - yp_x = -i\hbar\left\{x\,\frac{\partial}{\partial y} - y\,\frac{\partial}{\partial x}\right\} \qquad (6.29c)$$

Using the chain rule, we can write Eqs. (6.29) in spherical coordinates:

$$L_x = i\hbar\left\{\sin\phi\,\frac{\partial}{\partial\theta} + \cot\theta\cos\phi\,\frac{\partial}{\partial\phi}\right\} \qquad (6.30a)$$

$$L_y = -i\hbar\left\{\cos\phi\,\frac{\partial}{\partial\theta} - \cot\theta\sin\phi\,\frac{\partial}{\partial\phi}\right\} \qquad (6.30b)$$

$$L_z = -i\hbar\,\frac{\partial}{\partial\phi} \qquad (6.30c)$$

From Eqs. (6.30) we can now find the magnitude of the vector **L**. First we construct the L^2 operator as follows:

$$L^2 = L_x^2 + L_y^2 + L_z^2 \qquad (6.31)$$

with

$$L_x^2 = i\hbar\left\{\sin\phi\,\frac{\partial}{\partial\theta} + \cot\theta\cos\phi\,\frac{\partial}{\partial\phi}\right\}i\hbar\left\{\sin\phi\,\frac{\partial}{\partial\theta} + \cot\theta\cos\phi\,\frac{\partial}{\partial\phi}\right\}$$
$$(6.32a)$$

$$L_y^2 = i\hbar\left\{\cos\phi\,\frac{\partial}{\partial\theta} - \cot\theta\sin\phi\,\frac{\partial}{\partial\phi}\right\}i\hbar\left\{\cos\phi\,\frac{\partial}{\partial\theta} - \cot\theta\sin\phi\,\frac{\partial}{\partial\phi}\right\}$$
$$(6.32b)$$

$$L_z^2 = -\hbar^2\,\frac{\partial^2}{\partial\phi^2} \qquad (6.32c)$$

The result is

$$L^2 = -\hbar^2 \left\{ \frac{1}{\sin\theta} \frac{\partial}{\partial\theta} \left(\sin\theta \frac{\partial}{\partial\theta} \right) + \frac{1}{\sin^2\theta} \frac{\partial^2}{\partial\phi^2} \right\} \qquad (6.33)$$

You can see that the operator here is the same operator that appears in the Schrödinger equation [Eq. (6.27)]. We can rewrite Eq. (6.27) in terms of L^2 as

$$\frac{\partial^2}{\partial r^2}(ru) = -\frac{2m}{\hbar^2}\left\{ E - V(r) - \frac{L^2}{2mr^2} \right\}(ru) \qquad (6.34)$$

6.3.1 Separation of Variables in Spherical Coordinates

As we did with cartesian coordinates, let us now separate the variables. We write the wave function u as a product of three functions, one for each coordinate. In this case we have

$$u(r, \theta, \phi) = R(r)Y(\theta, \phi) = R(r)\Theta(\theta)\Phi(\phi) \qquad (6.35)$$

where $Y(\theta, \phi)$ is, as we shall see, a *spherical harmonic*, a function well known in solutions of Laplace's equation.[1] Equation (6.34) now becomes

$$Y(\theta, \phi)\frac{\partial^2}{\partial r^2}(rR(r)) = -\frac{2m}{\hbar^2}\left\{ E - V(r) - \frac{L^2}{2mr^2} \right\}(rR(r)Y(\theta, \phi)) \qquad (6.36)$$

Remembering that L^2 [defined in Eq. (6.33)] operates only on angular coordinates, we can now separate the variable r from the angular variables θ and ϕ. We divide both sides of (6.36) by $R(r)Y(\theta, \phi)$ to obtain

$$\frac{1}{R(r)}\frac{\partial^2}{\partial r^2}(rR(r)) = -\frac{1}{Y(\theta, \phi)}\frac{2m}{\hbar^2}\left\{ E - V(r) - \frac{L^2}{2mr^2} \right\}(Y(\theta, \phi)) \qquad (6.37)$$

[1] Laplace's equation, $\nabla^2 \Phi = 0$, arises in studies of electric potential and heat flow.

Multiplying both sides by r^2 and combining terms gives us

$$\frac{r^2}{R(r)} \frac{\partial^2}{\partial r^2} (rR(r)) = -\frac{2mr^2}{\hbar^2} \{E - V(r)\} + \frac{1}{Y(\theta, \phi)} \frac{L^2}{\hbar^2} (Y(\theta, \phi))$$

(6.38)

Following the same reasoning used before, we deduce that the last term on the right-hand side of Eq. (6.38) must be constant, because it varies only with the angular coordinates, and the other terms are independent of angle. Thus we may write

$$L^2 Y(\theta, \phi) = \alpha \hbar^2 Y(\theta, \phi)$$

(6.39)

where α is a dimensionless constant to be determined. Using Eq. (6.35), we can now further separate the variables by writing $Y(\theta, \phi) = \Theta(\theta)\Phi(\phi)$, so that Eq. (6.39) becomes

$$L^2 \Theta(\theta)\Phi(\phi) = \alpha \hbar^2 \Theta(\theta)\Phi(\phi)$$

(6.40)

or

$$\left\{ \frac{1}{\sin \theta} \frac{\partial}{\partial \theta} \left(\sin \theta \frac{\partial}{\partial \theta} \right) + \frac{1}{\sin^2 \theta} \frac{\partial^2}{\partial \phi^2} \right\} \Theta(\theta)\Phi(\phi) = \alpha \hbar^2 \Theta(\theta)\Phi(\phi)$$

(6.41)

As before, we can now obtain two separate equations. We divide both sides by $\Theta(\theta)\Phi(\phi)$ and multiply both sides by $\sin^2 \theta$, obtaining

$$\frac{\sin \theta}{\Theta(\theta)} \frac{\partial}{\partial \theta} \left(\sin \theta \frac{\partial}{\partial \theta} \right) \Theta(\theta) + \frac{1}{\Phi(\phi)} \frac{\partial^2}{\partial \phi^2} \Phi(\phi) = \alpha \hbar^2 \sin^2 \theta \quad (6.42)$$

The second term, the only term that is dependent on the variable ϕ, must be constant, so we may write

$$\frac{1}{\Phi} \frac{\partial^2 \Phi}{\partial \phi^2} = \text{constant} \quad \text{or} \quad \frac{\partial^2 \Phi}{\partial \phi^2} = \text{constant} \times \Phi$$

(6.43)

The form of the constant is dictated by a reasonable condition on the ϕ

coordinate. We require that

$$\Phi(\phi + 2\pi) = \Phi(\phi) \tag{6.44}$$

because the angle $\phi + 2\pi$ describes the same point in space as the angle ϕ, if r and θ are unchanged. If $\Phi(\phi + 2\pi)$ were not equal to $\Phi(\phi)$, then the wave function would be *double valued*. (Although there are certain conditions under which a double-valued wave function is acceptable, that is not possible in this case; see Problem 10.)

When condition (6.44) is satisfied, Φ is a periodic function with period 2π. Thus we can write Eq. (6.43) as

$$\frac{\partial^2 \Phi}{\partial \phi^2} = -m^2 \Phi \tag{6.45}$$

and Eq. (6.42) becomes

$$\frac{\sin\theta}{\Theta(\theta)} \frac{\partial}{\partial\theta} \left\{ \sin\theta \frac{\partial}{\partial\theta} \right\} \Theta(\theta) - m^2 = \alpha\hbar^2 \sin^2\theta \tag{6.46}$$

where m is an integer. The solution to Eq. (6.44) may be written in the form

$$\Phi + Ae^{im\phi} \tag{6.47}$$

where A is a normalization constant.

Notice that this solution is an eigenfunction of the L_z operator, for

$$L_z\Phi = -i\hbar \frac{\partial}{\partial\phi}\Phi = -i\hbar \frac{\partial}{\partial\phi}(Ae^{im\phi}) = m\hbar\Phi \tag{6.48}$$

The eigenvalues are $m\hbar$, and therefore the only possible results of a measurement of *any component* of angular momentum are integral multiples of $\hbar$. This is reminiscent of the Bohr condition, but with one highly significant difference: m can be zero. If $m = 0$, then Eq. (6.46) becomes

$$\frac{\sin\theta}{\Theta(\theta)} \frac{\partial}{\partial\theta} \left\{ \sin\theta \frac{\partial}{\partial\theta} \right\} \Theta(\theta) = \alpha\hbar^2 \sin^2\theta \tag{6.49}$$

Equation (6.49) is known as Legendre's equation, and the solutions are called *Legendre polynomials*. The Legendre polynomials can be written in the form

$$P_l(x) = \frac{(-1)^l}{2^l l!} \frac{d^l}{dx^l}(1 - x^2)^l \tag{6.50}$$

The solutions to Eq. (6.49) are these polynomials with the variable $\cos\theta$ instead of x. You should verify from Eq. (6.50) that the polynomials for $l = 0, 1, 2$, and 3 (with $x = \cos\theta$) are, respectively,

$$P_0 = 1, \quad P_1 = \cos\theta, \quad P_2 = \frac{3\cos^2\theta - 1}{2}, \quad P_3 = \frac{5\cos^3\theta - 3\cos\theta}{2}$$

$$(6.51)$$

Several features of these polynomials are significant.

- Each polynomial contains only even (odd) powers of $\cos\theta$ if l is even (odd).
- Each polynomial is normalized to be equal to 1 when $\theta = 0$.
- In each polynomial the highest power of $\cos\theta$ is equal to l.

Substitution of a Legendre polynomial into Eq. (6.49) now allows us to find the value of α, and thus to find the value of L^2. The first three results are:

- When $l = 0$, $\Theta(\theta) = 1$, and $\alpha = 0$; the total angular momentum is zero.
- When $l = 1$, $\Theta(\theta) = \cos\theta$; substitution into Eq. (6.49) shows that $\alpha = 2$.
- When $l = 2$, $\Theta(\theta) = \frac{1}{2}(3\cos^2\theta - 1)$; Eq. (6.49) shows that $\alpha = 6$.

These results were based upon the value $m = 0$. It is shown in Appendix D that, in general, $\alpha = l(l + 1)$; this is true for any value of m, as long as $|m| \leq l$. We can now summarize the results of Eqs. (6.39) as follows:

> The possible results of a measurement of L_z are $m\hbar$; $m = -l, -l + 1, \ldots, +l$.

> The possible results of a measurement of L^2 are $l(l + 1)\hbar^2$; $l = 0, 1, 2, \ldots$.

6.3.2 Eigenfunctions of L^2

Each eigenfunction of L^2 corresponds to an allowed combination of the quantum numbers l and m and may be written as a spherical harmonic:

$$Y_{l,m}(\theta, \phi) = P_l^{|m|}(\cos\theta)e^{im\phi} \quad (l \geq m) \quad (6.52)$$

where the functions $P_l^{|m|}(\cos\theta)$, found by solving Eq. (6.39a), are called *associated Legendre functions*. Notice that $P_l^{|m|}(\cos\theta)$ depends only on the absolute value of m, because only m^2 appears in Eq. (6.39a). A few normalized eigenfunctions are shown in Table 6.1.[2]

6.3.3 Probability and Spherical Coordinates

The normalization factors shown in Table 6.1 are based on the fact that the integral of the probability density $|u(r, \theta, \phi)|^2$ over all space must be equal to unity, or

$$\int_0^\infty \int_0^{2\pi} \int_0^\pi |R(r)Y_{l,m}(\theta, \phi)|^2 r^2 \sin\theta \, d\theta \, d\phi \, dr = 1 \qquad (6.53)$$

where $r^2 \sin\theta \, d\theta \, d\phi \, dr$ is the volume element in spherical coordinates.

It is convenient to normalize the radial and angular parts of the wave function separately. Thus we require that

$$\int_0^{2\pi} \int_0^\pi |Y_{l,m}(\theta, \phi)|^2 \sin\theta \, d\theta \, d\phi = 1 \qquad (6.54)$$

This requirement enables us to interpret $|Y_{l,m}|^2$ as a probability density for the angular coordinates of a particle. If the angular dependence of the wave function is $Y_{l,m}(\theta, \phi)$, then $\int_{\phi_1}^{\phi_2}\int_{\theta_1}^{\theta_2}|Y_{l,m}|^2 \sin\theta \, d\theta \, d\phi$ equals the probability that the particle will be observed where $\phi_1 < \phi < \phi_2$ and $\theta_1 < \theta < \theta_2$. Notice that $|Y_{l,m}|^2$ is a function of θ only, because $|e^{im\phi}|^2 = 1$.

Figure 6.3 shows polar plots of $|Y_{2,0}|^2$, $|Y_{2,1}|^2$, and $|Y_{2,2}|^2$. Each of these represents a particle with the same total angular momentum, but

TABLE 6.1 Some Spherical Harmonics

l:	0	1	2
$m = 2$			$\sqrt{15/32\pi}\,\sin^2\theta\, e^{2i\phi}$
$m = 1$		$-\sqrt{3/8\pi}\,\sin\theta\, e^{i\phi}$	$-\sqrt{15/8\pi}\,\sin\theta\cos\theta\, e^{i\phi}$
$m = 0$	$\sqrt{1/4\pi}$	$\sqrt{3/4\pi}\,\cos\theta$	$\sqrt{5/16\pi}\,(3\cos^2\theta - 1)$
$m = -1$		$\sqrt{3/8\pi}\,\sin\theta\, e^{-i\phi}$	$\sqrt{15/8\pi}\,\sin\theta\cos\theta\, e^{-i\phi}$
$m = -2$			$\sqrt{15/32\pi}\,\sin^2\theta\, e^{-2i\phi}$

---

[2] The minus signs in front of $Y_{1,1}$ and $Y_{2,1}$ are not essential, and they do not appear in some tables. They are included because they simplify the matrix representation of angular momentum operators, to be developed in Chapter 7.

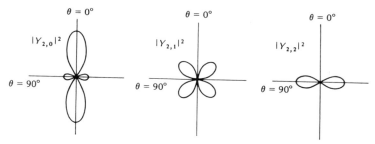

FIGURE 6.3 Polar plots of probability densities $|Y_{l,m}|^2$ as functions of θ.

with values of 0, $\hbar$, and $2\hbar$, respectively, for L_z. As L_z increases, we find that the maximum probability moves from the z axis ($\theta = 0$) to the xy plane ($\theta = \pi/2$). This behavior agrees with what we would expect classically; an orbit in the xy plane must correspond to a large value of L_z, and a particle on the z axis must have L_z equal to zero.

6.3.4 Summary

The functions $Y_{l,m}(\theta, \phi)$ are eigenfunctions of both the L_z operator and the L^2 operator, given by Eqs. (6.30) and (6.33), respectively. The L^2 operator is the sum of the three operators L_x^2, L_y^2, and L_z^2, given by the three equations (6.32). We have not yet considered the eigenfunctions of L_x and L_y, but logic dictates that, for a given value of l, the set of eigenvalues of L_x and L_y should be the same as those of L_z. Let us now see where this leads.

6.4 STATISTICAL ANALYSIS OF ANGULAR MOMENTUM EIGENVALUES

Section 6.3 poses two questions for the inquiring reader:

1. What happens when you measure L_x or L_y?
2. Is there a physical reason why L^2 is equal to $l(l + 1)\hbar^2$, when L_z^2 has a maximum value of only $l^2\hbar^2$?

The quick answer to question 1 is that the same thing happens when you measure L_x, L_y, or L_z, because x, y, and z are arbitrary labels. As we shall see in Section 6.5, we can only measure one component of **L** at a time, and the component that we have measured is customarily called the z component. The longer answer has to do with measuring a second component of **L**, say L_x, after measuring L_z. We deal with that in Section 6.5.

In answer to question 2, let us assume that we have a collection of atoms all having the same value of l, and consider what happens when we measure L_z. If we know nothing about the *direction* of the vector **L** for each atom, we must assume that each of the possible values of m is equally probable.[3] There are $2l + 1$ possible values of m; thus we can compute the expectation value of L_z^2 by adding up all of the possible results and dividing by $2l + 1$:

$$\langle L_z^2 \rangle = (2l + 1)^{-1}\hbar^2 \sum_{m=-l}^{l} m^2 = 2(2l + 1)^{-1}\hbar^2 \sum_{m=0}^{l} m^2 \quad (6.55)$$

It is easily shown by mathematical induction that the series sum is equal to $l(l + 1)(2l + 1)/6$. Substitution into (6.55) then shows that

$$\langle L_z^2 \rangle = l(l + 1)\hbar^2/3 \quad (6.56)$$

In the absence of information about the direction of **L**, we must assume that Eq. (6.56) holds for the other components of **L** as well; thus

$$\langle L_x^2 \rangle = \langle L_y^2 \rangle = \langle L_z^2 \rangle = l(l + 1)\hbar^2/3$$

and therefore

$$\langle L_x^2 \rangle = \langle L_y^2 \rangle = \langle L_z^2 \rangle = l(l + 1)\hbar^2 \quad (6.57)$$

Then, since each particle has the same value of L^2,

$$L = \langle L^2 \rangle = \langle L_x^2 \rangle + \langle L_y^2 \rangle + \langle L_z^2 \rangle = l(l + 1)\hbar^2 \quad \text{Q.E.D.} \quad (6.58)$$

After we have measured L^2 and L_z, we know that the vector **L** lies somewhere on a cone whose axis is the z axis and whose apex angle is η (Figure 6.4), where

$$\eta = \cos^{-1}(L_z/|\mathbf{L}|) = \cos^{-1}\left(m/\sqrt{l(l + 1)}\right) \quad (6.59)$$

All azimuthal angles ϕ are equally likely, because the probability density factor $|Y_{l,m}|^2$ is independent of ϕ. Thus it cannot tell us the value of either

---

[3]This is a valid assumption unless the particles' energy depends on m. See J. D. McGervey, *Am. J. Phys.* **49**, 494 (1981) for further discussion.

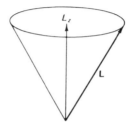

FIGURE 6.4 Visual representation of the angular momentum vector **L**, making an angle η with the z axis.

L_x or L_y. However, the values of L^2 and L_z do tell us something about L_x and L_y.

Example Problem 6.2 We have a collection of particles, all of which are known to have $L_z = 3\hbar$ and $L^2 = 12\hbar^2$. Find $\langle L_x \rangle$ and $\langle L_x^2 \rangle$ for these particles.

Solution. Since all allowed values of L_x, positive and negative, are equally likely, expectation value $\langle L_x \rangle$ must be zero. To find $\langle L_x^2 \rangle$, we rewrite Eq. (6.58) as $\langle L_x^2 \rangle + \langle L_y^2 \rangle = L^2 - \langle L_z^2 \rangle = 12\hbar^2 - 9\hbar^2 = 3\hbar^2$. Since all directions in the xy plane are equally unknown, there is no difference between $\langle L_x^2 \rangle$ and $\langle L_y^2 \rangle$, and $\langle L_x^2 \rangle = \langle L_y^2 \rangle$. Thus $\langle L_x^2 \rangle + \langle L_y^2 \rangle = 2\langle L_x^2 \rangle = 3\hbar^2$, and $\langle L_x^2 \rangle = 1.5\hbar^2$.

Example Problem 6.3 Determine the angular momentum of each eigenfunction u_{100}, u_{010}, and u_{001} of the three-dimensional harmonic oscillator.

Solution. Refer to Eqs. (6.21). The functions are

$$u_{100} = -2axe^{-ar^2/2}, \qquad u_{010} = -2aye^{-ar^2/2}, \qquad u_{001} = -2aze^{-ar^2/2}$$

The angular dependence of each function is contained in the factor x, y, or z. Table 6.1 shows that $Y_{1,0} = (3/4\pi)^{1/2} \cos\theta = (3/4\pi)^{1/2} z/r$. Therefore z is proportional to $rY_{1,0}$, making the angular dependence of u_{001} equal to $Y_{1,0}$. Thus for u_{001} we have $l = 1$ and $m = 0$, or $L^2 = 2\hbar^2$ and $L_z = 0$. We also can show from the table that $Y_{1,1} + Y_{1,-1} = (3/2\pi)^{1/2} x/r$, so that

$$x = (2\pi/3)^{1/2} r(Y_{1,1} + Y_{1,-1}) \quad \text{and} \quad u_{100} = Nr(Y_{1,1} + Y_{1,-1})e^{-ar^2/2}$$

where N is a normalization constant. Thus u_{100} is also an eigenfunction of

L^2 with $l = 1$, but it is a mixture of $m = +1$ and $m = -1$ with equal amplitudes. A measurement of L_z would yield $+\hbar$ and $-\hbar$ with equal probability. The same is true for u_{010}; it too is a mixture of $m = 1$ and $m = -1$.

6.5 EXPERIMENTAL TEST OF THEORY: DOUBLE STERN–GERLACH EXPERIMENT

6.5.1 Uncertainty in the Measurement of Angular Momentum Components

Figure 6.4 illustrates the fact that

> It is never possible to know the precise direction of the angular momentum.

Another way of saying this is:

> It is not possible to know two components of angular momentum simultaneously.

If we could know two components of **L** as well as the magnitude of **L**, then we would know the precise direction of **L**, so the two statements are equivalent. To test these statements, we must rely on experiment. How do we measure angular momentum on the atomic scale?

6.5.2 The Stern–Gerlach Experiment

The angular momentum of atomic states was first measured by Stern and Gerlach.[4] The measurement was based on the fact that an orbiting electron in an atom forms a current loop and thus has a magnetic dipole moment μ. This magnetic moment is proportional to the orbital angular momentum **L**; from standard electromagnetism[5] we know that $\mu = -e\mathbf{L}/2m$. We also know that if a beam of atoms passes through an *inhomogeneous* magnetic field **B** (Figure 6.5) the dipole moment of each

[4]O. Stern and W. Gerlach, Z. *Phys.* **8**, 110; **9**, 349 (1922).
[5]See, for example, E. M. Purcell, *Electricity and Magnetism*, Chapter 10, Vol. 2 of the Berkeley Physics Course, McGraw-Hill, New York (1965).

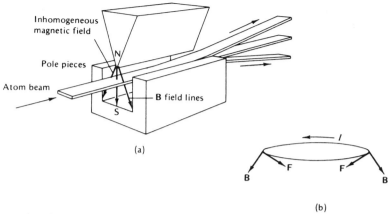

FIGURE 6.5 (a) Schematic view of Stern–Gerlach apparatus. Atoms travel along sharp pole piece, which produces a field that is stronger near the north pole (N). The resulting field lines converge and are not exactly vertical. (b) Edgewise view of a current loop in the **B** field of (a). Because the **B** field lines are in the directions shown, the forces **F** all have downward components, resulting in a net downward force on the loop.

atom results in a force whose z component is

$$F_z = \mu_z \frac{\partial B_z}{\partial z} = -\frac{eL_z}{2m} \frac{\partial B_z}{\partial z} \tag{6.60}$$

Figure 6.5b shows the origin of the net force, which acted on each atom which had nonzero L_z.

Because L_z can take on only a discrete set of values, only certain angles of deflection are seen in such an experiment. The figure illustrates the case $l = 1$, when three deflections are seen, corresponding to $L_z = \pm\hbar$ or $L_z =$ zero. Thus we know the value of L^2 and the value of L_z for each atom simply from its presence in one beam or another.

Let us now follow the beam in which the atoms have $L_z = +\hbar$, and try to determine the values of L_x for these atoms by sending them through another magnet in which the field is in the x direction. Presumably this beam will also split into three beams, corresponding to $L_x = \pm\hbar$ or $L_x =$ zero. Consider the atoms for which $L_x = -\hbar$. Can we not say that we now know the values of both L_x and L_z for these atoms? Where does the experimental difficulty lie?

Classically, there would be no difficulty. L_z would change when we measure L_x, but *it would change by a known amount*, and we would then know the values of both components, L_x and L_z. Analysis of the mechanism for this change in L_z will show why the change must be uncertain according to quantum mechanics.

When we send the beam through the magnet, the magnetic field **B** causes a torque of $\mu \times \mathbf{B}$ on each atom in the beam, and classical mechanics tells us that

$$\text{Torque} = \frac{d\mathbf{L}}{dt} = \mu \times \mathbf{B} = -\frac{e\mathbf{L}}{2m} \times \mathbf{B} \qquad (6.61)$$

To simplify the analysis, let us set the third angular momentum component, L_y, equal to zero. The first measurement told us that L_z was equal to $-\hbar$ when the atom entered the second magnet. When we measure L_x, **B** is in the x direction. Thus Eq. (6.61) becomes

$$\frac{d\mathbf{L}}{dt} = -\frac{e}{2m}L_z B_x \mathbf{j} = \frac{e\hbar}{2m}B_x \mathbf{j} \qquad (6.62)$$

where $\mathbf{j}$ is a unit vector in the y direction. Therefore, when the **L** vector lies in the xz plane as we have assumed, it must acquire a y component. In a short time δt, the change in **L** is

$$\delta \mathbf{L} = \frac{d\mathbf{L}}{dt}\delta t = \frac{e\hbar}{2m}B_x \delta t \mathbf{j} \qquad (6.63)$$

As time passes, the directions of **L** and $\delta \mathbf{L}$ change, and **L** precesses about the x axis (Figure 6.6). As the tip of the **L** vector describes a circle whose radius is $\hbar$, the angle ϕ_x through which it moves in time t is given

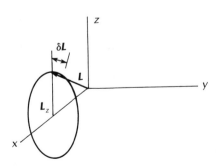

FIGURE 6.6 Precession of the tip of the **L** vector in a magnetic field along the x axis. **L** changes by $\delta \mathbf{L}$ in time δt.

by

$$\phi_x = \delta L/\hbar = (eB_x/2m)t \tag{6.64}$$

Thus classical theory tells us that L_z changes when we measure L_x, as you can see from Figure 6.6. But classical theory says that we still know the value of L_z, because *there is no classical limit on the precision with which the amount of change in L_z may be calculated.*

This recalls the Heisenberg microscope (Section 2.4), in which one could measure both x and p_x if we knew how much momentum the photon gave to the electron when x was measured the first time. We cannot know this, because we do not know the precise angle at which the photon was scattered from the electron.

The situation with angular momentum is similar. In this case, δL_z becomes uncertain because we do not know the precise angle through which the **L** vector precessed during the measurement of L_x. Why not?

Remember that the magnetic field cannot be uniform, because a field gradient is required to deflect the atoms. To know the precise angle of precession, we must know the magnetic field strength precisely, which means that we must know the precise x coordinate of the atoms as they pass through the magnet. If there is an uncertainty δx in this coordinate, then according to Eq. (2.31), the uncertainty in p_x must be

$$\delta p_x \geq \hbar/2\delta x \tag{6.65}$$

Any attempt to increase the precision in x will make δp_x larger. What if δp_x is larger than the momentum p_x that the magnetic field gives to the atoms that have $L_x = \hbar$? Then the $L_x = 0$ and $L_x = \hbar$ atoms will not be split into separate distinguishable beams. Thus we have two conflicting requirements:

1. To measure L_x, δp_x must be small. That means we need a sufficiently *broad* atomic beam; δx must be *large*.
2. To know the angle of precession (and thus to retain knowledge of L_z), we need a *narrow* beam; δx must be *small*.

The atoms with $L_x = \hbar$ acquire momentum $p_x = F_x t$, where t is the time over which the magnetic force F_x is applied, and $F_x = \mu_x \partial B_x/\partial x$. This momentum must be larger than the momentum uncertainty given by Eq. (6.65). Thus

$$\mu_x \frac{\partial B_x}{\partial x} t > \hbar/2\delta x \tag{6.66}$$

or, with $\mu_x = e\hbar/2m$,

$$(e\hbar/2m)\frac{\partial B_x}{\partial x}t\,\delta x > \hbar/2 \qquad (6.67)$$

In terms of δx, the uncertainty in the precession angle must be, according to Eq. (6.64),

$$\delta\phi_x = (e/2m)(\partial B_x/\partial x)t\,\delta x \qquad (6.68)$$

Substituting from Eq. (6.68), we have $\delta\phi_x > 1/2$ (radians). Thus we see that our precise knowledge of L_z has vanished as a result of measuring L_x and that this effect is directly related to the uncertainty relation between x and p_x.

This classical approach to the measurement serves only to make the quantum mechanical result plausible. We will explore the quantum mechanical states and operators in the next chapter, and we shall see that the knowledge of L_z vanishes completely. What we see experimentally is shown by the following steps. If we

1. split a beam into discrete components with different values of L_z (but the same l),
2. measure L_x for each of these atoms,
3. and measure L_z again,

we will find no correlation whatsoever between the two measurements of L_z. However, there will be a correlation between the measurement of L_x and either measurement of L_z. The results of Example Problem 6.2 illustrate this point. In that case, L_z had been measured, and we find an expectation value of $3\hbar^2$ for L_x. If we had known only L^2 and nothing about any component of L, the expectation value would have been $4\hbar^2$ for the square of each of the three components. Chapter 7 shows how to compute probabilities in cases like these.

ADDITIONAL READING

R. L. Liboff, *Introductory Quantum Mechanics*, Chapter 9, gives a thorough development of angular momentum operators and eigenvalues, with a list of the spherical harmonics through $l = 3$, plus many problems.

EXERCISES

1. A two-dimensional box is 3.0 nm long and 2.0 nm wide. An electron's potential energy V is zero inside the box and infinite outside.

(a) Find the electron's lowest three energy levels.

(b) Find the degeneracy of each level.

2. A three-dimensional box is 4.0 nm high, 3.0 nm long, and 2.0 nm wide. An electron's potential energy V is zero inside the box and infinite outside.

(a) Find the electron's lowest five energy levels.

(b) Find the degeneracy of each level.

3. An anisotropic three-dimensional harmonic oscillator has potential energy

$$V(x, y, z) = (m/2)\left(\omega_x^2 x^2 + \omega_y^2 y^2 + \omega_z^2 z^2 \right)$$

(a) Write a general expression for the energy levels, in terms of the frequencies ω_x, ω_y, and ω_z, and the quantum numbers n_x, n_y, and n_z.

(b) Find the four lowest energy levels if $\omega_x = \omega_y = \omega_z/2$, and determine the degeneracy of each level.

4. Verify that Eq. (6.19) follows from substitution into Eq. (6.3), using Eq. (6.17) and the corresponding equations for $\partial^2 u/\partial y^2$ and $\partial^2 u/\partial z^2$.

5. Verify by direct substitution that $Y_{1,1}$ is a solution of Eq. (6.41), with the expected eigenvalue for L^2 when $l = 1$.

6. Using Eqs. (6.29), verify by direct application of the operators that $L_x L_y - L_y L_x = i\hbar L_z$.

7. Use Eq. (6.50) to generate the Legendre polynomial $P_4(\cos \theta)$, and verify that it is an eigenfunction of Eq. (6.45) with the expected value of α.

8. Verify that the listed functions $Y_{1,1}(\theta, \phi)$ and $Y_{2,2}(\theta, \phi)$ satisfy the normalization condition (6.54).

9. From the form of the functions $Y_{1,1}(\theta, \phi)$ and $Y_{2,2}(\theta, \phi)$, make an educated guess to determine the function $Y_{3,3}(\theta, \phi)$. Verify that your guess is consistent with Eq. (6.30) and Eq. (6.39). Is the resulting value of α correct?

10. The requirement that m be an integer comes from the physical restriction that the wave function be single valued. Mathematically, $m = 1/2$ will yield a function that is a solution to Eq. (6.39); such a solution, with $l = 1/2$, can be written as

$$Y_{1/2, 1/2} = N(\sin \theta \, e^{i\phi})^{1/2} \quad \text{(compare with } Y_{1,1} \text{ and } Y_{2,2}.)$$

Verify by direct substitution into Eq. (6.39a) that $Y_{1/2, 1/2}$ is a solution to that equation when $m = 1/2$ and $\alpha = l(l + 1) = (1/2)(3/2) = 3/4$.

11. All of the spherical harmonics can be generated from the formula

$$Y_{l,m} = \left\{ \frac{(2l+1)(l+|m|)!}{4\pi(l-|m|)!} \right\}^{1/2} (2^l l!(\sin\theta)^{|m|})^{-1} \left\{ \frac{d}{d\cos\theta} \right\}^{l-|m|} (\sin\theta)^{2l} e^{im\phi}$$

Use this formula to generate $Y_{3,2}$ and verify that it is a solution of Eq. (6.39) with the correct value of α.

12. Show that the sum of the probability density functions

$$\sum_{m=-2}^{+2} |Y_{2,m}(\theta,\phi)|^2$$

is spherically symmetric. (In general,

$$\sum_{m=-l}^{+l} |Y_{l,m}(\theta,\phi)|^2$$

is spherically symmetric for any value of l.)[6]

[6] The general result was proven by A. Unsöld, *Annal. Phys.* **52**, 355 (1927).

CHAPTER 7

Angular Momentum and Superposition of States

We saw in Chapter 6 that the spherical harmonics $Y_{l,m}$ are eigenfunctions of the angular momentum operators L^2 and L_z, with eigenvalues $l(l + 1)\hbar^2$ and $m\hbar$, respectively. We also saw that we cannot simultaneously know more than one component of the vector $\mathbf{L}$ but a measurement of L_z does give information about the average values of L_x^2 and L_y^2. In this chapter we complete the picture by working out the eigenfunctions of L_x and L_y, and eventually compute the probabilities of various results of measuring any component of $\mathbf{L}$ after a different component has been measured.

In the process of doing this, we will see the power of the matrix representation of quantum mechanical operators.

7.1 EIGENFUNCTIONS OF ANGULAR MOMENTUM COMPONENTS

When we make a single measurement of an angular momentum component, that component is L_z, by definition. *The measurement defines the z axis.* If we now measure the component along another axis, we must give a different label to the second axis. If we call this axis the x axis, then the results of the measurement are dictated by the eigenfunctions of the L_x operator.

7.1.1 Rotation of Axes

Since we already know the eigenfunctions of the L_z operator, there is a simple way to find the eigenfunctions of the L_x operator: We write each eigenfunction of L_z in *rectangular* coordinates (instead of spherical coordinates), and then rotate the coordinate system. If we rotate the axes by 90°

139

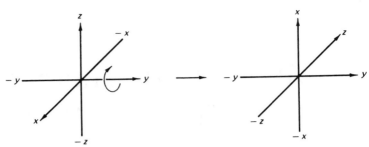

FIGURE 7.1 A rotation of 90° about the y axis causes the x axis to replace the z axis.

about the y axis, then the direction that was labeled z is now labeled x, and the direction previously labeled x becomes $-z$. (See Figure 7.1).

The effect of this rotation can be seen clearly if we write the eigenfunctions in rectangular coordinates rather than the spherical coordinates of Chapter 6. Thus we have, for $l = 1$,

$$Y_{1,1} = -\sqrt{3/8\pi}\,\sin\theta e^{i\phi} = -\sqrt{3/8\pi}\,\sin\theta(\cos\phi + i\sin\phi)$$

$$= -\sqrt{3/8\pi}\,\frac{x + iy}{r} \tag{7.1a}$$

$$Y_{1,0} = \sqrt{3/4\pi}\cos\theta = \sqrt{3/4\pi}\,\frac{z}{r} \tag{7.1b}$$

$$Y_{1,-1} = \sqrt{3/8\pi}\,\sin\theta e^{-i\phi} = \sqrt{3/8\pi}\,\sin\theta(\cos\phi - i\sin\phi) = \sqrt{3/8\pi}\,\frac{x - iy}{r} \tag{7.1c}$$

After rotation of the axes, we can consider the apparatus to be unchanged; only the coordinate labels are changed. Figure 7.1 shows that $x \to -z$, $y \to y$, and $z \to x$. The wave functions *relative to the apparatus* are unchanged, but they are described in terms of *the new labels* as eigenfunctions of L_x. We identify them by using the subscript l, mx, as follows:

$$Y_{1,1x} = -\sqrt{3/8\pi}\,\frac{z - iy}{r}, \qquad Y_{1,0x} = \sqrt{3/4\pi}\,\frac{x}{r},$$

$$Y_{1,1x} = \sqrt{3/8\pi}\,\frac{-z - iy}{r} \tag{7.2}$$

As an exercise, you may verify by direct application of the L_x and L^2 operators [Eqs. (6.27) and (6.31), respectively] that these functions are indeed eigenfunctions of L_x and L^2 with the indicated eigenvalues. Exercise 1 involves the application of this transformation to the five eigenfunctions for $l = 2$.

7.1.2 Superposition of Eigenfunctions of Angular Momentum

The functions of Eqs. (7.2) can now be used to solve the problem posed above, to determine the probabilities of the various possible results when measuring L_x after previously measuring L_z. Let us suppose that we have found that $l = 1$ and $m = 1$, so that the wave function's angular dependence is $Y_{1,1}$. To find the probabilities of the various results for L_x, we consider the function $Y_{1,1}$ to be a *superposition of the eigenfunctions of L_x*, just consider a wave to be a superposition of waves of different wavelengths when we do Fourier analysis. To see that this is possible, we simply do it; we write

$$Y_{1,1} = aY_{1,1x} + bY_{1,0x} + cY_{1,-1x} \tag{7.3}$$

where the coefficients a, b, and c are amplitudes that determine the probabilities of obtaining the results $+\hbar$, 0, and $-\hbar$, respectively, when L_x is measured. The probability is the square of the amplitude; $|a|^2$ is the probability of obtaining the result $L_x = +\hbar$, just as $|\phi(k)|^2$ equals the probability that p will be found to equal $\hbar k$ in a momentum measurement.

Substitution of the expressions from Eqs. (7.1) and (7.2) into Eq. (7.3) gives

$$-\sqrt{3/8\pi}\,\frac{x + iy}{r} = a\sqrt{3/8\pi}\,\frac{z - iy}{r} + b\sqrt{3/4\pi}\,\frac{x}{r} + c\sqrt{3/8\pi}\,\frac{-z - iy}{r} \tag{7.4}$$

or

$$-\frac{x + iy}{r} = a\frac{z - iy}{r} + b\sqrt{2}\,\frac{x}{r} + c\frac{-z - iy}{r} \tag{7.5}$$

Equation (7.5) must hold for all values of x, y, and z, so we equate the coefficients of x, y, and z in turn, obtaining

$$b\sqrt{2} = -1, \qquad a + c = -1, \qquad a - c = 0 \tag{7.6}$$

with the result that

$$a = -1/2, \quad b = -1/\sqrt{2}, \quad c = -1/2 \tag{7.7}$$

and

$$|a|^2 = 1/4, \quad |b|^2 = 1/2, \quad |c|^2 = 1/4 \tag{7.8}$$

How can we verify this result? First, we see that the sum of the probabilities is 1, as it should be. Second, we can determine the value of $\langle L_x^2 \rangle$ that follows from the result. We deduce from Eqs. (7.8) that $L_x^2 = 0$ half of the time, and $L_x^2 = \hbar^2$ half of the time, so that $\langle L_x^2 \rangle = \hbar^2/2$. Because there is no difference between the x axis and the y axis in this situation (after we have just measured L_z), we can also deduce the $\langle L_y^2 \rangle = \hbar^2/2$, and $\langle L_x^2 \rangle + \langle L_y^2 \rangle = \hbar^2$. This is consistent with the initial condition that $L^2 = 2\hbar^2$ and $L_z^2 = \hbar^2$; we require that $\langle L_x^2 \rangle + \langle L_y^2 \rangle = L^2 - L_z^2$.

Example Problem 7.1 Deduce the probabilities of the various results of measuring L_z for a system in which $l = 2$ and $L_x = 2\hbar$, given that the probability of finding $L_z = 0$ is 3/8.

Solution. From the reasoning above, $2\langle L_z^2 \rangle = L^2 - \langle L_x^2 \rangle = 6\hbar^2 - 4\hbar^2 = 2\hbar^2$, or $\langle L_z^2 \rangle = \hbar^2$. We know that the possible results are 0, $\pm\hbar$, and $\pm 2\hbar$. From symmetry we know that the probability that $L_z = +\hbar$ is equal to the probability that $L_z = -\hbar$; the same is true for $+2\hbar$ and $-2\hbar$. Then $\langle L_z^2 \rangle = \hbar^2 = \alpha\hbar^2 + \beta(4\hbar^2)$, where α is the probability that L_z will equal either $+\hbar$ or $-\hbar$, and β is the probability that L_z will equal either $+2\hbar$ or $-2\hbar$. Thus we can write $\alpha + 4\beta = 1$. Given that the total probability of a nonzero result is $\alpha + \beta = 1 - 3/8 = 5/8$, we can solve for α and β to obtain $\alpha = 1/2$ and $\beta = 1/8$.

7.2 THE SUPERPOSITION POSTULATE

The expansion of the spherical harmonics can be applied to many other functions in quantum mechanics, to obtain probabilities of experimental results. The procedure is of such general use that we have expressed it as the third fundamental postulate to accompany Postulates 1 and 2 of Section 3.3.

POSTULATE 3 Any acceptable wave function ψ can be expanded in a series of eigenfunctions of the operator corresponding to any dynamical variable.

As a postulate, this statement cannot be proved, but it underlies all of quantum mechanics. We have already used it implicitly in the previous section as well as in discussions of Fourier analysis of waves.[1] Postulate 3 has a useful corollary that concerns results of measurements:

COROLLARY TO POSTULATE 3 *The probability of finding a particular value of a dynamical variable as the result of a measurement on a system with normalized wave function ψ is equal to the sum of the squares of the coefficients of the corresponding normalized eigenfunctions in the expansion of the function ψ.*

The corollary refers to the *sum of the squares* of the coefficients. This is best explained with an example. If the angular dependence of the wave function is any function $f(\theta, \phi)$, then we can always write it as a sum of the functions $Y_{l,m}(\theta, \phi)$ with constant coefficients $c_{l,m}$:

$$f(\theta, \phi) = \sum_{l=0}^{\infty} \sum_{m=-l}^{m=l} c_{l,m} Y_{l,m}(\theta, \phi) \tag{7.9}$$

Then, for example, $|c_{3,2}|^2$ is the probability of finding $L^2 = 3(3 + 1)\hbar^2$ and $L_z = 2\hbar$ when these two variables are *both* measured. When L^2 *only* is measured, the probability that $l = 3$ is the sum of all seven possible values of $|c_{3,m}|^2$: $\sum_{m=-3}^{3} |c_{3,m}|^2$.

The identification of this sum with the probabilities is required if we are to obtain expectation values that are consistent with Postulate 2 (Section 3.3). For example, in analogy to Eq. (3.59), we write the expectation value of L^2 as

$$\langle L^2 \rangle = \int_0^{2\pi} \int_0^{\pi} f^*(\theta, \phi) L_{\mathrm{op}}^2 f(\theta, \phi) \sin \theta \, d\theta \, d\phi \tag{7.10}$$

where $f(\theta, \phi)$ is the *normalized* angular factor in the wavefunction and L_{op}^2 is the operator defined by Eq. (6.31). Substitution of the expansion (7.9) into the integral yields

$$\langle L^2 \rangle = \sum_{l,m} \sum_{l',m'} \int_0^{2\pi} \int_0^{\pi} c_{l',m'}^* Y_{l',m'}' L_{\mathrm{op}}^2 c_{l,m} Y_{l,m} \sin \theta \, d\theta \, d\phi \tag{7.11}$$

---

[1] One could always prove mathematically that any *particular* set of functions is a "complete set," having the property that by combining members of the set in various ways, one can form any other function that is reasonably well behaved. But the postulate is physical rather than mathematical; it concerns the completeness of those functions that are eigenfunctions of an operator representing a *dynamical variable*.

After performing the indicated operation with the L^2 operator, we obtain

$$\langle L^2 \rangle = \sum_{l,m} \sum_{l',m'} \int_0^{2\pi} \int_0^{\pi} c_{l',m'}^* Y_{l',m'}^* \hbar^2 l(l+1) c_{l,m} Y_{l,m} \sin \theta \, d\theta \, d\phi \quad (7.12)$$

To evaluate the integral, we use the fact that the spherical harmonics are orthogonal and normalized; that is

$$\int_0^{2\pi} \int_0^{\pi} Y_{l',m'}^* Y_{l,m} \sin \theta \, d\theta \, d\phi = \delta_{l'l} \delta_{m'm} \quad (7.13)$$

Thus every term of the infinite series vanishes except the terms for which $l' = l$ and $m' = m$. This reduces Eq. (7.12) to

$$\langle L^2 \rangle = \sum_{l,m} l(l+1)\hbar^2 |c_{l,m}|^2 = \sum_{l} l(l+1)\hbar^2 \sum_{m=-l}^{l} |c_{l,m}|^2 \quad (7.14)$$

But, by definition, $\langle L^2 \rangle = \sum_l l(l+1)\hbar^2 P(l)$, where $P(l)$ is the probability that L^2 will be equal to $l(l+1)$ when it is measured. Therefore we find that

$$P(l) = \sum_{m=-l}^{l} |c_{l,m}|^2 \quad (7.15)$$

in agreement with the corollary.

7.2.1 Calculation of the Coefficients $c_{l,m}$

To find any particular coefficient $c_{l',m'}$ in the expansion of any given function $f(\theta, \phi)$, we make use of the fact that the $Y_{l,m}$ are orthogonal and normalized. We multiply both sides of Eq. (7.9) by $Y_{l',m'}^*$, and then integrate both sides of the equation over the surface of a sphere, using the area element $\sin \theta \, d\theta \, d\phi$. Because of the orthogonality of the $Y_{l,m}$ [Eq. (7.13)], all terms except one drop out of the series, and we obtain

$$c_{l',m'} = \int_0^{2\pi} \int_0^{\pi} f(\theta, \phi) Y_{l',m'}^*(\theta, \phi) \sin \theta \, d\theta \, d\phi \quad (7.16)$$

Example. If $f(\theta, \phi) = N \sin \theta$, then

$$c_{1,0} = \sqrt{3/8\pi} \int_0^{2\pi} \int_0^\pi N \sin \theta [\cos \theta] \sin \theta \, d\theta \, d\phi$$

where $\sqrt{3/8\pi}$ is the normalizing factor for $Y_{1,1}$. The result is $c_{1,0} = 0$. The same procedure shows that all of the other coefficients are zero except $c_{1,1}$ and $c_{1,-1}$, which must be equal in magnitude, because $f(\theta, \phi)$ is independent of ϕ. If we use the correct normalizing factor N, then we will find that $|c_{1,1}|^2 + |c_{1,-1}|^2 = 1$. The normalizing factor N is given by the same condition used for normalizing the spherical harmonics: $\int_0^{2\pi} \int_0^\pi |Nf(\theta, \phi)|^2 \sin \theta \, d\theta \, d\phi = 1$

Example Problem 7.2 If $f(\theta, \phi) = N(3 + 4\sqrt{3} \cos \theta)$, where $N = \sqrt{1/100\pi}$ is a normalizing factor, find the probability that $l = 1$. What values of l are possible?

Solution. For $l = 1$, Eq. (7.16) shows that $c_{1,1} = c_{1,-1} = 0$, and

$$c_{1,0} = 2\pi/\sqrt{100\pi} \int_0^\pi (3 + 4\sqrt{3} \cos \theta)\sqrt{3/4\pi} \cos \theta \sin \theta \, d\theta = 4/5$$

Similarly,

$$c_{0,0} = 2\pi/\sqrt{100\pi} \int_0^\pi (3 + 4\sqrt{3} \cos \theta)\sqrt{1/4\pi} \sin \theta \, d\theta = 3/5$$

Thus the probability that $l = 1$ is $(4/5)^2$, or $16/25$, and the probability that $l = 0$ is $(3/5)^2$, or $9/25$. The sum of these probabilities is 1; therefore no other values of l are possible. You can verify this statement for $l = 2$ by referring to the spherical harmonics shown in Table 6.1. All of them are orthogonal to the given function $f(\theta, \phi)$.

7.2.2 Commuting Operators

In quantum mechanics there are many cases in which measurement of one dynamical variable disturbs the value of another variable, making it impossible to know the values of both variables simultaneously. There is a general rule for identifying such pairs of variables:

If it is possible to measure a dynamical variable A without losing the knowledge of another dynamical variable B, then the operators O_a and O_b representing the two variables must *commute.*

This means that $O_a O_b \psi = O_b O_a \psi$ for any function ψ, with the understanding that the operator on the right is applied first and then the other operator is applied to the result. The justification for this rule is as follows:

- If the value of a variable A is to be unchanged when variable A is measured, it must be possible to expand any eigenfunction of A in a series of eigenfunctions of *both* operators O_a and O_b.[2]
- Since any function whatever may be expanded in a series of eigenfunctions of operator A, and each of these may be expanded in a series of eigenfunctions of both variables, we see that any function may be expanded in a series of eigenfunctions of *both* operators O_a and O_b.
- Let ψ_{ij} be an eigenfunction of both O_a and O_b, with eigenvalues a_i and b_j, respectively. Then we may write

$$O_a \psi_{ij} = a_i \psi_{ij} \quad \text{and} \quad O_b \psi_{ij} = b_j \psi_{ij}$$

Applying both operators gives

$$O_a O_b \psi_{ij} = O_a b_j \psi_{ij} = a_i b_j \psi_{ij}$$

and

$$O_b O_a \psi_{ij} = O_b a_i \psi_{ij} = a_i b_j \psi_{ij}$$

Therefore

$$O_a O_b \psi_{ij} = O_b O_a \psi_{ij}$$

and these operators commute when applied to any of the functions ψ_{ij}. But this means that they must commute when applied to any wave function, because any wave function can be expanded in a series of the functions ψ_{ij}, according to Postulate 3. Thus the commutation of operators O_a and O_b is a necessary condition if variables A and B are to be simultaneously measurable.

For example, we have assumed that it is possible to measure E and L^2 simultaneously; this requires that the operator H (the energy operator) and the operator L^2 must commute. This is possible if the potential energy V is a function of r only, because in that case the Schrödinger equation [Eq. (6.34)] may be written in the form $H = p_r^2/2m + L^2/2mr^2 + V(r)$, where p_r operates on r only. Clearly L^2 commutes with H; the operator

---

[2]Let ψ be an eigenfunction of operator O_a. We can always write $\psi = \Sigma c_n \psi_{nb}$, where the functions ψ_{nb} are eigenfunctions of operator O_b. After we measure variable B for the state whose wave function is ψ, the new wave function must be one of the functions ψ_{nb}, according to the expansion postulate. But if we have measured B *without disturbing the value of variable A*, then the new wave function must be an eigenfunction of variable A as well as variable B. Thus *all* of the functions ψ_{nb} are eigenfunctions of both A and B.

L^2 must commute with each term in H, because L^2 does not operate on the variable r.

On the other hand, the operators L_x and L_z do not commute. You may verify from Eqs. (6.29) that

$$L_x L_y - L_y L_x = i\hbar L_z \qquad (7.17a)$$

Two additional commutation relations result from permuting the subscripts to obtain

$$L_y L_z - L_z L_y = i\hbar L_x \qquad (7.17b)$$

and

$$L_z L_x - L_x L_z = i\hbar L_y \qquad (7.17c)$$

7.3 MATRIX METHODS

We can express the methods of the previous section in a more manageable form by introducing the methods of matrix algebra. We can represent a quantum mechanical operator by a matrix that acts upon the quantum states, which are in turn represented as vectors in an abstract space. This space (called *Hilbert space* after mathematician David Hilbert) is constructed by representing each normalized eigenfunction by a unit vector. These unit vectors are orthogonal, as the eigenfunctions themselves are. Then any other function can be represented by a linear combination of these vectors, just as we have written the function as a superposition of the eigenfunctions.

Each dynamical system (e.g., the hydrogen atom or a harmonic oscillator) has its own Hilbert space, which has as many dimensions as there are eigenfunctions of the operators for the dynamical variables of that system. The angular momentum operators are uniquely suitable for illustrating how this works in practice.

Consider the three eigenfunctions for $l = 1$. The corresponding subspace of Hilbert space is three dimensional, and we can write the three eigenfunctions as unit vectors just as we would write an ordinary column vector:

$$Y_{1,1} \rightarrow \begin{pmatrix} 1 \\ 0 \\ 0 \end{pmatrix}, \qquad Y_{1,0} \rightarrow \begin{pmatrix} 0 \\ 1 \\ 0 \end{pmatrix}, \qquad Y_{1,-1} \rightarrow \begin{pmatrix} 0 \\ 0 \\ 1 \end{pmatrix} \qquad (7.18)$$

Then any function F for which $l = 1$ can be a three-dimensional vector:

$$\begin{pmatrix} a \\ b \\ c \end{pmatrix}, \text{ or } F \rightarrow a\begin{pmatrix} 1 \\ 0 \\ 0 \end{pmatrix} + b\begin{pmatrix} 0 \\ 1 \\ 0 \end{pmatrix} + c\begin{pmatrix} 0 \\ 0 \\ 1 \end{pmatrix}$$

representing the combination $aY_{1,1} + bY_{1,0} + cY_{1,-1}$. The angular momentum operators can now be represented as matrices that operate on these vectors. Knowing the results of applying these operators to the spherical harmonics, we can deduce that a correct representation is

$$L_x = \frac{\hbar}{\sqrt{2}}\begin{pmatrix} 0 & 1 & 0 \\ 1 & 0 & 1 \\ 0 & 1 & 0 \end{pmatrix}, \quad L_y = \frac{i\hbar}{\sqrt{2}}\begin{pmatrix} 0 & -1 & 0 \\ 1 & 0 & -1 \\ 0 & 1 & 0 \end{pmatrix},$$

$$L_z = \hbar\begin{pmatrix} 1 & 0 & 0 \\ 0 & 0 & 0 \\ 0 & 0 & -1 \end{pmatrix}$$

(7.19)

Using this representation, we can determine the result of applying operators L_x, L_y, or L_z to any function that can be expanded in spherical harmonics. For example, the operation $L_x Y_{1,1}$ may be represented as

$$\frac{\hbar}{\sqrt{2}}\begin{pmatrix} 0 & 1 & 0 \\ 1 & 0 & 1 \\ 0 & 1 & 0 \end{pmatrix}\begin{pmatrix} 1 \\ 0 \\ 0 \end{pmatrix}, \quad \text{with the result} \quad \frac{\hbar}{\sqrt{2}}\begin{pmatrix} 0 \\ 1 \\ 0 \end{pmatrix}$$

This means that $L_x Y_{1,1} = (\hbar/\sqrt{2})Y_{1,0}$, which you may verify by applying the differential operator for L_x to the function $Y_{1,1}$.

Example Problem 7.3 Using Eqs. (7.18) and (7.19), find a vector that is an eigenfunction of L_x, and express this function in terms of spherical harmonics.

Solution. Let us consider the eigenvalue zero. In that case the result of the matrix operation must be

$$L_x = \frac{\hbar}{\sqrt{2}}\begin{pmatrix} 0 & 1 & 0 \\ 1 & 0 & 1 \\ 0 & 1 & 0 \end{pmatrix}\begin{pmatrix} a \\ b \\ c \end{pmatrix} = 0$$

But the result of the indicated operation is the column vector

$$\frac{\hbar}{\sqrt{2}}\begin{pmatrix} b \\ a + c \\ b \end{pmatrix}$$

If this vector is zero, then each of its components must be zero, and thus we know that $b = 0$ and $a + c = 0$, or $a = -c$. The absolute values of a and c are given by normalization: $|a|^2 + |c|^2 = 1$. Thus $a = 1/\sqrt{2}$ and $c = -1/\sqrt{2}$, and the eigenfunction is $(1/\sqrt{2})(Y_{1,1} - Y_{1,-1})$. This can obviously be verified by application of the differential operator for L_x.

The same procedure can be applied to the set of five eigenfunctions for $l = 2$, using vectors in five dimensions, operated on by 5-by-5 matrices. For example, the matrix for L_z is

$$L_z = \hbar \begin{pmatrix} 2 & 0 & 0 & 0 & 0 \\ 0 & 1 & 0 & 0 & 0 \\ 0 & 0 & 0 & 0 & 0 \\ 0 & 0 & 0 & -1 & 0 \\ 0 & 0 & 0 & 0 & -2 \end{pmatrix} \tag{7.20}$$

with eigenvalues of $2\hbar$, $\hbar$, 0, $-\hbar$, and $-2\hbar$.

In each of these examples, the matrix operator is very convenient but is not necessary, because the eigenvalues have already been found by using the differential operators. But in some situations (for example, in the case of spin angular momentum), there are no space coordinates and no differential operator. In those cases we have no alternative but to use an abstract operator and to do the computations by matrix algebra.

But, you may ask, how do we know what the matrix is, when there is no differential operator to guide us? One way to proceed is to use the commutation relations, Eqs. (7.17). Later we shall see how these relations allow us to analyze spin, but here we illustrate the power of Eqs. (7.17) by using them to derive the eigenvalues of L^2 in a completely independent way. In doing this we shall introduce a powerful shorthand method, invented by P. A. M. Dirac, to write symbols for *eigenstates*, independently of the particular functions of space, time, or spin that might be involved.

7.4 GENERALIZED OPERATORS AND DIRAC NOTATION

Instead of dealing with the specific form of the wave function, Dirac[3] focused on the state of the system, and developed a set of symbols for *state vectors* in Hilbert space. We illustrate with the angular momentum states.

[3]P. A. M. Dirac, *Quantum Mechanics*, 4th edition, Oxford Science Publications, New York (1958), reprinted in paperback (1991).

The symbol for the state vector is $|l, m\rangle$, representing a state whose angular coordinates are given by $Y_{l,m}$. Operating on this state with operators L^2 and L_z gives the operator equations that we already know, now written

$$L^2|l, m\rangle = l(l + 1)\hbar^2|l, m\rangle \quad \text{and} \quad L_z|l, m\rangle = m\hbar|l, m\rangle \quad (7.21)$$

Can we derive the eigenvalue $l(l + 1)\hbar^2$, using only the eigenvalues of L_z, the definition of l as the maximum value of m, and Eqs. (7.17)? We can do it by using a trick similar to the one used to find the harmonic oscillator eigenfunctions. We define a new operator $L_+ = (L_x + iL_y)$, apply it to the state $|l, m\rangle$, and then demonstrate that the resulting state is also an eigenstate of L_z, as follows:

$$L_z L_+|l, m\rangle = (L_z L_x + iL_z L_y)|l, m\rangle \quad (7.22)$$

Using the commutation relations, we now make the substitutions

$$L_z L_x = L_x L_z + i\hbar L_y \quad \text{and} \quad L_z L_y = L_y L_z - i\hbar L_x \quad (7.23)$$

to obtain

$$L_z L_+|l, m\rangle = (L_x L_z + i\hbar L_y + iL_y L_z + \hbar L_x)|l, m\rangle \quad (7.24)$$

Using $L_z|l, m\rangle = m\hbar|l, m\rangle$, we then have

$$L_z L_+|l, m\rangle = (m\hbar L_x + i\hbar L_y + im\hbar L_y + \hbar L_x)|l, m\rangle \quad (7.25)$$

which can be written

$$L_z L_+|l, m\rangle = (m\hbar + \hbar)(L_x + iL_y)|l, m\rangle$$

or

$$L_z L_+|l, m\rangle = (m\hbar + \hbar)L_+|l, m\rangle \quad (7.26)$$

Thus we see that L_+ is a *raising operator*; $L_+|l, m\rangle$ is another eigenstate of L_z, with eigenvalue $m\hbar + \hbar$. We could write the state vector as $|l, m + 1\rangle$. In a similar way we can generate the state vector $|l, m - 1\rangle$ by applying the *lowering* operator $L_- = L_x - iL_y$, and by repeated use of these operators we can generate a whole set of eigenstates of L_z with eigenvalues $m\hbar, (m \pm \hbar), (m \pm 2\hbar)\ldots$

7.4.1 Eigenvalues of L^2

We can prove that each of these states is an eigenstate of L^2 as well as L_z; we do this by applying the operator $L^2 = L_x^2 + L_y^2 + L_z^2$ to the state $L_+|l, m\rangle$ and work it out with the aid of the commutation relations [Eqs. (7.21)]. But if all of these states have the same value of L^2, there must be an upper limit on L_z. By definition, that upper limit is $l\hbar$, and the corresponding state may be written $|l, l\rangle$.

This means that $L_+|l, l\rangle = (L_x + iL_y)|l, l\rangle = 0$, because there is no state $|l, l + 1\rangle$. If we apply the operator $(L_x - iL_y)$, we obtain

$$(L_x - iL_y)(L_x + iL_y)|l, l\rangle = 0 \tag{7.27}$$

or

$$\left[L_x^2 + L_y^2 + i(L_x L_y - L_y L_x) \right]|l, l\rangle = 0 \tag{7.28}$$

Using the commutation rule and the relation $L_x^2 + L_y^2 = L^2 - L_z^2$, we have

$$(L^2 - L_z^2 - \hbar L_z)|l, l\rangle = 0 \tag{7.29}$$

or

$$L^2|l, l\rangle = (L_z^2 + \hbar L_z)|l, l\rangle = (l^2\hbar^2 + l\hbar^2)|l, l\rangle = l(l + 1)\hbar^2|l, l\rangle \tag{7.30}$$

the same eigenvalue obtained in Chapter 6, but this time it is obtained from the commutation relations without reference to the form of the eigenfunctions. We shall encounter other uses of this method in later chapters.

7.4.2 Application: Decay of the Neutral K Meson

Next we investigate the superposition of states for which no coordinate representation even exists. The neutral K meson (K^0) is a short-lived particle that can decay into either two pions or into various combinations of three particles. Its antiparticle ($\overline{K}^0$), can decay in all of the same ways. For example:

$$K^0 \to \pi^+ + \pi^- \qquad \text{or} \quad K^0 \to \pi^0 + \pi^0 \tag{7.31}$$

or

$$K^0 \rightarrow \pi^+ + \pi^- + \pi^0 \quad \text{or} \quad K^0 \rightarrow \pi^0 + \pi^0 + \pi^0 \qquad (7.32)$$

Experiments showed that when a K^0 was first created, its decay was primarily into two, rather than three, particles, but after a few hundred picoseconds the decays were all into three particles.[4] (See Figure 7.2.) It appeared that there were two different particles, one of which lived longer and decayed in different ways, yet these two "particles" were created in exactly the same way.

M. Gell-Mann and A. Pais made the brilliant deduction that these two allegedly different particles were a single particle, the K^0, that could decay in two different ways.[5] They said that the K^0 exists as a superposition of two different *quantum states*, labeled $|K_1\rangle$ and $|K_2\rangle$, with equal amplitudes. Thus we can write the K^0 state as

$$|K^0\rangle = (|K_1\rangle + |K_2\rangle)/\sqrt{2} \qquad (7.33)$$

where $1/\sqrt{2}$ is a normalization factor.

The state $|K_1\rangle$ decays rapidly into two pions, but the state $|K_2\rangle$ decays into three particles [as shown in reaction (7.32)]. Eventually the $|K_1\rangle$

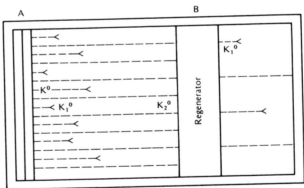

FIGURE 7.2   Decay of K^0 mesons. The symbol -< indicates decay of a K^0 into two pions, which is possible during a time interval of about 400 picoseconds. After that time the remaining K^0 produce no noticeable effect until they interact with a target, whereupon two-pion decay reappears. See text.

---

[4] The two charged particles always left a track in the detector. The neutral particle's presence could be deduced from its decay into two gamma rays.
[5] M. Gell-Mann and A. Pais, *Phys. Rev.* **97**, 1387 (1955).

component disappears, leaving only the $|K_2\rangle$ state. When the state $|K_1\rangle$ has disappeared, we no longer have a $|K^0\rangle$; we have a $|K_2\rangle$, a different "particle" with different properties. The $|K_2\rangle$ is much longer-lived than $|K_1\rangle$; its half life is about 35 nanoseconds.

Figure 7.3 shows these vectors in two dimensions; $|K_1\rangle$ and $|K_2\rangle$ are perpendicular unit vectors; $|K^0\rangle$ is a unit vector at an angle of 45° to both. When the state becomes $|K_2\rangle$, it can be written as a superposition of the two states $|K^0\rangle$ and $|\overline{K}^0\rangle$, which are orthogonal to each other:

$$|K_2\rangle = (|K^0\rangle - |\overline{K}^0\rangle)/\sqrt{2} \tag{7.34}$$

Similarly,

$$|K_1\rangle = (|K^0\rangle + |\overline{K}^0\rangle)/\sqrt{2} \tag{7.34a}$$

$|\overline{K}^0\rangle$ is the *anti* K^0 meson! Notice that each state is a unit vector; the sum of the squares of the coefficients in the superposition is always 1.

This identification of the $|\overline{K}^0\rangle$ is in agreement with experiment. The $|\overline{K}^0\rangle$ interacts with matter more strongly than the $|K^0\rangle$ does; it reacts with neutrons via $\overline{K}^0 + n \rightarrow K^- + p$. Thus as the K_2 beam passes through an absorber, the $|\overline{K}^0\rangle$ component is attenuated, leaving the mixed state vector

$$|K_{12}\rangle = a|K^0\rangle + b|\overline{K}^0\rangle \tag{7.35}$$

where initially $|K_{12}\rangle = |K_2\rangle$, $b = -a$, and normalization again requires that $|a|^2 + |b|^2 = 1$. This mixed state can also be expressed as a superposition of K_1 and K_2, as you can see from Figure 7.3. The decay into two pions reappears at this point. (See Figure 7.2.) The neutral K meson has other fascinating time-dependent features that will be better understood when we reach Chapter 14.

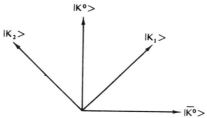

FIGURE 7.3 Vector diagram of relationships among state vectors $|K^0\rangle$, $|\overline{K}^0\rangle$, $|K_1\rangle$, and $|K_2\rangle$, all of which have unit length.

Example Problem 7.4 Ten thousand K_2 mesons enter an absorber; 4000 of them interact with neutrons, 2000 interact with protons (via $K^0 + p \rightarrow K^+ + n$), and the remaining 4000 emerge from the absorber in a mixed state.

(a) Write an expression for the resulting normalized mixed state in the form $|K_{12}\rangle = a|K^0\rangle + b|\overline{K}^0\rangle$, and find the values of a and b.

(b) Write an expression for the normalized mixed state in the form $|K_{12}\rangle = a'|K_1\rangle + b'|K_2\rangle$, and find the values of a' and b'. How many of the neutral K mesons emerging from the absorber will decay into two pions?

Solution. (a) If all of the K_2 mesons had interacted, 5000 would have had $\overline{K}^0$ interactions and 5000 K^0 interactions (in agreement with Eq. (7.32). We have seen 4000 $\overline{K}^0$ interactions and 2000 K^0 interactions already, so if the interactions could continue we would see 1000 more $\overline{K}^0$ events and 3000 more K^0 events. Thus the probabilities for a *given* meson are 3/4 that it acts as a K^0 and 1/4 that it acts as a $\overline{K}^0$, and $|K_{12}\rangle = \sqrt{3}/2|K^0\rangle - 1/2|\overline{K}^0\rangle$.

(b) We write

$$|K_{12}\rangle = a'|K_1\rangle + b'|K_2\rangle, \quad \text{with } |a'|^2 + |b'|^2 = 1.$$

We can say that a' is the length of the component of $|K_{12}\rangle$ along the $|K_1\rangle$ "axis." As with standard vectors, we can find this component by taking the "dot product" of the vectors $|K_{12}\rangle$ and $|K_1\rangle$. From Eqs. (7.32) we have

$$|K_1\rangle = (|K^0\rangle + |\overline{K}^0\rangle)/\sqrt{2} \quad \text{and} \quad |K_2\rangle = (|K^0\rangle - |\overline{K}^0\rangle)/\sqrt{2},$$

and we find the dot product in the usual way, by summing the products of the corresponding components; the result is $a' = (\sqrt{3}/2 - 1/2)/\sqrt{2} = 0.2588$. Squaring this coefficient we find that the probability is $|a'|^2 = 0.067$. The number of decays into two pions will then be $0.067 \times 4000 = 268$. (This result is of course subject to statistical fluctuations in any given test.)

Check. Find b' the same way; $b' = (\sqrt{3}/2 + 1/2)/\sqrt{2} = 0.9659$, and the corresponding probability is $|b'|^2 = 0.933$.

7.4.3 Conjugate Vectors

The dot product operation in Example Problem 7.4b is written in Dirac notation as $\langle K_{12}|K_1\rangle$, where $\langle K_{12}|$ is called the *conjugate vector* to $|K_{12}\rangle$; it can be written in terms of the conjugate vectors to $|K^0\rangle$ and $|\overline{K}^0\rangle$ as

$$\langle K_{12}| = \sqrt{3}/2\langle K^0| - 1/2\langle \overline{K}^0| \tag{7.36}$$

Then we can write the dot product in terms of the K^0 and anti-K^0 states as

$$\langle K_{12}|K_1\rangle = (\sqrt{3}/2\langle K^0| - 1/2\langle \overline{K}^0|)(|K^0\rangle + |\overline{K}^0\rangle)/\sqrt{2} \tag{7.37}$$

Since $|K^0\rangle$ is orthogonal to $\langle \overline{K}^0|$, we know that $\langle K^0|\overline{K}^0\rangle = \langle \overline{K}^0|K^0\rangle = 0$, and since both are normalized, $\langle K^0|K^0\rangle = \langle \overline{K}^0|\overline{K}^0\rangle = 1$. Equation (7.35) therefore yields

$$\langle K_{12}|K_1\rangle = \left(\sqrt{3}/2\langle K^0|K^0\rangle - 1/2\langle \overline{K}^0|\overline{K}^0\rangle/\sqrt{2}\right) = (\sqrt{3}/2 - 1/2)/\sqrt{2}$$

as in the example.

7.4.4 Complex Coefficients

The preceding example told only part of the story. In general, the coefficients a, b, a' and b' are complex numbers. In that case the coefficients of the conjugate vector $\langle K_1|$ are complex conjugates of the coefficients of the vector $|K_1\rangle$. Therefore,

$$\text{if } |K_{12}\rangle = a|K^0\rangle + b|\overline{K}^0\rangle, \quad \text{then } \langle K_{12}| = a^*\langle K^0| + b^*\langle \overline{K}^0| \tag{7.38}$$

The product $\langle K_{12}|K_{12}\rangle$ is equal to 1, because these are unit vectors. We know that $a^*a + b^*b$ must equal 1, because a^*a is the probability of finding K_{12} to be K^0, b^*b is the probability of finding K_{12} to be anti K^0, and there are no other possibilities. Dirac called any expression like $\langle K_{12}|K_{12}\rangle$ a "bracket," made up of two parts: $\langle K_{12}|$, called a "bra," and $|K_{12}\rangle$, called a "ket."

7.4.5 Expectation Values in Dirac Notation

Consider the problem of finding the expectation value of L_x^2 for a particle in the state denoted by $|l, m\rangle$, where $l = 1$ and $m = 1$. Using the form of Eq. (7.10), we could write

$$\langle L_x^2\rangle = \int_0^{2\pi}\int_0^{\pi} Y^*_{l,m} L_{x_{op}}^2 Y_{l,m} \sin\theta \, d\theta \, d\phi \tag{7.39}$$

then perform the $L_{x\,op}^2$ operation and integrate over all angles, eventually obtaining the result $\langle L_x^2 \rangle = \hbar^2/2$ in this case. But that is not necessary if we know the expansion of $Y_{l,m}$ in terms of eigenfunctions of L_x. In that case we simply use Eq. (7.11) and the orthogonality of the spherical harmonics to obtain the result without integration.

In the Dirac notation the equivalent definition of L_x^2 is

$$\langle L_x^2 \rangle = \langle l, m | L_x^2 | l, m \rangle \tag{7.40}$$

and we expand the bra and ket vectors in the same way that we expanded the spherical harmonics to obtain Eq. (7.11). With $l = 1$ and $m = 1$, the expansions are found from Eqs. (7.3) and (7.7) to be

$$|1, 1\rangle = a|1, 1_x\rangle + b|1, 0_x\rangle + c|1, -1_x\rangle \tag{7.41}$$

and

$$\langle 1, 1| = a^*\langle 1, 1_x| + b^*\langle 1, 0_x| + c^*\langle 1, -1_x| \tag{7.42}$$

with

$$a = -1/2, \qquad b = -1/\sqrt{2}, \qquad c = 1/2$$

We also know that

$$L_x^2|1, 1_x\rangle = \hbar^2|1, 1_x\rangle, \qquad L_x^2|1, 0_x\rangle = 0, \qquad L_x^2|1, -1_x\rangle = -\hbar^2 L_x^2|1, -1_x\rangle \tag{7.43}$$

Combining (7.38) through (7.41) and using the orthogonality and normalization of the vectors, we obtain $\langle L_x^2 \rangle = \hbar^2/4 + 0 + \hbar^2/4 = \hbar^2/2$, as we anticipated.

7.4.6 Hermitian Operators

We cannot measure imaginary quantities. This fact imposes a requirement on any operator O that can represent a dynamical variable. It is required that the expectation value $\langle O \rangle$ be a real quantity, or that for any acceptable wave function ψ, the integral $\int \psi^* F\psi \, d(\text{volume})$ must be real. In the notation of Dirac, $\langle \psi | O | \psi \rangle$ must be real.

Notice two important features of this bracket symbol:

- The form of the bracket is always interpreted to mean that the bra is the conjugate of the ket, so no asterisk is needed.
- The bracket can be evaluated by performing the operation O on

either the bra or the ket. If the operation is performed on the bra, then in the integral form it is equivalent to $\int (F\psi)^*\psi\, d(\text{volume})$. This requires that

$$\int(F\psi)^*\psi\,(d\text{volume}) = \int\psi^*F\psi\,d(\text{volume}) \qquad (7.44)$$

An operator that satisfies this requirement is called a *Hermitean* operator. Thus we have the requirement that

Any operator that represents a dynamical variable must be a Hermitean operator.

ADDITIONAL READING

D. Bohm, *Quantum Mechanics*, Prentice-Hall, New York (1951) has an excellent exposition of rotation of axes and transformation of angular momentum states, with some of the examples that are discussed here.

P. A. M. Dirac, *Quantum Mechanics*, 4th Edition, Oxford Science Publications, New York (1958, reprinted in paperback, 1991) approaches quantum physics in a completely abstract way, introducing state vectors, operators, and his notation first, before mentioning wave functions. He states, "The reason for this name (wave function) is that in the early days of quantum mechanics all the examples of these functions were in the form of waves. The name is not a descriptive one from the point of view of the modern general theory." His point of view is quite suited to the K mesons, whose only connection to conventional wave theory is in the relation between amplitudes and probabilities. Unfortunately, his book is lacking any such uncomplicated and enlightening example; the K mesons were analyzed after the last edition was published.

EXERCISES

1. Apply the transformation of Section 7.1 to the five functions $Y_{l,m}$ for $l = 2$, expressed in rectangular coordinates, to produce the five eigenfunctions of L_x for $l = 2$, expressed in rectangular coordinates.

2. Using the result of Exercise 1 and the algebraic method of Eqs. (7.4) through (7.7), find the five coefficients in the expansion of $Y_{2,2}$ in terms of eigenfunctions of L_x. Show that these coefficients give probabilities that agree with those found in Example Problem 7.1.

3. Solve Exercise 2 by using Eq. (7.16) instead of the algebraic method.

4. Using Eq. (7.16), with $f(\theta, \phi) = \sin\theta + \sin\theta\cos\theta$, find the probability that $l = 1$. What other value(s) of l is(are) possible?

5. (a) Using Eqs. (6.29), verify by direct application of the operators to the function $f(\theta, \phi) = \sin\theta\cos\theta$ that $L_xL_y - L_yL_x = i\hbar L_z$.

(b) Verify the same operator equation by means of matrix algebra, using Eqs. (7.19) and the appropriate column vector.

6. Using the method of Example Problem 7.3, find the eigenfunction of the L_x operator [from Eqs. (7.18)] with eigenvalue $-\hbar$.

7. By carrying out the matrix operations, find the results of the operations $L_xY_{1,0}$, $L_xY_{1,-1}$, $L_yY_{1,1}$, $L_yY_{1,0}$, and $L_yY_{1,-1}$, expressing them in terms of one or more of the normalized functions $Y_{1,m}$ with constant coefficients. Verify at least one of these results by application of the differential operator for L_x or L_y [Eqs. (6.29)].

8. Show that the state $L_+|l,m\rangle$ of Eq. (7.24) is an eigenstate of L^2, with the same eigenvalue as the state $|l,m\rangle$. (Apply the operator L^2 and use the commutation relations, as suggested in Section 7.4.)

9. Do Example Problem 7.4(b) graphically. Draw unit vector $|K_{12}\rangle$ on a copy of Figure 7.3, and find the length of the component of $|K_{12}\rangle$ along the $|K_1\rangle$ axis.

10. Do Example Problem 7.4(b) for the case that only 1000 of the K_2 mesons interact with protons, 3000 interact with neutrons, and 6000 emerge in a mixed state.

11. We have a beam of particles in which each particle has $L^2 = 2\hbar^2$ and $L_z = -\hbar$. For one of these particles we now measure the angular momentum component along the z' axis, which makes an angle of α with the z axis. In terms of α, write expressions for the probability that this component, $L_{z'}$, will be found to be equal to $\hbar$, $-\hbar$, or zero. [*Hint*: The eigenfunctions for these values are simply the eigenfunctions $Y_{1,m}$ expressed in terms of the primed coordinates instead of the unprimed coordinates; for example, we can rewrite the function $Y_{1,0}$ as $Y'_{1,0} = \sqrt{3/4\pi}(z'/r)$.] Then writing z' in terms of x, y, and z lets us find an expansion of $Y'_{1,0}$ in terms of $Y_{1,1}$, $Y_{1,0}$, and $Y_{1,-1}$.

12. Use Eq. (7.40) to obtain the value of $\langle L_x^2 \rangle$, by first performing the operation $L_x^2Y_{1,1}$, multiplying by $Y^*_{1,1}$, then integrating.

13. Use Eq. (7.40) to obtain the value of $\langle L_x^2 \rangle$ for the state $|1,0\rangle$, by expanding the bra and ket vectors in eigenvectors of L_x, then performing the L_x^2 operation on each ket and evaluating the sum of the resulting brackets. Is the result what you anticipated? Use the result of Exercise 7 instead of this expansion to arrive at this result.

14. Use the result of Exercise 1 to write the 5×5 matrix for the operator L_x with $l = 2$.

CHAPTER **8**

The Radial Schrödinger Equation

Equation (6.34), the time-independent Schrödinger equation in three dimensions, may be rewritten to eliminate the angular dependence, yielding

$$\frac{d^2}{dr^2}[rR(r)] = -\frac{2m}{\hbar^2}\left\{E - V(r) - \frac{l(l+1)\hbar^2}{2mr^2}\right\}[rR(r)] \qquad (8.1)$$

Consider the special case $l = 0$ first. Equation (8.1) can then be written

$$\frac{d^2}{dr^2}[rR(r)] = -\frac{2m}{\hbar^2}\{E - V(r)\}[rR(r)] \qquad (8.2)$$

This equation is identical to the time-independent one-dimensional Schrödinger equation [Eq. (3.5)], except that the variable is r rather than x and the eigenfunction is $rR(r)$ rather than $u(x)$. Therefore, the eigen*functions* are also identical when x is replaced by r and $u(x)$ is replaced by $rR(r)$.

8.1 SOLUTIONS FOR A FREE PARTICLE

For a free particle we can set $V(r) = 0$, and if $l = 0$ we have

$$\frac{d^2}{dr^2}[rR(r)] = -\frac{2m}{\hbar^2}E[rR(r)] \qquad (8.3)$$

159

whose solutions can be expressed as

$$rR = e^{\pm ikr} \quad \text{with } k = \sqrt{2mE}/\hbar \tag{8.4}$$

The complete solution, including time dependence, is therefore

$$\psi(r, \theta, \phi, t) = R(r)e^{i\omega t} = Ne^{\pm i(kr - \omega t)}/r \tag{8.5}$$

where N is a normalizing constant; ψ is independent of θ and ϕ, because $l = 0$.

The radial function for $l = 0$ may also be written in the form

$$R(r) = (A \cos kr + B \sin kr)/r \tag{8.6}$$

8.1.1 Boundary Condition at the Origin

Equation (8.6) describes a standing wave that cannot exist at the origin, because of the $1/r$ factor. However we can make this wave acceptable by setting A equal to zero, because $(\sin kr)/r$ is finite at $r = 0$. Thus, although Eq. (8.3) has the same form as the one-dimensional Schrödinger equation, the eigenfunctions $rR(r)$ that replace $u(x)$ in that equation must be zero at the origin, because they contain the factor r. This makes the solution for a *spherical* well significantly different from the solution for a one-dimensional well, even when the angular momentum is zero.

8.1.2 The Centrifugal Potential

Even if $l \neq 0$, we can write Eq. (8.1) in a form that resembles Eq. (8.2) by defining an *effective* potential V_{eff} given by

$$V_{\text{eff}} = V(r) + \frac{l(l + 1)\hbar^2}{2mr^2} \tag{8.7}$$

where the second term is called the *centrifugal potential*. This is not actually potential energy, but rather the kinetic energy associated with angular motion.[1] Equation (8.1) can now be written

$$\frac{d^2}{dr^2}(rR(r)) = -\frac{2m}{\hbar^2}[E - V_{\text{eff}}](rR(r)) \tag{8.8}$$

---

[1] This term can be written $L^2/2mr^2$, the classical expression for kinetic energy associated with the component of velocity perpendicular to the radial direction. Calling this component $v_\perp$, we know from the definition of angular momentum that $L^2 = (mv_\perp r)^2$. Then $v_\perp^2 = L^2/(mr)^2$ and the kinetic energy is $mv_\perp^2/2 = L^2/2mr^2$.

The quantity $E - V_{\text{eff}}$ is the energy that remains after we subtract the potential energy and the energy associated with angular motion. Thus this expression is the energy associated with radial motion, just as in one dimension the expression $E - V$ is the kinetic energy associated with motion along one axis.

8.1.3 The Radial Probability Density

The parallel between Eq. (8.8) and the one-dimensional Schrödinger equation [Eq. (3.5)] can be strengthened by considering probability densities. In Chapter 6 we saw that

$$\int_0^\infty \int_0^{2\pi} \int_0^\pi |R(r)Y_{l,m}(\theta, \phi)|^2 r^2 \sin \theta \, d\theta \, d\phi \, dr = 1 \qquad (6.53)$$

where $r^2 \sin \theta \, d\theta \, d\phi \, dr$ is the volume element in spherical coordinates, and we also saw that

$$\int_0^{2\pi} \int_0^\pi |Y_{l,m}(\theta, \phi)|^2 \sin \theta \, d\theta \, d\phi = 1 \qquad (6.54)$$

We can use Eq. (6.54) to eliminate the angular part and obtain the normalization condition for the radial part:

$$\int_0^\infty |R(r)|^2 r^2 \, dr = 1 \qquad (8.9)$$

The product $rR(r)$ is the probability amplitude for the radial coordinate, just as $u(x)$ is the probability amplitude for the x coordinate. That means that the probability $P(r_1, r_2)$ of finding the r coordinate of the particle to be in the range $r_1 < r < r_2$ is given by

$$P(r_1, r_2) = \int_{r_1}^{r_2} |rR(r)|^2 \, dr \qquad (8.10)$$

Equation (8.8) can now be expressed in words as

Operator $\times$ probability amplitude = kinetic energy $\times$ probability amplitude

where the kinetic energy is to be understood as the part of the energy that results from the component of the velocity along the r axis. This expression applies to motion along the x, y, or z axis in rectangular coordinates.

8.2 THE SPHERICAL POTENTIAL WELL

Let us compare the spherical potential well with the one-dimensional square well treated in Section 3.2. A "square" spherical well can be described by a potential $V(r)$ that has a sharp step (Figure 8.1):

$$V(r) = -V_0 \quad \text{for } r < a \qquad V(r) = 0 \quad \text{for } r > a \qquad (8.11)$$

When $l = 0$, the situation is very much like that of the one-dimensional well. For $r < a$, the radial Schrödinger equation is Eq. (8.3) and the solution is given by Eq. (8.6) with $A = 0$. That is,

$$u(r, \theta, \phi) = B(\sin kr)/r \qquad (8.12)$$

where, as before, the kinetic energy is $\hbar^2 k^2/2m$. But in this case the kinetic energy within the well is equal to $E + V_0$.

For $r > a$, the radial equation is, from Eq. (8.2)

$$\frac{d^2}{dr^2}[rR(r)] = -\frac{2mE}{\hbar^2}[rR(r)] \equiv -k^2 rR(r) \qquad (8.13)$$

and the solution for a bound state must go to zero as $r \to \infty$. Thus for $r > a$, we have a decaying exponential:

$$u(r) = Ce^{-\alpha r} \quad \text{with } \alpha = \sqrt{-2mE}/\hbar \qquad (8.14)$$

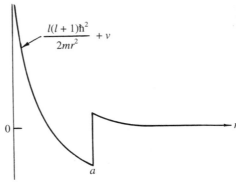

FIGURE 8.1 The effective potential well that appears in the radial equation for the square-well potential of Eq. (8.11). For $r < a$ the effective potential is simply the centrifugal potential.

Here α is real when E is negative,[2] just as in the one-dimensional well. We now apply the continuity condition at $r = a$ to find the allowed values of E as we did in Section 4.1. Because the mathematical functions are exactly the same, the energy levels must be the same as before, with one important exception:

The eigenfunction $rR(r)$ must be zero at $r = 0$. None of the even functions from Section 3.2 meets this requirement. Thus all of the solutions for the spherical well are odd functions of r.

Therefore the result for $l = 0$ must have the same form as the result found in Section 3.2 for the *odd-parity* solution; the condition that $rR = 0$ at $r = 0$ eliminates the even solution. The energy levels are thus found from Eq. (3.39b):

$$\sin ka = \pm ka/\beta a \qquad (\beta^2 = k^2 + \alpha^2 \text{ and } \cot ka < 0) \qquad (3.39b)$$

Because $\cot ka < 0$, the lowest-energy solution has $\pi/2 < ka < \pi$ (like the lowest odd solution in the one-dimensional case). But $ka \leq \beta a$. Therefore,

If $\beta a < \pi/2$ (that is, if $2mV_0a^2/h^2 < (\pi/2)^2$), there is no bound state.

There is one bound state if $\pi/2 < \beta a < 3\pi/2$, there are two if $3\pi/2 < \beta a < 5\pi/2$, and so on. The allowed energies are found by the same method followed in Section 3.2.

8.2.1 Example: Energy Level of the Deuteron

The deuteron, the nucleus of the ^{2}H atom, is a bound state of a neutron and a proton. It has only one energy level, at an energy of -2.2 MeV. (This means that an energy of at least 2.2 MeV is required to separate the neutron from the proton in this nucleus.) Experiments show that the potential energy $V(r)$ can be approximated by Eq. (8.11) with the value of a equal to 2 fm. From this information you can verify that the well depth V_0 is about 37 MeV. In spite of its depth, this narrow well has no excited state; the value of βa is not much larger than $\pi/2$. Curiously, there is no "dineutron" (a bound state of two neutrons in the absence of protons), even though there is a strong attractive force between neutrons.

---

[2]This is the same solution as for the one-dimensional well, but here we have a potential energy that is zero at infinity, rather than being equal to V_0 at infinity. Thus the energy E for a bound state must be negative, and the kinetic energy for $r < a$ is equal to $E - V(r)$, or $E + V_0$.

The Pauli exclusion principle (to be discussed in Chapters 10 and 11) provides the explanation for that. In a dineutron's ground state, both neutrons would be in the state of lowest energy. The exclusion principle does not permit this to happen for neutrons (and many other particles), and no excited state is bound. Thus there is no dineutron.

8.2.2 Solutions for Nonzero Angular Momentum

When $l > 0$, the radial equation for the spherical square well becomes

$$\frac{d^2}{dr^2}[rR(r)] = -\frac{2m}{\hbar^2}\left\{E + V_0 - \frac{l(l+1)\hbar^2}{2mr^2}\right\}[rR(r)] \qquad (r < a) \quad (8.15)$$

$$\frac{d^2}{dr^2}[rR(r)] = -\frac{2m}{\hbar^2}\left\{E - \frac{l(l+1)\hbar^2}{2mr^2}\right\}[rR(r)] \qquad (r > a) \quad (8.16)$$

With the substitutions $k^2 = 2m(E + V_0)/\hbar^2$ and $\alpha^2 = 2mE/\hbar^2$, these become

$$\frac{d^2}{dr^2}(rR) = -k^2(rR) + \frac{l(l+1)}{r^2}(rR) \qquad (r < a) \qquad (8.17)$$

$$\frac{d^2}{dr^2}(rR) = -i\alpha^2(rR) + \frac{l(l+1)}{r^2}(rR) \qquad (r > a) \qquad (8.18)$$

Solutions of Eq. (8.17) are called *spherical Bessel functions* $[j_l(kr)]$ and *spherical Neumann functions* $[n_l(kr)]$.[3] For Eq. (8.18) the argument of the functions is of course $i\alpha r$ rather than kr.

For each value of l there are two linearly independent solutions—a spherical Bessel function and a spherical Neumann function. The first

---

[3]The general formula for $j_l(x)$ is $j_l(x) = \sqrt{\pi/2x}\,J_{l+1/2}(x)$, where $J_{l+1/2}(x)$ is the ordinary Bessel function $J_n(x)$, defined by the infinite series

$$\frac{x^n}{2^n\Gamma(n+1)}\left(1 - \frac{x^2}{2(n+2)} + \frac{x^4}{2\times 4(2n+2)(2n+4)}\right.$$
$$\left. - \frac{x^6}{2\times 4\times 6(2n+2)(2n+4)(2n+6)} + \cdots\right)$$

three of each are

$$j_0(kr) = \frac{\sin kr}{kr} \qquad\qquad n_0(kr) = -\frac{\cos kr}{kr} \qquad (8.19a)$$

$$j_1(kr) = \frac{\sin kr}{(kr)^2} - \frac{\cos kr}{kr} \qquad n_1(kr) = -\frac{\cos kr}{(kr)^2} - \frac{\sin kr}{kr} \qquad (8.19b)$$

$$j_2(kr) = \left\{ \frac{3}{(kr)^3} - \frac{1}{kr} \right\} \sin kr - \frac{3\cos kr}{(kr)^2} \qquad (8.19c)$$

$$n_2(kr) = \left\{ -\frac{3}{(kr)^3} + \frac{1}{kr} \right\} \cos kr - \frac{3\cos kr}{(kr)^2}$$

We have seen that $j_0(kr)$ and $n_0(kr)$, as just given, are solutions of the radial equation (8.17) when $l = 0$. You may verify yourself that the other functions given above are solutions of Eq. (8.17) with the given values of l.

Figure 8.2 shows radial probability densities $|rR(r)|^2$, where $R(r)$ is the spherical Bessel function, for various values of l. Notice how $|rR(r)|^2$ is "pushed away" from the origin when $l > 0$. In that case $|rj_l(kr)|^2$ is maximum near $kr = l$ and is quite small for $kr < l$. A classical particle with momentum p and angular momentum L cannot be closer to the origin than $r = L/p$, because

$$L \equiv |\mathbf{L}| = |\mathbf{r} \times \mathbf{p}| \leq rp \quad \text{so that } r \geq L/p \qquad (8.20)$$

Using the values $L \approx l\hbar$ and $p = \hbar k$, we find that $kr \geq l$.

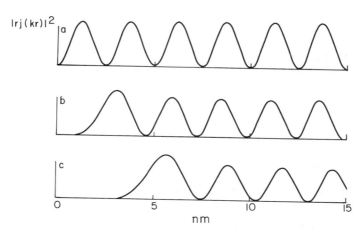

FIGURE 8.2 Radial probability densities $|rj(kr)|^2$ for $l = 0$, $l = 2$, and $l = 5$, generated by computer.

8.2.3 Calculation of Energy Levels for the Spherical Well

To find the energy levels, we now must apply the continuity conditions at $r = a$ to the solutions for $r < a$ and $r > a$. For $r < a$, the solution must be $j_l(kr)$, because it is the only solution that satisfies the condition that $rR(r)$ must be zero at $r = 0$. For $r > a$ there is no such condition, and $j_l(kr)$ does not go to zero as $r \to \infty$. Therefore we need to use the Neumann function n_l.

The general solution is a linear combination of j_l and n_l, and the combination that has the correct asymptotic behavior as $r \to \infty$ is called a *spherical Hankel function of the first kind*,[4] defined as

$$h_l^1(x) = j_l(x) + in_l(x) \qquad (8.21)$$

To use this function as a solution of Eq. (8.18) we let x equal $i\alpha r$. Then, for $l = 1$ and $l = 2$, the solutions are

$$h_1^1(i\alpha r) = i\left\{ \frac{1}{\alpha a} + \frac{1}{(\alpha a)^2} \right\} e^{-\alpha r} \qquad (8.22)$$

$$h_2^1(i\alpha r) = \left\{ \frac{1}{\alpha a} + \frac{3}{(\alpha a)^2} + \frac{3}{(\alpha a)^3} \right\} e^{-\alpha r} \qquad (8.23)$$

These functions clearly have the required behavior as $r \to \infty$, so we use them for the region $r > a$. The complete solution for $l = 1$ is therefore

$$R(r) = Aj_1(kr) \quad (r < a), \qquad Rr = Bh_1^1(i\alpha r) \quad (r > a) \qquad (8.24)$$

where A and B are constants to be determined by the boundary conditions at $r = a$ and by normalization. We may eliminate both A and B and solve for the energy level; we use the requirement that the ratio $(dR/dr)/R$ be continuous at $r = a$, as follows:

$$\frac{d/dr[j_1(kr)]_{r=a}}{j_1(ka)} = \frac{d/dr[h_1^1(i\alpha r)]_{r=a}}{h_1^1(i\alpha a)} \qquad (8.25)$$

Taking the derivatives and performing the division on each side leads (with

---

[4]The spherical Hankel function of the second kind is defined by $h_l^{(2)} = j_l - in_l$. We do not use it here, because as a function of $i\alpha r$ it contains the factor $e^{+\alpha r}$ and thus blows up as $r \to \infty$.

some rearrangement of terms) to

$$\frac{\cot ka}{ka} - \frac{1}{(ka)^2} = \frac{1}{\alpha a} + \frac{1}{(\alpha a)^2} \qquad (8.26)$$

This equation, with the definitions

$$k^2 = 2m(E + V_0)/\hbar^2 \quad \text{and} \quad \alpha^2 = -2mE/\hbar^2 \qquad (8.27)$$

may be solved numerically to find the possible values of E for $l = 1$.

Example Problem 8.1   Use Eqs. (8.26) and (8.27) to show that there is a bound state for $l = 1$ only if $V_0 a^2 \geq \pi^2 \hbar^2 / 2m$.

Solution. The right-hand side of Eq. (8.26) is never negative, but the left-hand side is negative when cot $ka < 1/ka$, which is true when $ka < \pi$. Thus we must have $ka \geq \pi$ if Eq. (8.26) is to have a solution. For a bound state in this well, E must be negative; therefore from Eq. (8.27), with $ka \geq \pi$, we have

$$\pi^2 \leq k^2 a^2 \leq 2mV_0 a^2/\hbar^2 \quad \text{or} \quad \pi^2 \leq 2mV_0 a^2/\hbar^2,$$

$$\text{and so} \quad V_0 a^2 \geq \pi^2 \hbar^2 / 2m \quad \text{Q.E.D.}$$

Notice that the required value of Va^2 for $l = 1$ is four times the value required for a bound state with $l = 0$.

8.3 EXAMPLE: THE SPHERICALLY SYMMETRIC HARMONIC OSCILLATOR REVISITED

Given the spherically symmetric harmonic oscillator potential [Eq. (6.16)]:

$$V(r) = Kr^2/2 \qquad (8.28)$$

We may write the radial Schrödinger equation [Eq. (8.8)] as

$$\frac{d^2}{dr^2}[rR(r)] = -\frac{2m}{\hbar^2}\left\{ E - Kr^2/2 - \frac{l(l+1)\hbar^2}{2mr^2} \right\}[rR(r)] \qquad (8.29)$$

For $l = 0$, Eq. (8.29) is identical to the one-dimensional equation, except that $u(x)$ is replaced by $rR(r)$. Therefore there is a solution whose energy eigenvalue is equal to $\hbar\omega/2$, as in one dimension. However, we found previously (Section 6.1) that, for the spherically symmetric harmonic oscillator, the energy levels are given by $E_n = (n + 3/2)\hbar\omega$, where n is a positive integer or zero. How can these results be reconciled?

Obviously, the function that gives an energy eigenvalue of $\hbar\omega/2$ must not be an acceptable solution. The reason is clear in the expression for $rR(r)$:

$$rR(r) = Ne^{-ar^2/2} \quad \text{where } N \text{ is a normalization factor} \quad (8.30)$$

This fails to satisfy the required boundary condition that $rR(r)$ must be zero for $r = 0$. On the other hand, the solution for energy level E_0, $E_0 = \frac{3}{2}\hbar\omega$, is

$$rR_0(r) = Nre^{-ar^2/2} \quad \text{or} \quad R_0(r) = Ne^{-ar^2/2} \quad (8.31)$$

This eigenfunction, having no angular dependence, must represent a state with zero angular momentum. (We might also say that its angular dependence can be expressed as the spherical harmonic $Y_{0,0}$, for which $l = 0$ and $m = 0$.) By applying the raising operator $(\partial/\partial x - ax)$ to $R_0(r)$, we can generate the eigenfunction

$$u_{100} = (\partial/\partial x - ax)Ne^{-ax^2/2}e^{-ay^2/2}e^{-az^2/2} = -2axNe^{-ar^2/2} \quad (8.32a)$$

Similarly, we can generate two other eigenfunctions with the same eigenvalue:

$$u_{010} = (\partial/\partial y - ay)Ne^{-ax^2/2}e^{-ay^2/2}e^{-az^2/2} = -2ayNe^{-ar^2/2} \quad (8.32b)$$

$$u_{001} = (\partial/\partial z - az)Ne^{-ax^2/2}e^{-ay^2/2}e^{-az^2/2} = -2azNe^{-ar^2/2} \quad (8.32c)$$

For each of these functions, $n = 1$ and the energy is $E = (n + 3/2)\hbar\omega = 5/2\hbar\omega$.

8.3.1 Connection with the Spherical Harmonics

We have already see that, in a spherically symmetric potential, each eigenfunction of the Schrödinger equation is the product of a purely radial function and a purely angular function and that the angular function must be a spherical harmonic or a linear superposition of spherical harmonics.

Let us show that this is true for the functions of Eqs. (8.32). The factor x [in Eq. (8.32a)] may be written in terms of the sum of the spherical harmonics $Y_{1,1}$ and $Y_{1,-1}$, because [as shown in Example Problem 6.3]

$$Y_{1,1} + Y_{1,-1} = (3/2\pi)^{1/2} x/r \quad \text{or} \quad x/r = (2\pi/3)^{1/2}(Y_{1,1} + Y_{1,-1}),$$

and therefore

$$u_{100} = N(Y_{1,1} + Y_{1,-1})re^{-ar^2/2} \tag{8.33}$$

where N is a normalization constant. Thus u_{100} is an eigenfunction of L^2 with $l = 1$, but it is a mixture of $m = +1$ and $m = -1$ with equal amplitudes. A measurement of L_z would yield $+\hbar$ and $-\hbar$ with equal probability.

From Eqs. (8.32) we can also deduce that u_{010} is also an equal mixture of $m = 1$ and $m = -1$, which can be written

$$u_{010} = N(Y_{1,1} - Y_{1,-1})re^{-ar^2/2} \tag{8.33a}$$

whereas u_{001} is an eigenfunction of L_z with $m = 0$:

$$u_{001} = NY_{1,0}re^{-ar^2/2} \tag{8.33b}$$

It is often convenient to use combinations of harmonic oscillator functions which are linearly independent and are eigenfunctions of L_z. These may be written, with $1/\sqrt{2}$ as a normalizing factor, as

$$u_a = (u_{100} + iu_{010})/\sqrt{2} \qquad (n = 1, l = 1, m = +1) \tag{8.34a}$$

$$u_b = u_{001} \qquad (n = 1, l = 1, m = 0) \tag{8.34b}$$

$$u_a = (u_{100} - iu_{010})/\sqrt{2} \qquad (n = 1, l = 1, m = -1) \tag{8.34c}$$

These functions are simultaneous eigenfunctions of energy ($n = 1$), of L^2 ($l = 1$), and of L_z (with $m = +1$, 0, and -1, respectively). It is interesting that, for any given value of n, a set of similar equations can be written. By application of the raising operators, you can verify that there are six linearly independent harmonic-oscillator eigenfunctions for $n = 2$, containing the respective factors x^2, y^2, z^2, xy, xz, and yz. We can construct six independent combinations of these factors by combining the five spherical harmonics for $l = 2$ with the spherical harmonic for $l = 0$. For example, the combination $u_{200} + u_{020} + u_{002}$ contains x, y, and z only in the combination $x^2 + y^2 + z^2$; thus it is spherically symmetric and is proportional to the spherical harmonic $Y_{0,0}$ with $l = 0$. Similarly, for

$n = 3$, the raising operators yield ten different factors with a combined exponent of 3: x^3, y^3, z^3, x^2y, xy^2, x^2z, xz^2, y^2z, yz^2, and xyz. These can be written as linear combinations of the seven spherical harmonics for $l = 3$ plus the three spherical harmonics for $l = 1$. The process is valid for any value of n. (See Problems 6 and 7.)

8.4 SCATTERING OF PARTICLES FROM A SPHERICALLY SYMMETRIC POTENTIAL

Let us now consider the three-dimensional counterpart of the transmission of particles past a potential barrier (Section 5.3). In this case the situation is obviously more complicated, because the particles can emerge in any direction (be "scattered") instead of simply being transmitted or reflected.

Suppose that a "beam" of particles—a plane wave of the form $\Psi_{in} = Ae^{i(kz-\omega t)}$—encounters a potential well (a "scattering center") where the potential energy is nonzero over a limited region ($r < a$) surrounding this scattering center. (See Figure 8.3.) The density of particles in the beam is $|\Psi_{in}|^2 = |A|^2$, and the intensity of the beam (the number of particles crossing a unit area in a unit of time) is the product of particle density and particle velocity, or $v|A^2|$, as discussed in Section 5.2.

Some fraction of the particles will interact with the scattering center to produce a wave that travels outward from the center with an amplitude that in general is a product of two functions: (1) a function $f(\theta, \phi)$ of the angular coordinates θ and ϕ and (2) a function $R(r)$ of the radial coordinate r.

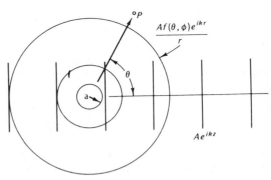

FIGURE 8.3 Scattering of a plane wave from a scattering center, producing a spherical scattered wave. The interaction that produces the scattered wave occurs only in the region $r < a$.

The function $f(\theta, \phi)$ enables us to find the probability that a particle will be scattered, as a function of the scattering angle. But before solving a wave equation, it may be helpful to investigate the scattering of classical particles.

8.4.1 Scattering Cross Section, Classical

Consider the scattering of classical particles from a sphere. Instead of a wave, we might have a stream of tiny pellets aimed toward the sphere. There are three possibilities; a pellet could

1. Be deflected by the sphere
2. Miss the sphere completely
3. Go through the sphere without deviation (if the sphere is porous)—a highly improbable classical result, but common in quantum mechanics.

If the beam intensity (the number of pellets per unit area per second in the beam) is I and the sphere is not porous, then the number N_{sc} of pellets that are scattered per second must be equal to $I\sigma$, where σ is the sphere's cross-sectional area, or

$$N_{sc} = I\sigma \quad \text{or} \quad \sigma = N_{sc}/I$$

and $\sigma = \pi R^2$, where R is the radius of the sphere. In general, the ratio N_{sc}/I is called the *scattering cross section* of the sphere for these particles, and it is denoted by the symbol σ. But what if the sphere is porous? In that case, σ is not defined as the geometrical cross section of the sphere; rather, it is defined as the ratio N_{sc}/I, using Eq. (8.34).

8.4.2 Differential Cross Section, Classical

We are often interested in the number of particles that are scattered into a specific range of angles. Classically, all pellets in a parallel beam will be scattered at the same angle θ if they have the same *impact parameter b*. By definition, b is the distance by which the pellet would miss the center of the sphere if it passed through the sphere in a straight line. If the beam is parallel to the z axis and the center of the sphere is at the origin, then we can relate b to R and the scattering angle θ. From Figure 8.4 we see that $b = R \sin \theta_i$, where θ_i is the angle between the vector $\mathbf{R}$ and the z axis in Figure 8.4.

When a pellet strikes a solid sphere and rebounds elastically, we can see from Figure 8.4 that the scattering angle θ, which is the angle between

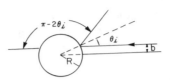

FIGURE 8.4   Classical elastic scattering of a pellet by a hard sphere. Each pellet is deflected through an angle of $\theta = \pi - 2\theta_i$.

the pellet's original direction and its final direction, is given by

$$\theta = \pi - 2\theta_i = \pi - 2\sin^{-1}(b/R) \tag{8.35}$$

where R is the sphere's radius.

The number of pellets that scatter into an angle between θ and $\theta + d\theta$ is proportional to the *differential cross section* $d\sigma/d\theta$, which is dependent on the angle θ; $d\sigma$ is simply the size of the area through which a pellet must go in order to be scattered into an angle in this range.

In terms of the impact parameter b, this is the area $\mathscr{A}$ of a ring of radius b and thickness db. We can write $\mathscr{A}$ in terms of θ and $d\theta$ by solving Eq. (8.35) for b, then differentiating, as follows:

$$b = R\sin[(\pi/2) - (\theta/2)] = R\cos(\theta/2) \tag{8.36}$$

$$db = -R/2\sin(\theta/2) \tag{8.37}$$

Therefore

$$d\sigma = 2\pi b\, db = 2\pi R\cos(\theta/2)(-R/2)\sin(\theta/2)\, d\theta \tag{8.38}$$

which can be written

$$d\sigma = -(\pi R^2/2)\sin\theta\, d\theta \tag{8.39}$$

Now the total cross section σ can be found by integration from $\theta = \pi$ to 0 (because $\theta = \pi$ when $b = 0$ and $\theta = 0$ when $b = R$):

$$\sigma = -\frac{\pi R^2}{2}\int_{\pi}^{0}\sin\theta\, d\theta = \frac{\pi R^2}{2}[\cos 0 - \cos\pi] = \pi R^2 \tag{8.40}$$

as it should be.

8.4.3 Scattering Cross Section, Quantum Mechanical

The scattered wave can be written as $\Psi_{sc} = Af(\theta, \phi)e^{i(kr - \omega t)}/r$, a wave traveling outward from the origin in the direction of increasing r (just as a function of $kx - \omega t$ travels in the $+x$ direction). The factor $1/r$ gives the proper $1/r^2$ dependence in the intensity of the wave, and the factor A expresses the fact that the scattered wave amplitude should be proportional to the amplitude of the incident wave [written before as $\Psi_{in} = Ae^{i(kz - \omega t)}$].

We can relate $f(\theta, \phi)$ to the scattering cross section as follows. If a perfectly efficient particle detector were placed at point P (see Figure 8.3), the number N_d of particles observed per unit time would be the product of the intensity of the scattered wave and the area $d\mathscr{A}$ of the detector. The intensity is $v|A|^2|f(\theta, \phi)|/r^2$, where v is the particle velocity; therefore

$$N_d = \left[v|A|^2|f(\theta, \phi)|^2/r^2 \right] d\mathscr{A} \tag{8.41}$$

But $d\mathscr{A}/r^2$ is the solid angle $d\Omega$ subtended by the detector at the scattering center, $v|A|^2$ is the intensity I_{inc} of the *incident* beam, and Eq. (8.41) gives

$$N_d = |f(\theta, \phi)|^2 I_{inc} \, d\Omega \tag{8.42}$$

where $d\Omega = \sin \theta \, d\theta \, d\phi$.

We can now use Eq. (8.34) to introduce the scattering cross section; by analogy to that equation, we must have

$$d\sigma = N_d/I_{inc} = |f(\theta, \phi)|^2 \, d\Omega \tag{8.43}$$

or

$$d\sigma/d\Omega = |f(\theta, \phi)|^2 \tag{8.44}$$

Notice that $d\sigma/d\Omega$, being a cross section, has the dimensions of an area. Therefore $f(\theta, \phi)$ must have the dimensions of a length. We shall now demonstrate that it is proportional to the wavelength of the incoming wave.

8.4.4 Partial Wave Analysis

We must now relate the scattering phenomenon to the Schrödinger equation solutions that we have seen. First, let us assume that the scattering potential is spherically symmetric, which implies that the scattered function has symmetry with respect to rotation about the z axis. In that case, there is no ϕ dependence, and $f(\theta, \phi)$ becomes simply $f(\theta)$. Next, we apply the conservation of angular momentum; we decompose the incoming wave into components called *partial waves*, each with a different value of l. We can then calculate the scattering of each component independently.

In the limit $r \rightarrow \infty$, the complete wave function approaches

$$\Psi \rightarrow A(e^{ikz} + f(\theta)e^{ikr}/r) \tag{8.45}$$

This is to be compared with the general solution of the Schrödinger equation, a linear combination of spherical harmonics, each multiplied by the appropriate radial factor $R_l(r)$. With no ϕ dependence, we have

$$\Psi = \sum_0^\infty b_l R_l(r) P_l(\cos \theta) \tag{8.46}$$

where $P_l(\cos \theta)$ is the Legendre polynomial of order l. In this and the following equations, the summation runs from $l = 0$ to infinity.

If the scattering potential goes to zero for $r > a$, the function R_l can be written as a superposition of $j_l(kr)$ and $n_l(kr)$ in that region:

$$R_l(r) = \cos \delta_l j_l(kr) - \sin \delta_l n_l(kr) \quad r > a \tag{8.47}$$

where the coefficients are written as cosine and sine to preserve the normalization of the solution (because $\cos^2 \delta_l + \sin^2 \delta_l = 1$ regardless of the value of δ_l). At this point we need only find the values of the phase angles δ_l (called *phase shifts*) in order to determine $f(\theta)$ and hence the scattering cross section.

To do this we match Eq. (8.45) to the limit of Eq. (8.46) for $r \rightarrow \infty$, eventually expressing $f(\theta)$ in terms of a series of Legendre polynomial with coefficients that are determined by the scattering potential. It is known that the limits of the functions $j_l(kr)$ and $n_l(kr)$ are, respectively,

$$j_l(kr) \rightarrow \frac{1}{kr}\sin(kr - l\pi/2) \quad \text{and} \quad n_l(kr) \rightarrow -\frac{1}{kr}\cos(kr - l\pi/2) \tag{8.48}$$

Consequently we may write

$$R_l(kr) \to \frac{1}{kr}\cos\delta_l \sin(kr - l\pi/2) + \frac{1}{kr}\sin\delta_l \cos(kr - l\pi/2) \quad (8.49)$$

$$R_l(kr) \to \frac{1}{kr}\sin(kr - (l\pi/2) + \delta_l) \quad (8.50)$$

Thus the only effect of the scattering center on each component of the radial function (each partial wave) is to shift its phase by the angle δ_l.

We now can find an expression for the cross section σ in terms of the phase shifts δ_l. We equate the right-hand side of (8.45) to the limit (as $r \to \infty$) of the right-hand side of (8.46), using (8.50) to eliminate R_l, obtaining

$$e^{ikz} + f(\theta)\frac{e^{ikr}}{r} = \sum_{l=0}^{\infty} b_l P_l(\cos\theta)\frac{1}{kr}\sin(kr - l\pi/2 + \delta_l) \quad (8.51)$$

To eliminate z as a variable, we can expand e^{ikz} as a series of spherical harmonics, as we could do for any function in three dimensions. (This is often a useful way to expand a plane wave, expressing it as a sum of component waves, each with a definite angular momentum. See Figure 8.5.) Since e^{ikz} has no ϕ dependence (z being equal to $r\cos\theta$), we write

$$e^{ikz} = \sum_{l=0}^{\infty} a_l P_l(\cos\theta) \quad (8.52)$$

and we evaluate the coefficients a_l in the same way that we find coefficients in a Fourier series, by using the fact that the P_l are orthogonal and normalized. The result is

$$e^{ikz} = \sum_{l=0}^{\infty} (2l + 1)i^l j_l(kr)P_l(\cos\theta) \quad (8.53)$$

Substitution of this expression into Eq. (8.51), and using the limit of $j_l(kr)$ as $r \to \infty$, yields

$$kf(\theta)e^{ikr} + \sum_{l=0}^{\infty} (2l + 1)i^l P_l(\cos\theta)\sin\left(kr - \frac{l\pi}{2}\right)$$

$$= \sum_{l=0}^{\infty} b_l P_l(\cos\theta)\sin\left(kr - \frac{l\pi}{2} + \delta_l\right) \quad (8.54)$$

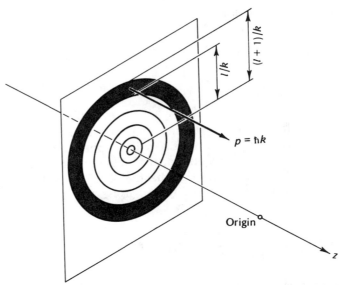

FIGURE 8.5 Analysis of a particle beam into distinct annular beams. A particle going through the shaded area has angular momentum between $l\hbar$ and $(l + 1)\hbar$.

or, writing the sine functions in exponential form [using the identity $\sin x \equiv (e^{ix} - e^{-ix})/2i$],

$$e^{ikr}\left(2ikf(\theta) + \sum (2l + 1)i^l P_l(\cos \theta)e^{-il\pi/2}\right)$$
$$- e^{-ikr}\sum (2l + 1)i^l e^{il\pi/2}P_l(\cos \theta)$$
$$= e^{ikr}\sum b_l P_l(\cos \theta)(e^{i\delta}l^{-il\pi/2}) - e^{-ikr}\sum b_l e^{il\pi/2-i\delta_l}P_l(\cos \theta)e^{-il\pi/2}$$

$$(8.55)$$

By equating the coefficients of e^{-ikr} on the two sides we find that the constant coefficients b_l are given by

$$b_l = (2l + 1)i^l e^{i\delta} \qquad (8.56)$$

By equating the coefficients of e^{ikr} we then find an expression for $f(\theta)$:

$$f(\theta) = (2ik)^{-1} \sum_{l=0}^{\infty} (2l + 1)(e^{2i\delta}l - 1)P_l(\cos \theta)$$

$$= \frac{1}{k} \sum_{l=0}^{\infty} (2l + 1)e^{i\delta}l \sin \delta_l P_l(\cos \theta) \qquad (8.57)$$

The total cross section is therefore

$$\sigma = \int \frac{d\sigma}{d\Omega} \, d\Omega = \int |f(\theta)|^2 \, d\Omega$$

$$= \frac{2\pi}{k^2} \int_0^\pi \left\{ \sum_{l=0}^\infty (2l + 1)\sin \delta_l P_l(\cos \theta)\sin \theta \, d\theta \right\}^2 \qquad (8.58)$$

You can verify Eq. (8.58) yourself. Notice that after squaring the series of Eq. (8.57), products of terms involving different values of l do not appear in the integral, because of the orthogonality of the Legendre polynomials. Therefore we can exchange the sum and integral signs. Then the integrals can be evaluated by use of the normalization of the Legendre polynomials, with the final result that

$$\boxed{\sigma = \frac{4\pi}{k^2} \sum_{l=0}^\infty (2l + 1)\sin^2 \delta_l} \qquad (8.59)$$

Again, the summation is over all values of l—all positive integers plus zero.

The problem now becomes the calculation of the phase shifts δ_l—and there are infinitely many of them! Fortunately, there are many situations in which we can deduce that only the waves with $l = 0$ or $l = 1$ make a significant contribution to the cross section.

Figure 8.5 may help us to visualize the decomposition of the incident plane waves into partial waves and to see why we can neglect waves with large values of l. The plane wave is a stream of particles of momentum $p = \hbar k$, moving parallel to the z axis. For a localized particle, the magnitude of the angular momentum about the origin is $pd = \hbar kd$, where d is the distance of that particle from the z axis. In this semiclassical view, particles with momentum $p = \hbar k$ and angular momentum between $i\hbar$ and $(l + 1)\hbar$ must pass through an annulus whose radius lies between l/k and $(l + 1)/k$.

If this radius is greater than the radius a at which the scattering potential becomes zero, then particles with this value of l (or greater) cannot be influenced by that potential, and the corresponding phase shift δ_l must be zero.

Figure 8.5 also can be related to the presence of the factor $2l + 1$ in each term of the series in Eq. (8.59). The particles of a given value of l must contribute to the cross section in proportion to their number, which in turn is proportional to the annular area through which they pass. From simple geometry we see that this area is proportional to $2l + 1$.

We can deduce from the above that when $ka \ll 1$ (or $\lambda \ll a$), the influence of the potential on particles with $l \geq 1$ is very small. In that case, σ is given with good accuracy by the first term in Eq. (8.59):

$$\sigma = (4\pi/k^2)\sin^2 \delta_0 \qquad (ka \ll 1) \qquad (8.60)$$

Figure 8.6, showing examples of radial wave functions, may further clarify the connections between l, a, and k in scattering from a negative potential well $V(r)$. The value of l determines the "centrifugal" potential V_{cent}; the value of a determines the region where the "effective" potential differs from V_{cent}, and of course the value of k determines the wavelength. The negative potential makes V_{eff} smaller than $V(r)$ for $r \leq a$. This in turn brings the point of inflection of the wave function in closer to the origin and therefore shifts the phase observed at large r. For sufficiently large values of l, V_{cent} is so large that the point of inflection is far outside the

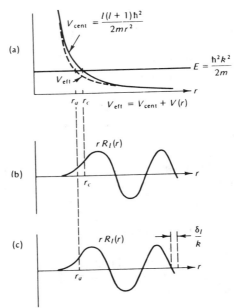

FIGURE 8.6 Connection between scattering potential and phase shift. (a) Centrifugal potential V_{cent} and effective potential V_{eff} (dashed line), for a negative scattering potential that extends to $r = a$. (b) Radial probability density $rR_0(r)$ for zero scattering potential (c) Radial probability density $rR(r)$ for this potential; at large r; $rR(r)$ is shifted by δ_l relative to $rR_0(r)$.

potential well, and the wave function is unaffected by the well. Thus there is no phase shift for such values of l.

Example Problem 8.2 Use QMVGA to plot the radial probability density $rR(r)$ for an electron with $E = 4.02$ eV and $l = 0$ in the potential

$$V(r) = 0 \quad \text{for } r < 2.4 \text{ nm}, \qquad V(r) = 4 \text{ eV} \quad \text{for } r \geq 2.4 \text{ nm}$$

and an electron with $E = 0.02$ eV and $l = 0$ in the potential $V(r) = 0$ for all r.

(a) From these graphs, find the phase shift δ_0 for an electron striking this potential well, if its kinetic energy outside the well is 0.02 eV.

(b) In the same way, find δ_1 and δ_2 for this electron energy, and compute the scattering cross section from three terms in the series of Eq. (8.59).

Solution. (a) Using option 4, part 2, we select $Z = 0$, $V = 4$, $E = 4.02$, and $l = 0$. Then we repeat with $V = .000001$ and $E = 0.02$. (Entering $V = 0$ gives the default value of V, which is $V = 4$.) The resulting graph is shown in Figure 8.7, along with the graph for $V = 4$ and $E = 4.02$. The graph shows that $\sin \delta_0 \cong 0.98$; we see that from any node of $rR(r)$ on the graph for $V = 0$. At each of those values of r, $rR(r)$ is about 98% of its maximum value on the graph for $V = 4$.

(b) In the same way, we observe the nodes on the graphs for $l = 1$ and $l = 2$, at points where the centrifugal potential is negligible, and we find that $\sin \delta_1 = 0.53$ and $\sin \delta_2 = 0.17$. Thus, from Eq. (8.59),

$$\sigma = (4\pi/k^2)(\sin^2 \delta_0 + 3 \sin^2 \delta_1 + 5 \sin^2 \delta_2 + \cdots)$$

$$= (4\pi/k^2)(0.96 + 0.84 + 0.15 \cdots) \qquad (8.61)$$

where $4\pi/k^2 = 4\pi\hbar^2/2mE = h^2/2mE = 75.4$ nm^2. Thus $\sigma = 75.4 \times 1.95 = 147$ nm^2, which is far more than the cross-sectional area of the well. This can be understood as a diffraction effect.

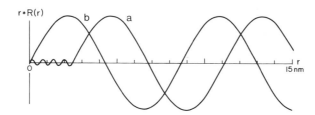

FIGURE 8.7 (a) Wave function for electrons of kinetic energy 0.02 eV scattering in a spherically symmetric potential step of depth 4.0 eV. (b) Wave function for 0.02 eV electrons when the step is zero.

Notice the large drop between the second and third terms in the series of Eq. (8.61). In general, a partial wave contributes significantly to the scattering cross section from a sphere of radius a only if its value of l is less than or equal to ka. In this example, since $4\pi/k^2 = 75.4$ nm^2 and $a = 2.4$ nm, we have $ka = 2.4\sqrt{4\pi/75.4}$, or $ka \cong 1$. Neglect of the $l = 2$ term in Eq. (8.61) results in an error of about 8% in the value of σ in this example. If you measure the graph for $l = 3$, you will find that $\sin \delta_3 \cong 0.02$, so the next term in the series is about 0.004.

ADDITIONAL READING

R. B. Leighton, *Principles of Modern Physics*, McGraw-Hill, New York (1959), pages 504–511, give a good coverage of scattering and phase shift analysis.

L. Pauling and E. B. Wilson, *Introduction to Modern Physics*, McGraw-Hill, New York (1935), pages 105–111, give an instructive treatment of the three-dimensional harmonic oscillator in cylindrical coordinates as well as cartesian coordinates.

EXERCISES

1. Use QMVGA to find the lowest three energy levels for an electron in a spherical well of radius 2.4 nm and depth 4.0 eV, given that the quantum number l is equal to 1. Then verify that these energies satisfy Eqs. (8.26) and (8.27). (The spherical square well is plotted by choosing option 4, then setting Z equal to zero. The program then plots the wave function for a well of radius 2.4 nm.)

2. Verify by direct substitution that $j_1(kr)$ and $j_2(kr)$ are solutions of Eq. (8.17).

3. (a) Find the lowest energy level for an electron trapped in a spherically symmetric shell whose boundaries are $r = 1.0$ nm and $r = 4.0$ nm. Inside the shell, $V = 0$; outside the shell, V is infinite.

   (b) Find the probability that this electron will be observed in that part of the shell lying between $r = 3.9$ nm and $r = 4.0$ nm.

4. Any spherical Bessel function can be generated by means of the relation

$$j_l(kr) = \left(-\frac{r}{k}\right)^l \left\{\frac{1}{r}\frac{d}{dr}\right\}^l j_0(kr)$$

   (a) Use this relation to find $j_1(kr)$, $j_2(kr)$, and $j_3(kr)$.

   (b) Verify that the $j_3(kr)$ that you found is a solution of Eq. (8.17).

5. From the same reasoning that led to Eq. (8.33), find
    (a) the possible values of L_z in the state whose wave function is u_{011}.
    (b) the possible values of L^2 in the state whose wave function is u_{322}.
6. (a) For a three-dimensional harmonic oscillator whose energy is 4.5 $\hbar\omega$, determine the possible values of the quantum number l.
    (b) For a three-dimensional harmonic oscillator with $L^2 = 30\hbar^2$, determine the two smallest possible energies, as multiples of $\hbar\omega$.
7. Construct the six linear combinations of u_{200}, u_{020}, u_{002}, u_{100}, u_{010}, and u_{001} that are eigenfunctions of L^2 and L_z.
8. (a) Write u_{200} as a linear combination of eigenfunctions of L^2 and L_z.
    (b) Find the probability of finding each possible values of L^2 and L_z when these variables are measured for a system with wave function u_{200}.
    (c) Compute $\langle L^2 \rangle$ and $\langle L_z^2 \rangle$ for this system. Do you expect to find that $\langle L^2 \rangle = 3\langle L_z^2 \rangle$?
9. (a) Write an expression for $d\sigma/d\theta$ for scattering of classical particles from vertical solid cylinder of height h and radius R. The particles are in a beam that travels in the horizontal plane.
    (b) Integrate this expression to find the cross section σ of this cylinder.
10. Determine the angular momentum in terms of k and a for a particle whose momentum is $p = \hbar k$ and whose impact parameter is equal to a. Show the connection between this result and the statement that terms for which $l > ka$ do not contribute significantly to the scattering cross section.
11. Use QMVGA to generate graphs and determine the value of $\sin \delta_3$ in the series solution for Example Problem 8.2, to verify the statement in the text concerning this term.
12. (a) Show that $\sigma \cong 4\pi a^2$ for scattering from a perfectly rigid sphere of radius a (that is, $V = +\infty$ for $r < a$; $V = 0$ for $r > a$), for a beam with $ka \ll 1$.
    Hint: For $r > a$, $R(r) = \sin(kr + \delta_0)/kr$ (because only the $l = 0$ partial wave is necessary), and $R(a) = 0$.
    (b) For the sphere of part (a), use two partial waves ($l = 0$ and $l = 1$) to show that for small values of ka, $\sigma \cong 4\pi a^2(1 + ka^4/3)$. *Hint:* Using Eq. (8.47) for the radius function for $l = 1$, you can compute the value of $\tan \delta_1$.
    (c) For the sphere of part (a), use Eqs. (8.44) and (8.47) with two partial waves to show that

$$d\sigma/d\Omega \cong a^2[1 + 2(ka)^2 \cos \theta]$$

If $ka = 0.1$, what is the percentage difference between $d\sigma/d\Omega$ at $\theta = 0$ and $d\sigma/d\Omega$ at $\theta = 180°$? How does this compare with the percentage difference between the σ for part (a) and the σ for part (b)? If we begin to bombard a scattering center with low-energy particles and then increase the energy, we see a difference between the forward and backward scattering intensity much sooner than we see an increase in the overall cross section.

13. (a) Use the partial waves $l = 0$, $l = 1$, and $l = 2$ to compute σ and $d\sigma/d\Omega$ for scattering from a rigid sphere of radius a, when $ka = 0.5$.

   (b) Compare this result with the result of using only the $l = 0$ wave.

14. A particle of mass m is scattered from the spherically symmetric potential

$$V(r) = -V_0 \quad (r < a), \qquad V(r) = -V_0/4 \quad (a \leq r < 2a),$$

$$V(r) = 0 \quad (r \geq 2a)$$

where $V_0 = h^2/8ma^2$.
Find σ and $d\sigma/d\Omega$. Assume that the particle's energy is $E \ll V_0$, so that the wave number for $r < a$ is $k_1 \cong \pi/2a$, and the wave number for $a < r < 2a$ is $k_2 \cong \pi/4a$. *Hint:* The radial wave function is a combination of j_0 and n_0, which can be written in the form

$$R(r) = \frac{A \sin k_1 r}{k_1 r} \qquad\qquad (r < a)$$

$$R(r) = \frac{B \sin(k_2 r + \delta')}{k_2 r} \quad (a < r < 2a)$$

$$R(r) = \frac{C \sin(kr + \delta)}{kr} \quad (r > 2a)$$

You can eliminate the arbitrary constants A, B, and C, and find δ, by use of the continuity conditions at $r = a$ and $r = 2a$.

15. If all the phase shifts except δ_0 are quite small, the total cross section σ is relatively small at the incident-particle energy that makes δ_0 equal to 180°, and σ becomes larger whether the energy increases or decreases. This effect has been observed in the scattering of electrons from rare-gas atoms; it is called the Ramsauer–Townsend effect. Show that this effect can occur with an attractive potential but not with a repulsive potential.

CHAPTER 9

The Hydrogen Atom

The solution of the problem of atomic spectra was a great triumph for the Schrödinger equation. Although this equation does not yield exact solutions for atoms containing more than one electron, it permits approximations which can be applied, in principle, to any problem and to any desired degree of accuracy. The ultimate result is enormous accuracy in theoretical calculations of hydrogen energy levels, taking into account such factors as the spin of the electron, a previously unknown quantity.

Subsequently, P. A. M. Dirac showed that electron spin emerges as a *natural* consequence of a relativistic wave equation. We cannot treat Dirac's equation in the detail with which we have treated the Schrödinger equation, but we will see in this chapter that two fundamental properties of matter—internal spin of particles and the existence of antimatter—follow directly from this equation.

Experimenters have responded to these theories with equally accurate experiments to test these calculations. This accuracy has not been sought simply to demonstrate the prowess of physics and physicists; proof of the existence of each small contribution to the energy in the amount predicted by the theory is an indication of the correctness of our fundamental ideas concerning the nature of matter.

9.1 WAVE FUNCTIONS FOR MORE THAN ONE PARTICLE

If the proton were of infinite mass, it would be a fixed center of force for the electron in the hydrogen atom, and we could solve the problem by the methods of Chapter 6. The wave function would be a function of the coordinates of a single particle, the electron. However, because the proton also moves, we must incorporate this fact into our wave function.

According to Postulates 1 and 2 (Section 3.3), each dynamical variable for each particle must be represented by an operator whose eigenvalues are the allowed values of the variable. As a logical extension of these postulates, we now assert that there must be a wave function for a *system* which is capable of generating all the dynamical variables of the system. Therefore, for a hydrogen atom the wave function may be written as $\Phi(x_p, y_p, z_p, x_e, y_e, z_e, t)$, where x_p is the x coordinate of the proton, x_e is the x coordinate of the electron, etc.

It is now clear that the wave function is simply a mathematical construction; there is no physical "wave" in the sense of a simple displacement that exists at each point of space and time, for the wave function is a function of *six* space coordinates in this case, rather than three. Our single-particle problems of the previous chapters enabled us to make useful analogies with conventional waves, but we must now go into a more abstract realm of theory. This does not mean that the problems are necessarily more difficult to solve. It simply means that we must beware of *visualizations* of the solutions; the limited experience that we have gained through our senses is not sufficient to permit this.

As usual, the wave function is an eigenfunction of the Schrödinger equation. The Schrödinger equation, in turn, contains the operator for the total energy of the system, as in Chapter 3. The total energy of the proton–electron system is

$$E_T = p_p^2/2m_p + p^2/2m_e + V(r) \qquad (9.1)$$

or, in operator form

$$E_T = -\left(\hbar^2/2m_p\right)\nabla_p^2 - \left(\hbar^2/2m_e\right)\nabla_e^2 + V(r) \qquad (9.2)$$

where r is the proton–electron distance, m_p is the proton's mass, m_e is the electron's mass, $V(r) = -e^2/4\pi\varepsilon_0 r$ is the Coulomb potential, and the symbol ∇^2 represents $\partial^2/\partial x^2 + \partial^2/\partial y^2 + \partial^2/\partial z^2$ or the equivalent expression in spherical coordinates; ∇_p^2 operates on the proton's coordinates, and ∇_e^2 operates on the electron's coordinates in the wave function.

The time-independent Schrödinger equation for the hydrogen atom is therefore

$$\left[-\left(\hbar^2/2m_p\right)\nabla_p^2 - \left(\hbar^2/2m_e\right)\nabla_e^2 + V(r)\right]\psi = E_T\psi \qquad (9.3)$$

but when the equation is written in this form, the energy eigenvalue E_T includes the kinetic energy of translation of the center of mass of the whole atom, a quantity in which we are not interested at the moment. The states that tell us about the hydrogen spectrum are the *internal* energy states—states of the *relative* motion of proton an electron. Fortunately, the potential energy is a function only of the relative coordinates of proton and electron, and we can rewrite Eq. (9.3) in terms of the coordinates X, Y, and Z of the center of mass and the coordinates x, y, and z of the electron relative to the proton. These coordinates are related to the coordinates of the individual particles as follows:

$$X = \frac{m_e x_e + m_p x_p}{m_e + m_p}, \qquad x = x_e - x_p \qquad (9.4a)$$

$$Y = \frac{m_e y_e + m_p y_p}{m_e + m_p}, \qquad y = y_e - y_p \qquad (9.4b)$$

$$Z = \frac{m_e z_e + m_p z_p}{m_e + m_p}, \qquad z = z_e - z_p \qquad (9.4c)$$

It is not difficult to use Eqs. (9.4) to write the kinetic energy in terms of the relative velocity and the velocity of the center of mass; the result is

Total kinetic energy

$$= (m_e + m_p)(\dot{X}^2 + \dot{Y}^2 + \dot{Z}^2)/2 + m_r(\dot{x}^2 + \dot{y}^2 + \dot{z}^2)/2 \quad (9.5)$$

where $\dot{x}$, $\dot{y}$, and $\dot{z}$ are the x, y, and z components of the relative velocity, and $m_r = m_e m_p / (m_e + m_p)$ is the reduced mass, which we previously encountered in discussing the Bohr model of hydrogen (Chapter 3).
    In terms of momentum variables defined as $P_x = (m_e + m_p)\dot{X}$, $p_x = m_r \dot{x}$, etc., the kinetic energy may be written as

$$\frac{P_x^2 + P_y^2 + P_z^2}{2(m_e + m_p)} + \frac{p_x^2 + p_y^2 + p_z^2}{2m_r} \qquad (9.6)$$

so that, according to the rules for writing the energy and momentum operators, the Schrödinger equation becomes

$$\left\{ \frac{\hbar^2}{2(m_e + m_p)} \nabla_c^2 + \frac{\hbar^2}{2m_r} \nabla^2 + V(r) \right\} \psi = -i\hbar \frac{\partial \psi}{\partial t} \qquad (9.7)$$

where the operators ∇_c^2 and ∇^2 operate on center of mass and relative coordinates, respectively.

Equation (9.7) could also have been obtained by direct transformation of the partial derivatives in Eq. (9.3), making use of Eqs. (9.4) to find the derivatives with respect to the new variables.

9.1.1 Separation of Variables

The wave function ψ is now a function of X, Y, Z, x, y, z, and t. Let us assume that it is possible to write the space dependence of Ψ as a function of x, y, and z multiplied by another function of X, Y, and Z. We write

$$\psi(x, y, z, X, Y, Z, t) = u(x, y, z)U(X, Y, Z)e^{-i(E+E')t/\hbar} \qquad (9.8)$$

where E is the energy of the relative motion and E' the energy of translation of the center of mass. From Eq. (9.7) we may now extract the two time-independent equations

$$-\frac{\hbar^2}{2(m_e + m_p)}\nabla_c^2 U = E'U \qquad (9.9)$$

$$\left\{-\frac{\hbar^2}{2m_r}\nabla^2 + V(r)\right\}u = Eu \qquad (9.10)$$

Equation (9.9) is simply the equation of motion of the center of mass of the whole hydrogen atom; it tells us nothing about the atom's internal energy levels. Equation (9.10) is the Schrödinger equation for the motion of the electron relative to the proton. It is identical to the equation for a single particle of mass m_r moving under the influence of a fixed potential energy $V(r)$.

As in the analysis of the Bohr atom, the fact that the proton and electron both move is completely accounted for by using the reduced mass m_r instead of the actual mass of the moving electron (or proton). The eigenvalues E are the energy levels of the hydrogen atom in the frame of reference in which the center of mass of the atom is at rest.

9.2 ENERGY LEVELS OF THE HYDROGEN ATOM

9.2.1 Coulomb Potential and Effective Potential

Because Eq. (9.10) is identical to the one-particle equation treated in Chapter 6, we already know that this equation can be separated into an angular equation and a radial equation and that the function u can be written in spherical coordinates as $u(r, \theta, \phi) = R(r)Y_{l,m}(\theta, \phi)$, where $Y_{l,m}(\theta, \phi)$ is a spherical harmonic, a solution of the angular equation.

To find the energy levels, we need to solve only the radial equation, whose general form is Eq. (8.8):

$$\frac{d^2}{dr^2}[rR(r)] = -\frac{2m}{\hbar^2}\{E - V_{\text{eff}}\}[rR(r)] \qquad (9.11\text{a})$$

or, in this case

$$\frac{d^2}{dr^2}[rR(r)] = -\frac{2m_r}{\hbar^2}\left\{E - V(r) - \frac{l(l+1)\hbar^2}{2m_r r^2}\right\}[rR(r)] \qquad (9.11\text{b})$$

where $l(l+1)\hbar^2/2m_r r^2$ is the "centrifugal" potential, whose introduction, as we saw (Chapter 8), results from eliminating the angular dependence in the equation.

Equation (9.11) is identical to the equation for one-dimensional motion of a particle in the potential field $V_{\text{eff}} = V(r) + l(l+1)\hbar^2/2m_r r^2$. This effective potential, the sum of the Coulomb potential and the centrifugal potential V_{cent}, is sketched in Figure 9.1.

In hydrogen, an electron of negative total energy is trapped in the "potential well" formed by the effective potential. Classically, the particle would describe an elliptical orbit under these conditions; it would oscillate between the two values of r at which its total energy would be equal to the

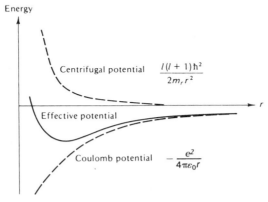

FIGURE 9.1 The effective potential V_{eff} (solid line), the sum of the centrifugal potential V_{cent}, and the Coulomb potential $V(r) = -e^2/4\pi\varepsilon_0 r$ for the hydrogen atom.

potential energy. In quantum theory, we may find a wave function for this well just as we did for the one-dimensional wells considered in Chapters 3 and 4.

The general method of solution of Eq. (9.11) begins with the assumption that the solution is the product of a polynomial and an exponential. Details of this solution are given in Appendix E, which treats the general case of a one-electron ion with nuclear charge Ze. The polynomial contains $n - l$ terms, where l is the angular momentum quantum number and n is a new quantum number, the radial quantum number. The statement that $R(r)$ contains $n - l$ terms may be considered to be the definition of n.

9.2.2 Degeneracy of Solutions

It is a curious feature of the solutions that the energy depends only on n, not on l. For example, the energy is the same for the $l = 1$ solution with a two-term radial solution and for the $l = 2$ solution with a one-term radial solution; in both cases $n = 3$. The energy levels which result are identical to the levels predicted by Bohr for the hydrogen atom or any one-electron ion:

$$E_n = -\frac{m_r c^2 \alpha^2 Z^2}{2n^2} = -13.60(Z^2/n^2) \text{ eV} \tag{9.12}$$

although n now has a completely different interpretation from that of Bohr.

Figure 9.2 shows the effective potentials for $l = 0$, 1, and 2 and the energy levels for $n = 1$, 2, and 3. Because the energy depends only on n and not on l, the same levels which are allowed for any given l are also allowed for *all lower* values of l. The only effect of l on the levels appears through the condition that $n \geq l + 1$, so that lower energy levels are possible for smaller l, because smaller n values are then possible. For example, E_3 is an energy eigenvalue for all three effective wells shown in Figure 9.2a, but E_2 is an eigenvalue only for $l = 1$ and $l = 0$. The fact that all of these different wells have the same set of energy levels is a remarkable property peculiar to the Coulomb potential.

Because Eq. (9.11) is identical to the one-dimensional Schrödinger equation, we can use a graphical analysis, as in Chapter 5, to gain more understanding of the form of the eigenfunctions. As we saw in Chapter 8, the product $rR(r)$ is the probability amplitude for the radial coordinate. Figure 9.2 also shows graphs of the probability amplitudes $rR_{nl}(r)$ for the radial probability amplitudes rR_{21} and rR_{31}, whose energy eigenvalues are

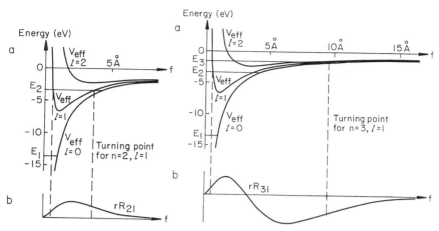

FIGURE 9.2   (a) Effective potential and energy levels of the hydrogen atom for $l = 0, l = 1$, and $l = 2$. The lowest two levels are shown; the number of levels is infinite. (b) Radial probability amplitudes $rR_{nl}(r)$. Notice the points of inflection where $E = V_{eff}$, at the classical turning points of the motion.

shown on the vertical axis as E_2 and E_3, respectively. [You can draw additional graphs yourself by using the QMVGA program, which will plot $rR_{nl}(r)$ for any combination of n and l with $n \leq 100$.]

Notice that, as in the one-dimensional examples of Chapters 3 and 4, the probability amplitude curves away from the axis in the region where $E < V_{eff}$ (the classically forbidden region) and it curves toward the axis, tending to oscillate, where $E > V_{eff}$. The classical turning point, where $E = V_{eff}$, is a point of inflection for the eigenfunction. You may also notice that the function rR_{31} curves more rapidly than rR_{21} in the allowed region, because the curvature is proportional to the kinetic energy. As usual, each eigenfunction contains one more node than the eigenfunction immediately lower in energy. This point is reflected in the fact that the polynomial factor in $R(r)$ has $n - l$ terms, and thus there are $n - l$ roots to the equation $R(r) = 0$.

Because each of the functions $rR(r)$ is a probability amplitude, its absolute square, $|rR(r)|^2$, is the probability density $P(r)$ for finding the electron at a radius between r and $r + dr$. (See Section 8.1 for details.) Thus Figure 9.2 shows where the electron is likely to be, as far as the r coordinate is concerned. It indicates how the average radius of the "orbit" increases as n increases; it is evident from the figure that this average radius must be close to the value given by the original Bohr theory.

TABLE 9.1 Normalized Wave Functions for Hydrogen Atoms and Hydrogenlike Ions

u_{nlm}	$u(r, \theta, \phi)$
u_{100}	$\dfrac{1}{\sqrt{\pi}}\left(\dfrac{Z}{a_0'}\right)^{3/2} e^{-\rho}$
u_{200}	$\dfrac{1}{\sqrt{32\pi}}\left(\dfrac{Z}{a_0'}\right)^{3/2} (2-\rho)e^{-\rho/2}$
u_{210}	$\dfrac{1}{\sqrt{32\pi}}\left(\dfrac{Z}{a_0'}\right)^{3/2} \rho\cos\theta\, e^{-\rho/2}$
$u_{21\pm1}$	$\dfrac{1}{\sqrt{64\pi}}\left(\dfrac{Z}{a_0'}\right)^{3/2} \rho\sin\theta\, e^{-\rho/2} e^{\pm i\phi}$
u_{300}	$\dfrac{1}{81\sqrt{3\pi}}\left(\dfrac{Z}{a_0'}\right)^{3/2} (27-18\rho+2\rho^2)e^{-\rho/3}$
u_{310}	$\dfrac{1}{81\sqrt{\pi}}\left(\dfrac{Z}{a_0'}\right)^{3/2} (6-\rho)\rho\cos\theta\, e^{-\rho/3}$
$u_{31\pm1}$	$\dfrac{1}{81\sqrt{\pi}}\left(\dfrac{Z}{a_0'}\right)^{3/2} (6-\rho)\rho\sin\theta\, e^{-\rho/3} e^{\pm i\phi}$
u_{320}	$\dfrac{1}{81\sqrt{6\pi}}\left(\dfrac{Z}{a_0'}\right)^{3/2} \rho^2(3\cos^2\theta-1)e^{-\rho/3}$
$u_{32\pm1}$	$\dfrac{1}{81\sqrt{\pi}}\left(\dfrac{Z}{a_0'}\right)^{3/2} \rho^2\sin\theta\cos\theta\, e^{-\rho/3} e^{\pm i\phi}$
$u_{32\pm2}$	$\dfrac{1}{162\sqrt{\pi}}\left(\dfrac{Z}{a_0'}\right)^{3/2} \rho^2\sin^2\theta\, e^{-\rho/3} e^{\pm 2i\phi}$

$\rho \equiv Zr/a_0'$, where a_0 = first Bohr radius = 5.29 nm.

$a_0' = a_0\left(\dfrac{m_e + M}{M}\right)$, where M = mass of nucleus.

9.2.3 Table of Wave Functions

The complete normalized wave functions $u_{nlm}(r, \theta, \phi)$ for the lowest energy states of the hydrogen atom are given in Table 9.1. These wave functions also apply to any one-electron ion, if one uses the appropriate values for the atomic number Z and the nuclear mass M. The probability densities associated with some of these wave functions are shown in Figure 9.3.

9.2.4 Comparison with the Bohr Model: Correspondence Principle and Orbits in Hydrogen

Figure 1.7 shows elliptical orbits in the Bohr model of the hydrogen atom, for $n = 4$. The major axis of each ellipse has a length of $16a_0$. If you

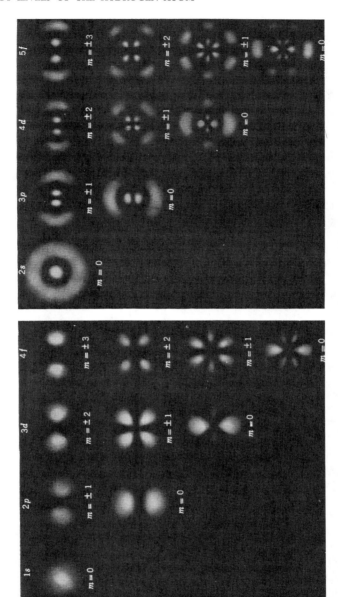

FIGURE 9.3 Photographic representation of the probability density $|rR(r)|^2$ for various states of the hydrogen atom. The bright areas may be regarded as cross-sectional views of each distribution in the plane of the paper with the $\theta = 0$ axis vertical. The scale varies from one distribution to another. (From R. B. Leighton, *Principles of Modern Physics*. Used by permission.)

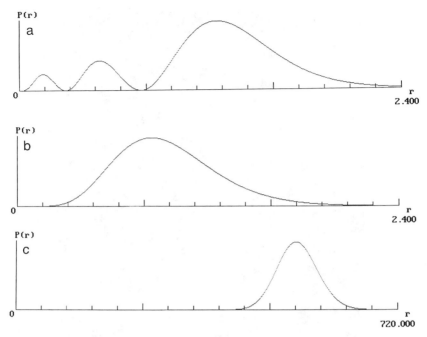

FIGURE 9.4 Radial probability densities for three different states of the hydrogen atom. (a) For $n = 4$, $l = 1$, the allowed region stretches from $r \cong 0.1$ nm to $r \cong 1.6$ nm. (b) For $n = 4$, $l = 3$, the center of the allowed region is unchanged, but the region is narrower, extending from $r \cong 0.6$ to $r \cong 1.1$ (c) For $n = 100$, $l = 99$, the allowed region is centered on $r = 530$ nm, the radius of the Bohr orbit for $n = 100$. It extends from $r \cong 490$ to $r \cong 570$. This is the narrowest possible allowed region for this value of n, and it is much narrower than curve (b), relative to the value of $\langle r \rangle$.

measure the orbits, you see that the most eccentric of these has a minimum value of r that is less than a_0 (0.053 nm), and the maximum value of r in that orbit is greater than $31a_0$, or about 1.64 nm. In that orbit the angular momentum is equal to $\hbar$, so this would correspond to $n = 4$ and $l = 1$ in the Schrödinger equation.

The graph of the probability density for this pair of quantum numbers is shown in Figure 9.4a. You can see that the classically allowed region does indeed extend from about 0.1 nm to about 1.6 nm. In Figure 9.4b, with the same n but $l = 3$, the allowed region is much narrower, because the larger value of l means a less eccentric orbit.

For another example, you see in Figure 9.4c that the allowed region for $n = 100$, $l = 99$ is centered on approximately 530 nm, just the Bohr radius for $n = 100$. You recall that this state has the maximum angular momentum of all the states with $n = 100$, and as such the function has no nodes. Notice how much narrower the classically allowed region appears with $n = 100$. It is actually broader, but as a fraction of the radius it has become smaller. (See Exercise 9.)

9.2.5 Spectroscopic Symbols

For historical reasons associated with observation of the various series of lines in atomic spectra, the l value of each state is designated by a letter, as follows:

Letter:	s	p	d	f	g	h	i
l:	0	1	2	3	4	5	6

The letters go in alphabetical order for $l > 3$. Each state is then identified by the number for n followed by the letter for l: for example, $3d$ for $n = 3$, $l = 2$.[1]

ADDITIONAL READING

Robert Eisberg and Robert Resnick, *Quantum Physics of Atoms, Molecules, Solids, Nuclei, and Particles*, Wiley, New York (1974). Chapter 7 covers many of the topics of this chapter.

R. B. Leighton, *Principles of Modern Physics*, McGraw-Hill, New York (1959). Sections 5-1 to 5-4 give a detailed account of the topics of this chapter.

Linus Pauling and E. B. Wilson, *Introduction to Quantum Mechanics*, McGraw-Hill, New York (1935). Chapter 5 contains a wealth of information about the hydrogen atom wave functions, quantum numbers, and probability densities.

EXERCISES

1. By direct substitution into the Schrödinger equation for the hydrogen atom [Eq. (9.11)], verify that several of the wave functions given in Table 9.1 are solutions of that equation, with the appropriate energy eigenvalue in each case.

---

[1] There is no such "code" to distinguish between states with different values of the m quantum number. The value of m indicates only the orientation of a state in space, relative to the z axis. But in the absence of a magnetic field, the direction of that axis is arbitrary, and thus states that differ only in the value of m all have the same energy. Therefore there was no need to indicate the value of m in the label of a state.

2. By reference to Appendix E, construct the unnormalized functions u_{400}, u_{410}, and u_{411}.

3. By deducing the values of n, $n - l$, and m, determine which one of the following indicates an acceptable set of values and thus could be an (unnormalized) eigenfunction of the Schrödinger equation for the hydrogen atom:

$$(1 - \rho/3 + \rho^2/8 - \rho^3)e^{-\rho/4}, \qquad (1 - \rho/3 + \rho^2/8 - \rho^3)e^{-\rho/5}e^{i\phi}$$

$$(1 - \rho/3 + \rho^2/8 - \rho^3)e^{-\rho/4}e^{-4i\phi}, \qquad (1 - \rho/3 + \rho^2/8 - \rho^3)e^{-\rho/5}$$

4. Compute $\langle r \rangle$ (Section 3.3) for hydrogen atoms whose wave functions are u_{100}, u_{210}, and u_{320}, respectively. (See Table 9.1.) Compare your result with the orbit radius given by the Bohr formula for these values of n. (Notice that each of the orbits is as close to circular as it can be for its value of n.)

5. A proton has a radius of about 10^{-13} cm. Find the probability that an electron whose wave function is u_{200} will be found inside the proton. Compute the same probability for the wave function u_{210}, and comment on the difference in terms of classical orbits and angular momentum.

6. The radial probability amplitude $rR(r)$ for an $n = 5$ state of the hydrogen atom is shown below.
   (a) Deduce the value of L^2 for this state.
   (b) Write an expression, as an integral on r and θ only, for the probability that the electron will be found within a distance of $2a_0'$ from the proton.
   (c) Determine from the accompanying graph the values of r at the boundaries of the classically allowed region. Compute the value of V_{eff} for each of these values of r and show that at these locations the kinetic energy from radial motion is zero.

7. For the state shown in the preceding exercise,
   (a) What are the possible results of a measurement of the z component of its angular momentum?
   (b) After L_z has been measured and found to be equal to $-\hbar$, what are the values of $\langle L_x \rangle$ and $\langle L_x^2 \rangle$ and the probabilities of the various results of a measurement of L_x?

8. Find the probability that a negative muon will be found inside a deuteron (a nucleus of ^{2}H) when the muon–deuteron system is in a $2s$ state. Assume that the radius of a deuteron is 1.6 fm. Compare this with the corresponding probability for an electron. (Muon mass equals 207 times the electron mass. See Exercise 5, Chapter 1.)
   For the following four exercises, use the program QMVGA.

9. Plot the radial probability densities $|rR(r)|^2$ for a series of states for which $n = l + 1$, as follows: $n = 10$, $l = 9$; $n = 20$, $l = 19$; $n = 50$, $l = 49$; and $n = 100$, $l = 99$.
   (a) Compare the energy with the energy given by Eq. (9.12).
   (b) From your plots, estimate the value of $\langle r \rangle$ in each case, and draw a graph of $\langle r \rangle$ vs. n.
   (c) From your graph of $\langle r \rangle$ vs. n determine the value of n that you would need to plot this function for $\langle r \rangle \cong 1$ cm. How is this related to the Bohr atom?

10. Observe the dependence of one-electron atomic or ionic energy levels on
    (a) the value of the atomic number Z.
    (b) the mass of the negative particle. In particular, what would be the lowest energy level of a bound state of a proton and a negative tau lepton? (Tau mass is 3500 times the electron mass.)

11. Determine the points of inflection for all of the hydrogen radial functions with $n = 4$. Is each result consistent with the boundaries of the classically allowed region, as determined by the effective potential?

12. From the graph of probability density, estimate the probability that an electron will be found at $r < 0.05$ nm when a hydrogen atom is in a $2p$ state. Compare with theory by evaluating the appropriate integral.

CHAPTER 10

Spin

It is not strictly true that the energy levels of the hydrogen atom depend only on the quantum number n, as implied by Eq. (9.12). We saw in Section 1.5 that certain levels must be split into two or more sublevels, because we see splitting of lines in the hydrogen spectrum. We also saw that Sommerfeld introduced a second quantum number to explain this splitting. The effect was thought to be relativistic, arising because the electron travels at a much greater speed as it approaches the nucleus when the orbit is eccentric. This idea explained fine structure in the spectrum of the hydrogen atom, but inconsistencies arose when it was applied to alkali metals, whose spectra (Chapter 13), should be quite similar to the hydrogen spectrum. The mystery was resolved by Goudsmit and Uhlenbeck,[1] who attributed the results to an intrinsic property of the electron, called spin. Electron spin gives each electron an intrinsic magnetic dipole moment, which is independent of the electron's state of motion. This hypothesis, which accounted for the observed line splitting, was also supported by the results of the Stern–Gerlach experiment (Section 6.4).

In the Stern–Gerlach experiment, a beam of neutral silver atoms was split into *two* components after passing through an inhomogeneous magnetic field. If the magnetic moment of the silver atom resulted from *orbital* angular momentum of its electrons, the beam would have split into an *odd* number of components, because L_z has $2l + 1$ possible values (from $+l\hbar$ to $-l\hbar$ in steps of $\hbar$). But two components are possible if there are only two possible values for the z component of this *new kind* of angular momentum—*spin* angular momentum. As long as the *difference* between the two possible values is $\hbar$, there would be no contradiction with our previous ideas concerning angular momentum in general. The two possible values for this z component must therefore be $+\hbar/2$ and $-\hbar/2$. (Clearly,

---

[1]S. Goudsmit and G. E. Uhlenbeck, *Nature* **117**, 264 (1926).

for each possible positive value, there must be a negative value of equal magnitude, because space is isotropic, so there is no other way to allow two values which differ by $\hbar$.)

10.1 THE SPIN OPERATORS

Following the method of Section 7.3, we represent the eigenstates of the spin operator as two-dimensional vectors. In the case of electron spin, there are only two eigenfunctions instead of three or more. The Hilbert spin space is therefore two-dimensional, and we can write the eigenfunctions as column vectors just as we did for orbital angular momentum; the unit vectors that are eigenfunctions of S_z are the two-component vectors

$$| + \rangle_z = \begin{pmatrix} 1 \\ 0 \end{pmatrix}, \text{ the eigenstate for } S_z = +\hbar/2 \tag{10.1a}$$

$$| - \rangle_z = \begin{pmatrix} 0 \\ 1 \end{pmatrix}, \text{ the eigenstate for } S_z = -\hbar/2 \tag{10.1b}$$

Compare with Eq. (7.18) for the three eigenstates of L_z with $l = 1$. Following the reasoning given there, we say that any electron-spin function F_s can be a represented by a superposition of $| + \rangle_z$ and $| - \rangle_z$:

$$F_s = a| + \rangle_z + b| - \rangle_z = a\begin{pmatrix} 1 \\ 0 \end{pmatrix} + b\begin{pmatrix} 0 \\ 1 \end{pmatrix} \equiv \begin{pmatrix} a \\ b \end{pmatrix} \tag{10.2}$$

Continuing this line of reasoning, we can represent the spin operators S_x, S_y, and S_z as the following 2×2 matrices that operate on these vectors:

$$S_x = \frac{\hbar}{2}\begin{pmatrix} 0 & 1 \\ 1 & 0 \end{pmatrix}, \quad S_y = \frac{\hbar}{2}\begin{pmatrix} 0 & -i \\ i & 0 \end{pmatrix}, \quad S_z = \frac{\hbar}{2}\begin{pmatrix} 1 & 0 \\ 0 & -1 \end{pmatrix} \tag{10.3}$$

These may be written

$$S_x = \frac{\hbar}{2}\sigma_x, \quad S_y = \frac{\hbar}{2}\sigma_y, \quad S_z = \frac{\hbar}{2}\sigma_z \tag{10.4}$$

where the dimensionless operators σ_x, σ_y, and σ_z (the *Pauli spin matrices*) are expressed by the matrices in Eq. (10.3). These matrices can be derived by using commutation rules analogous to those of Eqs. (7.17):

$$S_xS_y - S_yS_x = i\hbar S_z, \qquad S_yS_z - S_zS_y = i\hbar S_x, \qquad S_zS_x - S_xS_z = i\hbar S_y$$

$$(10.5)$$

It is left as an exercise to verify that these equations are indeed satisfied by the matrices given. You can also verify the *anticommutation relation* $\sigma_x\sigma_y + \sigma_y\sigma_x = 0$, and the corresponding relations involving σ_z.

Example Problem 10.1 Find values of the constants a and b that yield normalized eigenfunctions of S_x, by applying this operator to the general expression for a spin state: $a| + \rangle_z + b| - \rangle_z$.

Solution. We know that the eigenvalues of S_x must be the same as those of S_z.

The statement that $\begin{pmatrix} a \\ b \end{pmatrix}$ is an eigenfunction of S_x with eigenvalue $\hbar/2$

is expressed by the equation $S_x\begin{pmatrix} a \\ b \end{pmatrix} = \dfrac{\hbar}{2}\begin{pmatrix} a \\ b \end{pmatrix}$. Substituting the matrix

for S_x and performing the indicated matrix operation, we have $S_x\begin{pmatrix} a \\ b \end{pmatrix} = \dfrac{\hbar}{2}\begin{pmatrix} 0 & 1 \\ 1 & 0 \end{pmatrix}\begin{pmatrix} a \\ b \end{pmatrix} = \dfrac{\hbar}{2}\begin{pmatrix} b \\ a \end{pmatrix}$. Therefore $\begin{pmatrix} a \\ b \end{pmatrix} = \begin{pmatrix} b \\ a \end{pmatrix}$, or $b = a$.

Normalization requires that $|a|^2 + |b|^2 = 1$, so that $a = b = 1/\sqrt{2}$.

Similarly, if $\begin{pmatrix} a \\ b \end{pmatrix}$ is an eigenfunction with the eigenvalue $-\hbar/2$, then

$S_x\begin{pmatrix} a \\ b \end{pmatrix} = \dfrac{\hbar}{2}\begin{pmatrix} 0 & 1 \\ 1 & 0 \end{pmatrix}\begin{pmatrix} a \\ b \end{pmatrix} = -\dfrac{\hbar}{2}\begin{pmatrix} b \\ a \end{pmatrix}$. In this case $\begin{pmatrix} a \\ b \end{pmatrix} = -\begin{pmatrix} b \\ a \end{pmatrix}$, or

$b = -a = -1/\sqrt{2}$. In conclusion, the eigenvectors are $\begin{pmatrix} 1/\sqrt{2} \\ 1/\sqrt{2} \end{pmatrix}$ and

$\begin{pmatrix} 1/\sqrt{2} \\ -1/\sqrt{2} \end{pmatrix}$, respectively.

10.2 FINE STRUCTURE IN THE HYDROGEN SPECTRUM

10.2.1 Effect of Spin on Energy Levels

The intrinsic magnetic moment of the electron, which results from its spin, leads to a splitting of the energy levels of hydrogen into sublevels, *provided that there is a magnetic field in the atom*, because in each of the states determined by the Schrödinger equation the electron would then have two possible values for its magnetic potential energy, according to the value of the z component of its magnetic moment.

But why should there be a magnetic field in the hydrogen atom? The only other particle present is the proton. One might expect the proton, like the electron, to have internal spin and magnetic moment, and it does, but it turns out that the proton's intrinsic magnetic moment produces a field which is too small to account for the observed fine structure in the hydrogen spectrum. (The proton's magnetic moment leads to *hyperfine structure*, to be discussed later in this section.) However, the proton's *electric charge* leads to the existence of a magnetic field of sufficient strength, because *in the electron's rest frame* the *proton* moves in an orbit around the electron. To put in another way, a body moving through an *electric* field also "sees" a magnetic field, according to the theory of relativity.

10.2.2 Energy of the Spin–Orbit Interaction

Let us use the preceding idea to estimate the magnitude of the splitting of the hydrogen levels. First we must know the magnetic moment of the electron, which may be deduced from the magnitude of the splitting of the atomic beam in the Stern–Gerlach experiment. This moment is very close to

$$\mu = -(e/m_e)\mathbf{S}, \quad (\text{or } \mu = -(e/m_e c)\mathbf{S} \text{ in cgs units}) \quad (10.6)$$

where $\mathbf{S}$ is the spin angular momentum of the electron, and e is the magnitude of the electronic charge. Since $S_z = \pm\hbar/2$, the z component of μ is (in mks units)

$$\mu_z = \pm e\hbar/2m_e = \pm 1 \text{ Bohr magneton}$$

[The Bohr magneton is often denoted by the symbol μ_B and simply called the magnetic moment of the electron. Strictly speaking, μ_B is not the electron's magnetic moment; it is only one component of that moment. However, one component is all that is observed in an experimental measurement of the magnetic moment of the electron.]

To determine the energy attributable to the existence of the electron's magnetic moment, we write the energy of a magnetic moment μ in a magnetic field $\mathbf{B}$ as $W = -\mu \cdot \mathbf{B}$, and from Eq. (10.6) we have

$$W = -(e/m_e)\mathbf{S} \cdot \mathbf{B} \quad (10.7)$$

$\mathbf{B}$ may be related to the electron's *orbital* angular momentum in the following way: If the velocity of the electron is $\mathbf{v}$, the velocity of the proton relative to the electron[2] is $-\mathbf{v}$. The magnetic field $\mathbf{B}$ produced by the proton's charge e is therefore

$$\mathbf{B} = \frac{\mu_0 e(-\mathbf{v}) \times \mathbf{r}}{4\pi r^3} = e\frac{-\mathbf{v} \times \mathbf{r}}{4\pi r^3 \varepsilon_0 c^2}, \quad \text{where } \mu_0 \equiv 1/\varepsilon_0 c^2 = 4\pi \times 10^{-7} \text{ H/m}$$

$$(10.8)$$

This may be written in terms of the electron's angular momentum $\mathbf{L} = \mathbf{r} \times m\mathbf{v}$ as[3]

$$\mathbf{B} = \frac{e\mathbf{L}}{4\pi r^3 m\varepsilon_0 c^2} \tag{10.9}$$

The interaction energy is then, in the electron's rest frame

$$W = -\boldsymbol{\mu} \cdot \mathbf{B} = -\frac{e^2}{4\pi\varepsilon_0 r^3 m^2 c^2} \mathbf{S} \cdot \mathbf{L} \tag{10.10}$$

Expression (10.10) is not the correct one to use for W, however, because the energy splitting seen in the lab is not the same as that seen in the electron's rest frame. The reason for this is rather complicated, involving a phenomenon known as the Thomas precession,[4] which is related to the fact that the time scale is not the same in the laboratory as in the rest frame of the electron. The proper relativistic treatment shows that the energy shift seen in the laboratory is just one-half of the value given in Eq. (10.10), or

$$W = -\boldsymbol{\mu} \cdot \mathbf{B} = -\frac{e^2}{8\pi\varepsilon_0 r^3 m^2 c^2} \mathbf{S} \cdot \mathbf{L} \tag{10.11}$$

We can now estimate the value of W. We set $|\mathbf{L}| \cong \hbar$. For $\mathbf{S}$, we know that the component in any direction, such as the direction of $\mathbf{L}$, must be $\hbar/2$ in magnitude. If we then let $r \cong a_0$, we find W to be of the order of 10^{-4} eV, which is of the order of magnitude required by the observed fine structure. We shall see in Chapter 12 that effects of this size may be calculated by the methods of perturbation theory.

[2] We neglect the slight motion of the proton in the center-of-mass frame.
[3] For cgs units, simply remove $4\pi\varepsilon_0$ from the denominator.
[4] For details, see R. Eisberg, Fundamentals of Modern Physics, pp. 341–346, Wiley, New York (1961).

It should come as no surprise to find that as a first approximation this energy can be assumed to be the expectation value of $-\boldsymbol{\mu} \cdot \mathbf{B}$, computing this value by using the eigenfunctions already found for the hydrogen atom; that is

$$W = \langle \mathbf{S} \cdot \mathbf{L} \rangle = \int \int \int u^* \frac{e^2}{8\pi\varepsilon_0 r^3 m^2 c^2} (\mathbf{S} \cdot \mathbf{L})_{\text{op}} u \, dx \, dy \, dz \quad (10.12)$$

where u is an eigenfunction of Eq. (9.10). We should therefore be able to express this energy in terms of the quantum numbers n, l, and m for these eigenfunctions, if we can evaluate the $\mathbf{S} \cdot \mathbf{L}$ operator; that is, we need to know the result of the operation $(\mathbf{S} \cdot \mathbf{L})_{\text{op}} u$.

10.2.3 Combining Spin and Orbital Angular Momentum

The $\mathbf{S} \cdot \mathbf{L}$ operator can be included in the operator for the total energy as it appears in the Schrödinger equation. The energy is then found by finding the eigenvalues of the Schrödinger equation in this form, with the variables $\mathbf{S}$ and $\mathbf{L}$ represented as operators. Thus

$$-\frac{\hbar^2}{2m_{\text{r}}} \nabla^2 u_{\text{s}} - \frac{e^2}{4\pi\varepsilon_0 r} u_{\text{s}} + \frac{e^2}{8\pi\varepsilon_0 r^3 m^2 c^2} (\mathbf{S} \cdot \mathbf{L})_{\text{op}} u_{\text{s}} = E u_{\text{s}} \quad (10.13)$$

where the function is labeled u_{s} to show that it is a function of the spin coordinate as well as the three space coordinates, because $\mathbf{S} \cdot \mathbf{L}$ must operate on it.

We may account for this spin dependence by assuming that u_{s} is some superposition of *two wave functions* ψ_1 and ψ_2 that differ in the value of the spin coordinate, as follows:

$$\psi_1 = e^{-i\omega t} u_1(x, y, z) | + \rangle_{\text{s}} \quad (10.14a)$$

$$\psi_2 = e^{-i\omega t} u_2(x, y, z) | + \rangle_{\text{s}} \quad (10.14b)$$

where the state vectors $| + \rangle_{\text{s}}$ and $| - \rangle_{\text{s}}$ are the two eigenfunctions introduced in Eqs. (10.1). There we saw that the spin operator S_z operates on these functions to yield the two possible eigenvalues of S_z:

$$S_z | + \rangle_{\text{s}} = +(\hbar/2)| + \rangle_{\text{s}} \quad \text{and} \quad S_z | - \rangle_{\text{s}} = -(\hbar/2)| - \rangle_{\text{s}} \quad (10.15)$$

Now we must examine the effect of all three components of $\mathbf{S}$, as they appear in the $\mathbf{S} \cdot \mathbf{L}$ operator. To proceed, we must introduce the *total angular momentum* vector $\mathbf{J} = \mathbf{L} + \mathbf{S}$. Because there are no *external* torques on the system the *total* angular momentum is constant during the motion, but neither $\mathbf{L}$ nor $\mathbf{S}$ is necessarily constant. The $\mathbf{B}$ field caused by the electron's orbital motion *exerts a torque* on the *spin* magnetic moment; when this magnetic moment changes direction, $\mathbf{L}$ must also change direction, in order that $\mathbf{L} + \mathbf{S}$ remain constant.

In quantum mechanical terms, this means that the wave function cannot be either of the functions of Eqs. (10.14), which have fixed values of S. Rather, the wave function must be a superposition of these two functions. The vector $\mathbf{S}$, of course, remains constant in *magnitude*, because this is a fundamental property of the electron, but the *direction* of $\mathbf{S}$ is not determined.

Having introduced $\mathbf{J}$, we now can express $\mathbf{S} \cdot \mathbf{L}$ in terms of $\mathbf{J}$, $\mathbf{L}$, and $\mathbf{S}$, using the equation

$$J^2 \equiv \mathbf{J} \cdot \mathbf{J} = (\mathbf{L} + \mathbf{S}) \cdot (\mathbf{L} + \mathbf{S}) = S^2 + L^2 + (\mathbf{S} \cdot \mathbf{L}) + (\mathbf{L} \cdot \mathbf{S}) \quad (10.16)$$

But the $\mathbf{L}$ operator must commute with the $\mathbf{S}$ operator, because $\mathbf{L}_{op}$ and $\mathbf{S}_{op}$ operate on different coordinates. Therefore $(\mathbf{S} \cdot \mathbf{L}) = (\mathbf{L} \cdot \mathbf{S})$, and we can solve for $\mathbf{S} \cdot \mathbf{L}$, obtaining

$$\mathbf{S} \cdot \mathbf{L} = (J^2 - S^2 - L^2)/2 \quad (10.17)$$

The problem of evaluating $(\mathbf{S} \cdot \mathbf{L})$ is now reduced to evaluating $(J^2 - S^2 - L^2)$.

We have already found [Eq. (6.58)] that the eigenvalues of L^2 can be written in the form

$$L^2 = l(l + 1)\hbar^2 \quad (10.18)$$

where l is defined as the maximum value of m, or of $L_z/\hbar$ (Section 6.3). By analogy we might suspect that we could also write

$$S^2 = s(s + 1)\hbar^2 \quad (10.19)$$

and

$$J^2 = j(j + 1)\hbar^2 \quad (10.20)$$

where s is defined as the maximum value of $S_z/\hbar$ and j is defined as the maximum value of $J_z/\hbar$. Because $S_z = \pm\hbar/2$, the value of s must be 1/2. This leads to a value for S^2 that is consistent with what we know about orbital angular momentum: From the isotropy of space, $S_x^2 = S_y^2 = S_z^2 =$

$\hbar^2/2$, and by definition,

$$S^2 = S_x^2 + S_y^2 + S_z^2 = 3\hbar^2/4 \tag{10.21}$$

the same result that we find from Eq. (10.19).

Notice the similarity of this discussion to the one that led to Eq. (6.58). The only difference is that Eq. (6.58) involved expectation values (e.g., $\langle L_x^2 \rangle$). In the case of spin, S_x^2 has only one possible value, $\langle S_x^2 \rangle = S_x^2$ in all cases.

10.3 STEPPING OPERATORS AND EIGENVALUES OF J^2

The eigenvalues of J^2 follow a similar pattern; we only have to determine the maximum value of J_z. We know that $J_z = L_z + S_z$ and that $S_z = \pm\hbar/2$. Thus it would seem clear that the maximum value of J_z must equal the maximum value of L_z plus the maximum value of S_z. Apparently, therefore, $j = (l\hbar + \hbar/2)/\hbar = l + 1/2$, by definition. This is a possible value for j, but there is more to it.

The value of j is related to the eigenvalue of J^2, which in turn is related to the value of $\mathbf{L} \cdot \mathbf{S}$. For a given value of l, there must be two possible values of $\mathbf{L} \cdot \mathbf{S}$, because the spin vector always has two possible directions relative to any other direction (including the direction of the $\mathbf{L}$ vector). One of these directions yields a positive value for $\mathbf{L} \cdot \mathbf{S}$, the other a negative value of equal magnitude. This tells us that there must be two possible values of j.

10.3.1 Stepping Operators for Angular Momentum States

To put the discussion on a firmer basis, we now generate the eigenvalues of J^2 by using the stepping-operator method that was introduced in Section 4.2 for harmonic-oscillator eigenfunctions. In doing this, it is convenient to introduce the symbol $[L_x, L_y]$ to represent the so-called *commutator* of L_x and L_y: the operator $L_x L_y - L_y L_x$. It is easy to show (Exercise 6, Chapter 6) that the three operators of this type obey the *commutation relations*

$$[L_x, L_y] = i\hbar L_z, \qquad [L_y, L_z] = i\hbar L_x, \qquad [L_z, L_x] = i\hbar L_y \tag{10.22}$$

Each relation can be generated from the previous one by the replacement of x by y, y by z, and z by x.

We can now derive the eigenvalues of J^2 by making two reasonable assumptions:

1. Similar commutation relations apply to all forms of angular momentum, so that $[S_x, S_y] = i\hbar S_x$, $[J_x, J_y] = i\hbar J_x$, and so on.
2. We can find a function that is an eigenfunction of both J^2 and J_z, with the eigenvalue $a\hbar^2$ for J^2 and eigenvalue $b\hbar$ for J_z, where a and b are dimensionless.

Denoting this function by the symbol u_{ab}, we write

$$J^2 u_{ab} = a\hbar^2 u_{ab} \tag{10.23}$$

$$J_z u_{ab} = b\hbar u_{ab} \tag{10.24}$$

Let us now introduce the operator $J_+ = J_x + iJ_y$ and demonstrate that the function $J_+ u_{ab}$ is also an eigenfunction of J_z. Applying J_z to $J_+ u_{ab}$ gives us

$$J_z J_+ u_{ab} = J_z(J_x + iJ_y)u_{ab} = (J_z J_x + iJ_z J_y)u_{ab} \tag{10.25}$$

We can apply J_z directly to u_{ab} by using the commutation relations in the form

$$J_z J_x = J_x J_z + i\hbar J_y \quad \text{and} \quad J_z J_y = J_y J_z - i\hbar J_x \tag{10.26}$$

Substituting from Eq. (10.26) into Eq. (10.25) produces

$$J_z J_+ u_{ab} = (J_x J_z + i\hbar J_y + iJ_y J_z + \hbar J_x)u_{ab} \tag{10.27}$$

The substitution of $J_z u_{ab} = b\hbar u_{ab}$ from Eq. (10.24) now yields

$$J_z J_+ u_{ab} = (J_x b\hbar + i\hbar J_y + iJ_y b\hbar + \hbar J_x)u_{ab} \tag{10.28}$$

which reduces to

$$J_z J_+ u_{ab} = (J_x + iJ_y)(b\hbar + \hbar)u_{ab} = (b + 1)\hbar J_+ u_{ab} \tag{10.29}$$

Thus we have proven that the function $J_+ u_{ab}$ is an eigenfunction of J_z, with eigenvalue $(b + 1)\hbar$. Following the same procedure, we could prove that $J_- u_{ab}$, where $J_- = J_x - iJ_y$, is an eigenfunction of J_z with eigenvalue $(b - 1)\hbar$. Therefore we can generate from u_{ab} a set of eigenfunctions whose eigenvalues are $\hbar$ multiplied by $b, b \pm 1, b \pm 2, \ldots$. Each of these functions is also an eigenfunction of J^2, as you may verify by applying the J^2 operator to the function J_+u_{ab}, then applying the commutation relations and Eqs. (10.23) and (10.24).

This set of eigenfunctions cannot be infinite, because we know that no eigenvalue of J_z^2 can be greater than the eigenvalue of J^2, which equals $a\hbar^2$ for every eigenfunction in the set. Thus there must be a largest eigenvalue of J_z, which we denote as $j\hbar$. We write the eigenfunction corresponding to this eigenvalue as u_{aj}; that means that $J_zu_{aj} = j\hbar u_{aj}$. It is clear that

$$J_+u_{aj} = 0 \tag{10.30}$$

for otherwise J_+u_{aj} would be an eigenfunction of J_z with eigenvalue $(j + 1)\hbar$, in contradiction to the definition of $j\hbar$ as the largest eigenvalue. Let us now apply the operator $(J_x - iJ_y)$ to this function, to obtain

$$(J_x - iJ_y)(J_x + iJ_y)u_{aj} = 0 \tag{10.31}$$

or

$$\left[J_x^2 + J_y^2 + i(J_xJ_y - J_yJ_x)\right]u_{aj} = 0 \tag{10.32}$$

The commutation rule for J_x and J_y shows that this can be written as

$$\left(J_x^2 + J_y^2 - \hbar J_z\right)u_{aj} = 0 \tag{10.33}$$

or, since $J_zu_{aj} = j\hbar u_{aj}$,

$$\left(J_x^2 + J_y^2 - j\hbar^2\right)u_{aj} = 0 \tag{10.34}$$

Adding $(J_z^2 + j\hbar^2)u_{aj}$ to both sides, we have

$$\left(J_x^2 + J_y^2 + J_z^2\right)u_{aj} = (J_z^2 + j\hbar^2)u_{aj} \tag{10.35}$$

which reduces to

$$\boxed{J^2u_{aj} = j(j + 1)\hbar^2u_{aj}} \tag{10.36}$$

In summary, we have shown *directly from the assumed commutation relations* that the eigenvalues of J^2 are $j(j + 1)\hbar^2$ and that the states with a given value of j form a set for which the eigenvalue of J_z is one of the sequence $+j\hbar, +(j - 1)\hbar, \ldots, -j\hbar$. Thus we may write $J_z = m_j\hbar$, where

m_j has a maximum value of j and a minimum value of $-j$. Since the difference between any two values of m_j must be an integer and the difference between j and $-j$ is $2j$, the value of j must be either an integer or a half-integer.

This derivation is based solely upon the commutation relations; nothing in the derivation depends upon the nature of the particle involved. For a single electron, the fact that $s = 1/2$ makes j a half-integer; for a particle whose spin is an integer, j would also be an integer.

Now we must determine the general relation between the value of j and the value of s. We do this by referring to the z components of the vectors $\mathbf{J}$, $\mathbf{L}$, and $\mathbf{S}$. Since $\mathbf{J} = \mathbf{L} + \mathbf{S}$ we know that $J_z = L_z + S_z$. The eigenvalue relation is $m_j\hbar = m_l\hbar + m_s\hbar$, so that $m_j = m_l + m_s$, or

$$m_j = m_l \pm 1/2 \tag{10.37}$$

The maximum value of m_l is j, so

$$m_j \leq l + 1/2 \tag{10.38}$$

and there must be a set of states for which m_j takes on the values

$$m_j = l + \tfrac{1}{2}, l - \tfrac{1}{2}, l - \tfrac{3}{2}, \ldots, -(l + \tfrac{1}{2}) \qquad (j = l + \tfrac{1}{2}) \tag{10.39}$$

and for every state in this set the value of j is $l + \tfrac{1}{2}$. This means that the magnitude of $\mathbf{J}$ is greater than the magnitude of $\mathbf{L}$ for each of these states, and we deduce that $\mathbf{S} \cdot \mathbf{L}$ is positive for each of these states. There must also be states for which $\mathbf{S} \cdot \mathbf{L}$ is negative; these form a set for which j equals $l - \tfrac{1}{2}$, and in this set

$$m_j = l - \tfrac{1}{2}, l - \tfrac{3}{2}, \ldots, -(l - \tfrac{1}{2}) \qquad (j = l - \tfrac{1}{2}) \tag{10.40}$$

In both of these sets we find m_j values in the range $l - \tfrac{1}{2} \geq m_j \geq -(l - \tfrac{1}{2})$. Thus there is only one state for which $m_j = l + \tfrac{1}{2}$, only one state for which $m_j = l - \tfrac{1}{2}$, and two states for each of the other possible values of m_j. Let us now examine in detail the distinction between two states with different j and the same m_j. Consider the case $m_j = l - \tfrac{1}{2}$. This value of m_j can be produced in two ways; either $m_l = l - 1$ and the spin is positive, or $m_l = l$ and the state is negative. In Dirac notation (as presented in Section 7.4) we would write these states as $|l, l - 1\rangle|+\rangle_s$ and $|l, l\rangle|-\rangle_s$, respectively. These two states are of course orthogonal to each other. Neither of them is an eigenstate of the J^2 operator, but we can form two linear combinations of them, such that one combination is an eigenstate of the J^2 operator with eigenvalue $j = l + \frac{1}{2}$, and the other combination is an eigenstate of the J^2 operator with eigenvalue $j = l - \frac{1}{2}$. As illustrated in the following example, $\mathbf{S} \cdot \mathbf{L}$ is positive when $j = l + \frac{1}{2}$, and $\mathbf{S} \cdot \mathbf{L}$ is negative when $j = l - \frac{1}{2}$.

Example Problem 10.2   Find the eigenvalue of J^2 for each of the two independent linear combinations of $|1, 0\rangle| + \rangle_s$ and $|1, 1\rangle| - \rangle_s$ that are eigenfunctions of J^2.

Solution. Applying the $\mathbf{S} \cdot \mathbf{L}$ operator to the general expression

$$F = a|1, 0\rangle| + \rangle_s + b|1, 1\rangle| - \rangle_s$$

gives us the expression $(S_x L_x + S_y L_y + S_z L_z)[a|1, 0\rangle| + \rangle_s + b|1, 1\rangle| - \rangle_s]$.
Using the spin operators [Eqs. (10.3)] combined with the operators for the components of $\mathbf{L}$ [Eqs. (7.19)] we find that the individual terms are

$$S_x L_x [a|1, 0\rangle| + \rangle_s + b|1, 1\rangle| - \rangle_s]$$
$$= \hbar^2/2\sqrt{2}\,[a|1, 1\rangle| - \rangle_s + a|1, -1\rangle| - \rangle_s + b|1, 0\rangle| + \rangle_s]$$
$$S_y L_y [a|1, 0\rangle| + \rangle_s + b|1, 1\rangle| - \rangle_s]$$
$$= \hbar^2/2\sqrt{2}\,[a|1, 1\rangle| - \rangle_s - a|1, -1\rangle| - \rangle_s + b|1, 0\rangle| + \rangle_s]$$
$$S_z L_z [a|1, 0\rangle| + \rangle_s + b|1, 1\rangle| - \rangle_s]$$
$$= \hbar^2/2[-b|1, 1\rangle| - \rangle_s]$$

This becomes, after adding and grouping terms:

$$\mathbf{S} \cdot \mathbf{L}F = \hbar^2/2\sqrt{2}\,[a|1, 1\rangle| - \rangle_s + b|1, 0\rangle| + \rangle_s] - \hbar^2/2[b|1, 1\rangle| - \rangle_s]$$
$$= A\hbar^2 [a|1, 0\rangle| + \rangle_s + b|1, 1|\rangle| - \rangle_s]$$

where $A\hbar^2$ is the assumed eigenvalue, A being a dimensionless number. Equating corresponding coefficients we have

$$a/\sqrt{2} - b/2 = Ab \quad \text{and} \quad b/\sqrt{2} = Aa \qquad (10.41)$$

We can solve these to find two roots for A: $A = +1/2$ or $A = -1$.
Therefore the eigenvalues of $\mathbf{S} \cdot \mathbf{L}$ are $+\hbar^2/2$ and $-\hbar^2$. Since $L^2 = 2\hbar^2$ and $S^2 = 3\hbar^2/4$, $J^2 = L^2 + S^2 + 2\mathbf{S} \cdot \mathbf{L} = 15\hbar^2/4$ (when $\mathbf{S} \cdot \mathbf{L} = +\hbar^2/2$), and $J^2 = 3\hbar^2/4$ (when $\mathbf{S} \cdot \mathbf{L} = -\hbar^2$).

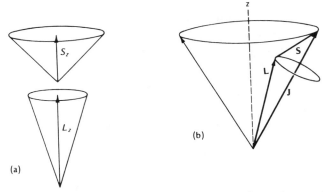

FIGURE 10.1 (a) State in which L_z and S_z are known; J^2 is unknown, because the direction of **L** relative to **S** is unknown. (b) State in which J^2 is known. The values of L_z and S_z are not known, because **S** and **L** rotate around **J**.

Notice that these values agree with the formula $J^2 = j(j + 1)\hbar^2$ with $j = 3/2$ in the first case and $j = 1/2$ in the second case. To summarize:

1. There are *two sets* of states for a given value of l, when $s = \frac{1}{2}$; these have the values of m_l given in Eqs. (10.39) and (10.40).
2. There are two independent states for each value of m_j from $m_j = l - (1/2)$ down to $m_j = -[l - (1/2)]$.
3. If one of these states is an eigenstate of J^2, that state is not an eigenstate of either L_z or S_z, but it is a linear combination of the two eigenstates $|l, m_l\rangle| - \rangle_s$ and $|l, m_l - 1\rangle| + \rangle_s$ (as illustrated in the preceding example).[5]
4. In general, the quantities L_z, S_z, and J^2 are not simultaneously observable, because the J^2 operator does not commute with the L_z and S_z operators.

Figure 10.1 illustrates this point, Figure 10.1a shows the case in which L^2, L_z, S^2, and S_z are all known. Each of the vectors **S** and **L** lies on a cone whose axis is the z axis, but the position of the vector on its cone is completely undetermined. Therefore the direction of **S** relative to **L** is unknown, **S · L** is unknown, and J^2 must be unknown. In contrast, Figure 10.1b shows the situation in which J^2, L^2, S^2, and J_z are known. In this case **S · L** must be known; thus we know that the angle between **S** and **L**

[5]The analogy of states with vectors is helpful here. The two eigenfunctions of J^2 are analogous to orthogonal unit vectors in a plane. The two states for which L_z and S_z are known are orthogonal unit vectors in the same plane. Each of the latter two can be written as a linear combination of the other two, and vice versa.

must be fixed, making **S** lie on a cone whose axis is the direction of **L**. Observe that the triangle formed by **S**, **L**, and **J** is completely determined in size and shape, but this triangle rotates about the direction of **J**. Thus L_z and S_z are undetermined, but J^2 and J_z are known.

10.3.2 Evaluation of the Spin–Orbit Energy

We now have the necessary information to evaluate the spin–orbit energy. Let us generalize Eq. (10.12) to the case of a nuclear charge of Ze in a one-electron ion. Then we have

$$W = \langle \mathbf{S} \cdot \mathbf{L} \rangle = \int \int \int u^* \, \frac{Ze^2}{8\pi\varepsilon_0 r^3 m^2 c^2} (\mathbf{S} \cdot \mathbf{L})_{op} u \, dx \, dy \, dz \quad (10.42)$$

Using Eq. (10.17) for the operator, we have

$$W = \langle \mathbf{S} \cdot \mathbf{L} \rangle = \int \int \int u^* \, \frac{Ze^2}{16\pi\varepsilon_0 r^3 m^2 c^2} (J^2 - S^2 - L^2)_{op} u \, dx \, dy \, dz$$

$$(10.43)$$

If u is an eigenfunction of J^2, S^2, and L^2, we may write

$$(J^2 - S^2 - L^2)_{op} u = \hbar^2 [j(j + 1) - l(l + 1) - s(s + 1)]u \quad (10.44)$$

And after setting s equal to $1/2$, we have

$$W = \frac{Ze^2 \hbar^2}{16\pi\varepsilon_0 m^2 c^2} [j(j + 1) - l(l + 1) - \tfrac{3}{4}] \int \int \int \frac{u^* u}{r} \sin \theta \, dr \, d\theta \, d\phi$$

$$(10.45)$$

The angular part of this integral is equal to 1, because of the normalization of the spherical harmonics. It is a simple matter to evaluate the radial part for any *particular* state, by plugging in the appropriate radial function (see Table 9.1 or Appendix E). Evaluating the integral for the general case is more difficult; the result (for $l \neq 0$) is[6]

$$\int \int \int \frac{u^* u}{r} \sin \theta \, dr \, d\theta \, d\phi = \frac{2Z^3}{a_0' n^3 l(2l + 1)(l + 1)} \quad (10.46)$$

where a_0' is defined as in Table 9.1.

[6] Details are given in L. Pauling and E. B. Wilson, *Introduction to Quantum Mechanics*, Appendix VII, McGraw-Hill, New York (1935).

It is instructive to write W in terms of the energy of the unperturbed state $[E_n = -m_r c^2 Z^2 \alpha^2 / 2n^2$, where $\alpha = e^2 / 4\pi\varepsilon_0 \hbar c \cong 1/137$ (the fine-structure constant)]:

$$W = \frac{Z^2 |E_n| \alpha^2 [j(j+1) - l(l+1) - 3/4]}{nl(2l+1)(l+1)} \tag{10.47}$$

Since j equals either $l + 1/2$ or $l - 1/2$, we can write this as two expressions:

$$W = \frac{Z^2 |E_n| \alpha^2}{n(2l+1)(l+1)} \quad \text{if } j = l + 1/2 \tag{10.48a}$$

$$W = -\frac{Z^2 |E_n| \alpha^2}{nl(2l+1)} \quad \text{if } j = l - 1/2 \tag{10.48b}$$

Thus $|W|$ is proportional to $|E_n| \alpha^2$, while $|E_n|$ itself is proportional to $mc^2\alpha^2$. Thus we could say that we have the first three terms in an expansion of the total energy in a power series in α^2. However, there is another effect, a relativistic contribution to the energy, that is comparable to the effect just computed. This effect is a consequence of the fact that the kinetic energy of a particle of mass m and momentum p is not accurately given by the expression $p^2/2m$, as had been assumed in constructing the Schrödinger equation.

With both effects included, the energy shift ΔE is given by a single formula

$$W = -\frac{Z^2 |E_n| \alpha^2}{n^2} \left\{ \frac{3}{4} - \frac{2n}{2j+1} \right\} \tag{10.49}$$

Notice that the total energy depends on n and j only, and not on l. (The calculation involves an approximation that we will discuss in Chapter 12.) For this reason, the labeling of states includes a subscript giving the value of j. For example, $2s_{1/2}$ is the state with $n = 2$, $l = 0$, and $j = 1/2$, and $2p_{1/2}$ is the state with $n = 2$, $l = 1$, and $j = 1/2$.

10.3.3 Hyperfine Structure

The *proton* and *neutron* also have intrinsic spin and magnetic moment. These are observable because of the energy shift resulting from the interaction of this magnetic moment and the orbital and spin magnetic moments of the electron. Although the energy involved is tiny, compared to the $\mathbf{S} \cdot \mathbf{L}$ effect, it has clearly been observed. The magnitude of this "hyperfine" energy splitting can be estimated theoretically by an extension of the methods used above.

In general, an atom has a *total* angular momentum that can be defined as the vector $\mathbf{F} = \mathbf{J} + \mathbf{I}$, where $\mathbf{I}$ is the internal angular momentum of the nucleus (analogous to $\mathbf{S}$ for the electron). In the ^{1}H atom, $\mathbf{I}$ is simply the spin angular momentum of the proton, which is equal to the spin angular momentum of the electron, with two possible values, $\pm\hbar/2$, of the component along any specified direction. In all other nuclei, $\mathbf{I}$ is the vector sum of the angular momenta of all of the constituent protons and neutrons. These vectors include spin angular momentum as well as angular momentum resulting from orbital motions of these particles within the nucleus.

The vectors $\mathbf{F}$ and $\mathbf{I}$ obey the same rules that apply to all angular momentum vectors. The z component of the vector $\mathbf{I}$ is $I\hbar$, and the possible values of $\mathbf{I}^2$ are $I(I + 1)\hbar^2$. The z component of the vector $\mathbf{F}$ is $F\hbar$, and the possible values of $\mathbf{F}^2$ are $F(F + 1)\hbar^2$. The quantum number F takes on integrally spaced values between $J + I$ and $|J - I|$. The magnetic energy resulting from the magnetic moment of the nucleus is proportional to $\mathbf{I} \cdot \mathbf{J}$, and this energy has as many possible values as there are values of F—namely, either $2I + 1$ or $2J + 1$, whichever is smaller. (You may verify this for yourself.)

The energy splitting is so tiny (or hyperfine) because the magnetic moment μ_p of the proton is very small compared to that of the electron; we find that

$$\mu_p = 2.79 \, e\hbar/2M_p \quad \text{(whereas } \mu_B = e\hbar/2m_e \gg e\hbar/2M_p) \quad (10.50)$$

The quantity $e\hbar/2M_p$ is called the *nuclear magneton* (μ_N).[7]

[The magnetic moment of the neutron is smaller than the proton's; its magnitude is 1.93 μ_N. You might wonder why a neutron, being neutral, would have any magnetic moment at all. But the neutron is not uniformly neutral; it is composed of other particles (three quarks) that are charged—two of them with a charge of $+e/3$, one with a charge of $-2e/3$. It is the motion and spin of these charges that produces the neutron's magnetic moment.]

Because the nuclear magneton is so small relative to the Bohr magneton, hyperfine structure is much more difficult to resolve than fine struc-

---

[7]Remember that the magnetic moment is a vector and that what we conventionally call the magnetic moment is only one component of that vector.

ture. However, the nuclear magnetic moment is responsible for the splitting of the $1s$ level of hydrogen into two levels which differ in energy by about 6×10^{-6} eV (considerably smaller than the fine-structure splitting, which is of order 10^{-4} eV). The wavelength of photons of this energy is 21 cm; this wavelength is quite prominent in radiation from outer space, and it has made it possible to map the distribution of atomic hydrogen in our galaxy.[8]

The agreement between theory and experiment over the energy range from 10 eV to less than 10^{-5} eV is impressive, but even more impressive results are obtained by including effects of relativity and quantization of the energy levels of the electromagnetic field.

Calculations have shown that the $2s_{1/2}$ and $2p_{1/2}$ levels are not equal, as Eq. (10.48) implies. Instead, they should be split by 4.3743×10^{-6} eV; this result (the *Lamb shift*) has been confirmed by experiment.[9]

10.4 SPIN AND RELATIVITY: THE DIRAC EQUATION AND ANTIMATTER

Schrödinger recognized the need for a relativistic form of his equation, and he developed one, but his equation applied only to a particle without spin. The equation developed (in 1930) by P. Dirac does not have this shortcoming. On the contrary, the equation *automatically* includes spin, as an inevitable result of the relativistic energy operator chosen by Dirac. The full implications of the Dirac equation are beyond the scope of this text, but two interesting facets—spin and antimatter—are suitable for a brief discussion.

10.4.1 The Dirac Equation

Dirac's equation for a free particle differs from the Schrödinger equation in having an operator for the energy $E = (p^2c^2 + m^2c^4)^{1/2}$, rather than an operator for E^2. To make a Hamiltonian operator H_{op} out of this square root, he assumed that the operator must be a *linear* operator in the momentum components, or

$$H_{op} = (p^2c^2 + m^2c^4)^{1/2} = \boldsymbol{\alpha} \cdot \mathbf{p}c + \beta mc^2 \qquad (10.51)$$

---

[8] This radiation was first obseved by H. I. Ewen and E. M. Purcell, *Phys. Rev.* **83**, 881 (1951).

[9] W. E. Lamb and R. C. Retherford, *Phys. Rev.* **72**, 241 (1947).

where α must be a vector operator and β a scalar operator, as we shall see in the following. Constants c and mc^2 are included to make these operators dimensionless.

It is significant that *neither α nor β can depend on the time or the position of the particle*, because a free particle's behavior must be invariant with respect to translation or rotation of the coordinate system. Consequently, both of the these operators must commute with the **p** operator; i.e., the differential operator $-i\hbar\nabla$.

We can now square both sides of Eq. (10.51), and equate the coefficients of p_x^2, p_y^2, p_z^2, and mc^2, to find that

$$\alpha_x^2 = \alpha_y^2 = \alpha_y^2 = \beta^2 \tag{10.52}$$

Similarly, we can show that (see Exercise 14)

$$\alpha_x\alpha_y + \alpha_y\alpha_x = 0 \qquad \alpha_x\alpha_z + \alpha_z\alpha_x = 0 \qquad \alpha_y\alpha_z + \alpha_z\alpha_y = 0 \tag{10.53a}$$

$$\alpha_x\beta + \beta\alpha_x = 0 \qquad \alpha_y\beta = \beta\alpha_y = 0 \qquad \alpha_z\beta + \beta\alpha_z = 0 \tag{10.53b}$$

We see from these equations that the four Dirac operators each have unit magnitude, and they *anticommute* in pairs. We have previously found that the three spin operators also anticommute and that the commutation relations can be applied to derive all the important properties of angular momentum.

From these facts and from the fact that these operators are not dependent on the space or time variable, we have reason to suspect that the operators α_x, α_y, α_z, and β must be related to the spin of the electron. However, there can be no more than three commuting operators that operate only on the spin function, as the spin operators do. Since the above four operators all anticommute with one another, another variable must be involved. This variable turns out to be the *algebraic sign of the total energy* (including rest energy) of the particle.

We may thus express each wave function as a product of three factors: a space–time factor $\psi(x, y, z, t)$, a spin factor, which is a linear combination of the spin functions $|+\rangle_s$ and $|-\rangle_s$, and an energy factor, which is a linear combination of the energy functions $|+\rangle_E$ and $|-\rangle_E$. The energy functions are eigenfunctions for the two algebraic signs of the energy eigenvalue of the particle. We shall see in a moment that the possibility of a negative *total energy* (which implies a *negative rest mass!*) provides a theoretical basis for the existence of antimatter.

Let us now show how the Dirac equation (if it is valid) forces us to conclude that the electron has spin. First, we show that the Dirac energy operator H_{op}, unlike the Schrödinger energy operator, *does not commute with the operator L_z*. Thus there are no functions that are simultaneously eigenfunctions of L_z and the energy. Thus in the relativistic case, for constant (known) energy, the value of L_z cannot be uniquely determined. To prove this, we begin with the definitions:

$$L_z = -i\hbar \left\{ x \frac{\partial}{\partial y} - y \frac{\partial}{\partial x} \right\} \tag{10.54}$$

$$H_{op} = c(\alpha_x p_x + \alpha_y p_y + \alpha_z p_z + \beta mc) =$$

$$= -ci\hbar \left\{ \alpha_x \frac{\partial}{\partial x} + \alpha_y \frac{\partial}{\partial y} + \alpha_z \frac{\partial}{\partial z} + \frac{i\beta mc}{\hbar} \right\} \tag{10.55}$$

Because the α and β operators are independent of x, y, and z, and the L_z operator is independent of z, L_z must commute with the third and fourth terms of H_{op}. Thus the commutator of L_z and H_{op} may be written

$$L_z H_{op} - H_{op} L_z = -c\hbar^2 \left\{ x \frac{\partial}{\partial y} - y \frac{\partial}{\partial x} \right\} \left\{ \alpha_x \frac{\partial}{\partial x} + \alpha_y \frac{\partial}{\partial y} \right\} +$$

$$+ c\hbar^2 \left\{ \alpha_x \frac{\partial}{\partial x} + \alpha_y \frac{\partial}{\partial y} \right\} \left\{ x \frac{\partial}{\partial y} - y \frac{\partial}{\partial x} \right\} \tag{10.56}$$

which eventually reduces to

$$L_z H_{op} - H_{op} L_z = -c\hbar^2 \left\{ \alpha_y \frac{\partial}{\partial x} - \alpha_x \frac{\partial}{\partial y} \right\} \neq 0 \tag{10.57}$$

This is a highly significant result. Given that energy is conserved, this result shows that the value of L_z is uncertain, and consequently L_z *cannot be a conserved quantity*. How then can the law of conservation of angular momentum be rescued?

The only way to rescue it is to assume that any change in orbital angular momentum leaves the *total* angular momentum unchanged, because spin angular momentum is also present. In that case both L_z and S_z can change, but their sum, J_z, is constant. With a slight further development of the algebra of the α and β operators, it is not hard to show that J_z does indeed commute with H_{op}. (See Exercise 16.)

10.4.2 Antimatter

The introduction of a negative-energy function is a radical innovation. The possibility that the total energy, including rest energy, can be negative was never considered before the Dirac equation was developed. Obviously, one can take the negative square root in the energy equation $E = \sqrt{p^2c^2 + m^2c^4}$, but this had been considered to be an unphysical solution. It means that the positive energies range from $+mc^2$ to infinity and the negative energies range from minus infinity to mc^2, leaving a gap of impossible energies between $-mc^2$ to $+mc^2$. Classically, with energy changing in a continuous manner, there would be no way for a particle to cross this gap in either direction. If there were negative-energy particles, their existence would have no observable consequences under these circumstances.

The situation is different for quantum mechanics, where discrete changes in energy are not only permitted, they are often required. If a negative energy state exists, a particle with a positive energy could emit a photon and fall into that state. Conversely, a particle with negative energy could absorb a photon and acquire a positive energy. However, it is hard to imagine how a particle with negative energy could be observed. If a particle dropped into a negative energy state, would it simply disappear? Could it then reappear suddenly out of nowhere, by absorbing a sufficiently energetic photon? Has anything like this ever been observed?

It certainly does not occur regularly in everyday life. What could prevent it from happening? We could simply invent a rule that transitions between positive and negative energy states are not allowed. However, we have found that such transitions do indeed occur, but not often enough to wipe out our existence. The solution to the puzzle involves another fundamental principle of quantum mechanics, the *Pauli exclusion principle* (to be discussed in detail in Section 11.1), *applicable to particles of half-integer spin* (called *fermions*) which can be stated as follows:

> No two identical fermions can exist in the same quantum state.

As we shall see in Chapters 11 and 13, this principle is well documented in its effect on atomic structure. Only one electron can have a given set of values of the quantum numbers n, l, m_l, and m_s. Therefore, if electrons are added to a positive ion, each additional electron must go into a different state. Dirac applied this principle to suggest that particles cannot make transitions into negative energy levels because *all of the negative-energy states are normally occupied*. This suggestion is not so preposterous as it appears. Strange as it seems, we can actually test it and observe the consequences.

We cannot detect a negative-energy particle in ordinary life, because we cannot give one enough energy to change to an unoccupied state; an energy of twice its rest energy would be required. Thus you can walk right through this infinite sea of negative-energy particles with no hindrance, because you cannot lose energy to these particles. (The exclusion principle does not prevent identical particles from being in the same *place*; it simply prevents them from being in the same *quantum state*.)

Nevertheless, we can excite a negative-energy particle of mass m if we can increase its energy by $2mc^2$ or more, to bring it from an energy of $-mc^2$ (or less) to an energy of $+mc^2$ (or more). This is routinely observed in cosmic radiation. When two particles collide inelastically, losing an energy ΔE, a particle of mass m can be excited from a negative-energy state if $\Delta E > 2mc^2$. It is also possible for a negative-energy particle to be excited via the photoelectric effect, by a gamma-ray or X ray whose energy is $2mc^2$ or greater. (As in the normal photoelectric effect, a third body must be present in order that momentum and energy be conserved. A gamma ray can collide with a proton; the recoiling proton acquires almost all of the gamma ray's momentum, while a negative-energy electron acquires almost all of its energy.)

What do we *observe* when the electron comes out of the negative-energy sea? We see an electron, of course. We also see the *unoccupied negative-energy state* that the electron has vacated. Because charge must be conserved, the unoccupied state appears positively charged. (The absence of a negative is a positive.) Other electrons are attracted to it; the sea *becomes capable of interacting with the rest of the world*. When this state is filled by a negative-energy electron, and that electron moves, the state itself changes its position in the opposite direction, as you would expect a positive charge to do.

Thus we see a positive "particle" with the properties of an electron; this particle is called the *positron*. If we apply an electric field, that field exerts a force on the positron, in the same direction as the force that would be exerted on any other positively charged particle.

The existence of this particle, called an antielectron by Dirac, was suggested in 1931,[10] and it was seen in cosmic rays a year later.[11]

---

[10] It was not immediately obvious that the existence of positrons is implied by the Dirac equation. Dirac at first said that the empty states might be protons, but Oppenheimer pointed out that electrons and protons would annihilate one another too rapidly if that were true. Herman Weyl then proved that an empty negative-energy state would behave like a particle with the *same mass* as an electron, and Dirac then suggested that such particles could be produced by high-energy gamma rays. [*Proc. Roy. Soc.* **A133**, 60 (1931)].

[11] C. D. Anderson, *Science* **76**, 238 (1932).

When a positron encounters an electron, both are *annihilated* as the electron falls into the empty state, and their total energy ($2mc^2$ plus kinetic energy) is emitted in the form of gamma rays. Dirac pointed out that gamma rays of sufficiently high energy could also produce antiprotons, which were eventually observed in 1955. It is now clear that every particle of nonzero spin has an antiparticle.

Further development of the Dirac equation would take us beyond the level of this text, but it is hoped that you now see a connection between relativity and the phenomena of spin and antimatter.

ADDITIONAL READING

David Bohm, *Quantum Theory*, Prentice-Hall, New York (1951). Chapter 16 is a good exposition of matrices in quantum mechanics. Chapter 17 gives detailed information and good insights into spin and angular momentum operators.

Richard L. Liboff, *Introductory Quantum Mechanics*, Holden-Day, San Francisco (1990). Pages 442–455 give a thorough summary of angular momentum matrices.

EXERCISES

1. Operator $S_{z'}$ is the component of operator $\mathbf{S}$ along the z' axis. We define the z' axis as lying in the xz plane, making an angle of 45° with the z axis.
   (a) Write $S_{z'}$ as a linear combination of the operators S_x and S_z.
   (b) Write the eigenfunctions of $S_{z'}$ as a linear combination of the eigenfunctions of S_z.
   (c) For an electron whose spin component along the z' axis is positive, find the probability that a measurement of the spin component along the z axis will yield a positive result (that is, that the result will be $S_z = +\hbar/2$).

2. $S_w = 0.8S_x + 0.6S_z$ is the operator for the spin component along the "w" axis
   (a) Draw a figure showing the directions of the x, z, and w axes.
   (b) Apply S_w to the general spin state $a|+\rangle_s + b|-\rangle_s$ to find the two sets of values of a and b that make this state a normalized eigenstate of S_w.
   (c) For a particle in the state that has a positive eigenvalue for S_w, what is the probability that a measurement of S_z will yield a positive result?
   (d) After a measurement of S_z gives a negative value, S_z is immediately measured again for the same particle. What is the probability that the value will be positive?

3. As we saw in Chapter 7, states with $l = 2$ can be represented by 5-component column vectors, and the operators for L_x and L_y can be written as 5×5 matrices. For example,

$$|2,2\rangle = \begin{pmatrix} 1 \\ 0 \\ 0 \\ 0 \\ 0 \end{pmatrix}, \quad |2,1\rangle = \begin{pmatrix} 0 \\ 1 \\ 0 \\ 0 \\ 0 \end{pmatrix}, \dots, \quad |2,-2\rangle = \begin{pmatrix} 0 \\ 0 \\ 0 \\ 0 \\ 1 \end{pmatrix},$$

$$\text{and} \quad L_x = \frac{\hbar}{2} \begin{pmatrix} 0 & 2 & 0 & 0 & 0 \\ 2 & 0 & \sqrt{6} & 0 & 0 \\ 0 & \sqrt{6} & 0 & \sqrt{6} & 0 \\ 0 & 0 & \sqrt{6} & 0 & 2 \\ 0 & 0 & 0 & 2 & 0 \end{pmatrix}$$

Refer to the numbers in Example Problem 7.1. Write the column vector that is an eigenvector of the L_x operator, and show that the eigenvalue is $+2\hbar$.

4. (a) Use the commutation relation $L_x L_y - L_y L_x = i\hbar L_z$ to find the elements of the L_y matrix. *Hint:* Each element of L_y has the same magnitude as the corresponding element of L_x; remember that L_z is diagonal.
   (b) Apply the stepping operators $L_x \pm iL_y$ to each of the five eigenvectors of L_z for these $l = 2$ states, to verify that the appropriate eigenvectors are produced. (See also Problem 7.14.)

5. (a) Write the matrices for L_z^2 and L^2 for the case $l = 2$.
   (b) By squaring the matrix of Exercise 3, find the matrix for L_x^2 for $l = 2$.
   (c) Find the matrix for L_y^2 for $l = 2$ by combining your results from (a) and (b). Is your result consistent with the result of Exercise 4(a)?

6. Find the values of a and b in the eigenfunctions

$$[a|1,0\rangle| + \rangle_s + b|1,1\rangle| - \rangle_s]$$

of Example Problem 10.2.

7. Following the method of Example Problem 10.2, find the values of a and b in the functions $[a|1,-1\rangle| + \rangle_s + b|1,0\rangle| - \rangle_s]$ that yield eigenfunctions of J^2, and find the eigenvalues.

8. F_1 and F_2 are eigenstates of J^2; F_1 has the larger J^2 eigenvalue.

$$F_1 = \sqrt{7}|4, 2\rangle|+\rangle + \sqrt{2}|4, 3\rangle|-\rangle, \qquad F_2 = -\sqrt{2}|4, 2\rangle|+\rangle + \sqrt{7}|4, 3\rangle|-\rangle$$

(a) What are the eigenvalues of J^2 for states F_1 and F_2, respectively?

(b) F_1 is also an eigenstate of which of the following: J_z, L^2, L_z, S_z, S^2, $\mathbf{L} \cdot \mathbf{S}$?

(c) Suppose that we measure S_z for an electron in the state F_1, and we find that S_z is positive. What result will be find when we measure L_z for the same particle immediately after that? Why?

9. F_1 and F_2 are normalized eigenstates of J^2; F_1 has the larger J^2 eigenvalue. N_1 and N_2 are normalizing constants.

$$F_1 = N_1\{\sqrt{6}|3, 2\rangle|+\rangle + |3, m\rangle|-\rangle\},$$

$$F_2 = N_2\{-|3, 2\rangle|+\rangle + \sqrt{6}|3, m\rangle|-\rangle\}$$

(a) Find the values of m, N_1, and N_2.

(b) For state F_1, find the probability of finding each possible value of L_z as the result of a measurement.

10. Write the matrices for J_z and J^2 for $j = 5/2$.

11. For $l = 4$, write the (a) matrices for L_z and L_z^2 and (b) column eigenvectors of L_z.

12. Verify that Eq. (10.47) correctly gives the value of the integral when u is the function u_{210} given in Table 9.1.

13. Use Eqs. (10.48) to compute the splitting, in eV, of the $2p$ level of the hydrogen atom. Repeat for the $3d$ level.

14. Verify that Eqs. (10.52) and (10.53) follow from Eq. (10.51), by squaring both sides of Eq. (10.51), as suggested in the text.

15. Calculate classically the maximum interaction energy $\boldsymbol{\mu}_B \cdot \mathbf{B}_N$ between a Bohr magneton and a nuclear magneton when they are separated by a distance of one Bohr radius. Compare this energy with the energy of the hyperfine splitting of the ground state of hydrogen (5.9×10^{-6} eV). The energy may be written as

$$\boldsymbol{\mu}_B \cdot \mathbf{B}_N = -\frac{\mu_N \mu_B}{4\pi\varepsilon_0 c^2} \cdot \nabla \frac{\cos\theta}{r^2}$$

where θ is the direction between the $\boldsymbol{\mu}_N$ vector and the radius vector between the two dipoles. The vector operator ∇ is, in spherical coordinates,

$$\nabla = \hat{\mathbf{r}} \frac{\partial}{\partial r} + \frac{\hat{\theta}}{r} \frac{\partial}{\partial \theta}$$

16. Each of the four Dirac operators may be written as a product of two operators, one of which is one of the Pauli spin operators σ_x, σ_y, or σ_z, and the other an analogous operator ρ_x, ρ_y, or ρ_z, which operates on the functions $|+\rangle_E$ and $|-\rangle_E$ in the same way that the spin operators operate on the spin functions $|+\rangle_s$ and $|-\rangle_s$. Thus ρ_x, ρ_y, and ρ_z must satisfy the same commutation relations as σ_x, σ_y, and σ_z; for example, $\rho_x \rho_y + \rho_y \rho_x = 0$.

    (a) Show that Eqs. (10.53) are satisfied if

$$\alpha_x \rho_x = \sigma_x, \qquad \alpha_y \rho_x = \sigma_y, \qquad \alpha_z \rho_x = \sigma_z, \qquad \beta = \rho_z$$

    (b) Using the above results plus the commutation rules $[S_x, S_y] = i\hbar S_z$, $[S_y, S_z] = i\hbar S_x$, and $[S_z, S_x] = i\hbar S_y$, show that

$$S_z H_{op} - H_{op} S_z = c\hbar^2 \left(\alpha_y \frac{\partial}{\partial x} - \alpha_x \frac{\partial}{\partial y} \right)$$

    (c) Combine the result of (b) with Eq. (10.57) to conclude that $[J_z, H_{op}] = 0$, and therefore total angular momentum can be conserved.

17. An electron is in a state whose *unnormalized* spin factor is $-|+\rangle + 3|-\rangle$.

    (a) The normalized state is $a|+\rangle + b|-\rangle$. Find the values of a and b.

    (b) If the x component of this electron's spin is measured, what is the probability that the result will be positive?

CHAPTER 11

Identical
Particles

We have seen in Section 10.4 that the states available to identical particles of *half-integer spin* (fermions) are restricted by the Pauli exclusion principle:

> No two identical fermions can exist in the same quantum state.

We now explore a multitude of consequences of this principle.

11.1 IDENTICAL PARTICLES AND SYMMETRY OF WAVE FUNCTIONS

11.1.1 Indistinguishability

To analyze systems containing a number of identical electrons (e.g., any atom except hydrogen) we must express our equations in a way that *makes no distinction between one electron and another*. It is sometimes difficult to grasp this fact, because in writing equations we have become accustomed to identifying each particle in a collection by a separate label. If the particles are truly indistinguishable, we must be very careful in using labels.

For example, suppose that two electrons come together, so that their wave functions overlap. When they fly apart again, there can be no way, after they have interacted, to determine which one came in from the left and which one came from the right. *If there were a way, this would mean that electrons are not identical.*

221

11.1.2 Symmetry of Wave Functions

How can we write a wave function for two identical particles, if we cannot use labels to describe the coordinates of each electron? The answer is clear when we develop the general form of the wave function of a system of any number of particles. (This is an extension of the discussion in Section 9.1.)

The spatial part of the wave function for a system of N particles is a function of $3N$ coordinates, $u_T(x_1, y_1, x_1, \ldots, x_N, y_N, z_N)$, which we abbreviate as $u_T(1, 2, \ldots N)$. What happens if we interchange the coordinates of particles 1 and 2? After this interchange, the function u_T depends on the coordinates of particle 1 in the same way that it formerly depended on the coordinates of particle 2, and vice versa. Thus we have formally interchanged the particles. But according to the definition of indistinguishability, if the particles are identical, the state resulting from this interchange cannot be distinguished from the original state. This means that

$$u_T(2, 1, \ldots N) = Au_T(1, 2, \ldots N) \tag{11.1}$$

where A is a constant. Interchanging particles 1 and 2 again must have the same effect on the wave function, yielding

$$u_T(1, 2, \ldots N) = Au_T(2, 1, \ldots N) \tag{11.2}$$

Combining Eqs. (11.1) and (11.2) then yields

$$u_T(2, 1, \ldots N) = A^2 u_T(2, 1, \ldots N) \tag{11.3}$$

and thus $A^2 = 1$. If $A = +1$, we say that the wave function is *symmetric* with respect to interchange of the two particles; if $A = -1$, the function is *antisymmetric*. Figure 11.1 illustrates these two situations graphically for the one-dimensional case, where the function u is a function of two variables only: the coordinate x_1 for particle 1, and the coordinate x_2 for particle 2. Notice that the line $x_1 = x_2$ is a line of symmetry. Interchanging x_1 and x_2 is equivalent to reflecting the figure along this line.

Figure 11.1 shows graphically the indistinguishability of the two particles. Although we put different labels on the two axes, any physical result must be independent of the label. For example, we can find the probability that at least one of the two particles has x coordinate between $+a$ and $+b$, by using the fact that u^*u is the probability density for *both* particles. That means that u^*u is the probability density for a particle at each of two coordinates. To find the probability that one or both of the particles is

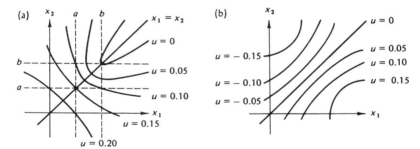

FIGURE 11.1 Contour map, showing values of possible (a) symmetric and (b) antisymmetric wave functions u as functions of the x coordinates x_1 and x_2 of two identical particles. Contours connect points at which u has a constant value. On the line $x_1 = x_2$, u must be zero in the antisymmetric case.

located between $x = a$ and $x = b$, we integrate u^*u over the entire region for which *either* $a < x_1 < b$ or $a < x_2 < b$. This region is enclosed by dashed lines in Figure 11.1.

Figure 11.1(b) illustrates an important general feature of antisymmetric functions, namely that $u = 0$ when $x_1 = x_2$. To prove that this must be so, let $x_1 = x_2 = c$. Then, from Eq. (11.1), we may write

$$u(x_2, x_1) = -u(x_1, x_2) \tag{11.4}$$

or

$$u(c, c) = -u(c, c) \tag{11.5}$$

which can be true only if $u(c, c) = 0$.

11.1.3 Separation of Variables

To determine the wave function for a system of two or more particles is a formidable task. Therefore we start with the approximation that there is *no interaction between the particles*. That is, we assume that each particle moves in a known external potential that is independent of the position(s) of the other particle(s). Thus for two particles we write the Schrödinger equation as

$$\left[-(\hbar^2/2m)(\nabla_1^2 + \nabla_2^2) + V(1) + V(2)\right]u(1, 2) = E_T u(1, 2) \tag{11.6}$$

where ∇_1 operates on the coordinates of particle 1, and $V(1)$ is the potential energy of particle 1 and is a function of the coordinates of particle 1 only.[1]

We next assume that the particles are distinguishable, and we separate the variables by writing $u(1, 2) = u_a(1)u_b(2)$, where u_a and u_b may be different functions. Substitution into Eq. (11.6) and regrouping terms yields

$$\left[-(\hbar 2/2m)\nabla_1^2 + V(1)\right]u_a(1)u_b(2)$$
$$+\left[-(\hbar^2/2m)\nabla_2^2 + V(2)\right]u_a(1)u_b(2)$$
$$= E_T u_a(1)u_b(2) \tag{11.7}$$

As before, we now separate the variables by dividing all terms by the wave function $u_a(1)u_b(2)$, to obtain

$$\frac{1}{u_a(1)}\left[-(\hbar^2/2m)\nabla_1^2 + V(1)\right]u_a(1)$$

$$+ \frac{1}{u_b(2)}\left[-(\hbar^2/2m)\nabla_2^2 + V(2)\right]u_b(2) = E_T \tag{11.8}$$

We now conclude, as in Section 9.1, that each term must equal a constant. Labeling these constants E_a and E_b, we have

$$\frac{1}{u_a(1)}\left[-(\hbar^2/2m)\nabla_1^2 + V(1)\right]u_a(1) = E_a$$

$$\frac{1}{u_b(2)}\left[-(\hbar^2/2m)\nabla_2^2 + V(2)\right]u_b(2) = E_b$$

or

$$\left[-(\hbar^2/2m)\nabla_1^2 + V(1)\right]u_a(1) = E_a u_a(1) \tag{11.9a}$$
$$\left[-(\hbar^2/2m)\nabla_2^2 + V(2)\right]u_b(2) = E_b u_b(2) \tag{11.9b}$$

where $E_a + E_b = E_T$.

Except for the labels, the two equations (11.9) are really the same equation, the single-particle Schrödinger equation. Therefore, if both particles are subject to the same potential, u_a and u_b belong to the same set of eigenfunctions. For example, as a first approximation we can assume that each of the two electrons in a helium atom separately occupies one of the states of a helium *ion*, whose wave functions are given in Section 9.2

---

[1]This treatment is analogous to that of Eq. (9.3); it differs in having external potentials instead of a potential dependent on the distance between the particles.

(Table 9.1 with $Z = 2$). The energy of each electron is therefore -54.4 eV [from Eq. (9.12)], and the total energy is -108.8 according to Eq. (11.8). This is far from the measured energy for the helium atom, because we have neglected the potential energy of repulsion between the two electrons. However, Chapter 12 shows how to approximate this energy by a method that gives great agreement with experiment.

11.2 SYMMETRY OF STATES FOR TWO IDENTICAL PARTICLES

The function $u_a(1)u_b(2)$ of Eq. (11.7) is in general neither antisymmetric nor symmetric; thus it is unacceptable as a wave function for a state of two identical particles. But we can use this function to construct acceptable wave functions.

The symmetric function is the sum

$$u_S(1,2) = [u_a(1)u_b(2) + u_a(2)u_b(1)]/\sqrt{2} \qquad (11.10)$$

The antisymmetric function is

$$u_A(1,2) = [u_a(1)u_b(2) - u_a(2)u_b(1)]/\sqrt{2} \qquad (11.11)$$

The divisor $\sqrt{2}$ is needed to preserve the normalization of the wave function, on the assumption that u_a and u_b are individually normalized. For each of these wave functions there is one particle in the single-particle state whose wave function is u_a, and one particle in the single-particle state whose wave function is u_b, but we have no way to say which particle is in which state. You may verify that

$$u_S(1,2) = u_S(2,1) \qquad (11.12)$$

$$u_A(1,2) = -u_A(2,1) \qquad (11.13)$$

in agreement with Eq. (11.3).

11.2.1 Spin States and Symmetry

Study of many phenomena has shown that *all particles with half-integer spin* (electrons, protons, neutrons, muons, neutrinos, and many others) *must have antisymmetric wave functions.* Particles with integer spin (photons, pions, and others) must have symmetric wave functions. These facts are directly connected with the Pauli exclusion principle, because the antisymmetric wave function (Eq. 11.10) vanishes when both particles are in the same state:

$$u_a(1)u_a(2) - u_a(2)u_a(1) \equiv 0 \qquad (11.14)$$

Therefore no two electrons (or other spin-1/2 particles) can simultaneously occupy the same quantum state. The antisymmetry requirement applies to all coordinates, including spin. It is convenient to separate the wave function into two factors, a spin function and a space function. When the space factor is symmetric, the spin factor must be antisymmetric, and vice versa. For example, there are four independent completely antisymmetric state functions for two electrons in states a and b. In Dirac notation, the normalized states are

$$[|a\rangle_1|b\rangle_2 + |a\rangle_2|b\rangle_1][|+\rangle_1|-\rangle_2 - |+\rangle_2|-\rangle_1]/2 \qquad (11.15a)$$

$$[|a\rangle_1|b\rangle_2 - |a\rangle_2|b\rangle_1]|+\rangle_1|+\rangle_2/\sqrt{2} \qquad (11.15b)$$

$$[|a\rangle_1|b\rangle_2 - |a\rangle_2|b\rangle_1][|+\rangle_1|-\rangle_2 + |+\rangle_2|-\rangle_1]/2 \qquad (11.15c)$$

$$[|a\rangle_1|b\rangle_2 - |a\rangle_2|b\rangle_1]|-\rangle_1|-\rangle_2/\sqrt{2} \qquad (11.15d)$$

The first of these, called the *singlet state*, is symmetric in the space functions $|a\rangle$ and $|b\rangle$ but antisymmetric in the spin functions $|+\rangle$ and $|-\rangle$. Thus it is antisymmetric with respect to the exchange of *all* coordinates of the two particles. The other three, called *triplet states*, are antisymmetric in the space functions but symmetric in the spin functions.

For the triplet states, the z component of the spin angular momentum is $+\hbar$, 0, and $-\hbar$, respectively. From the general rules for angular momentum [Eqs. (10.18)–(10.20)], we thus deduce that this set of states has a total spin quantum number of $S = 1$. The square of the total spin angular momentum is, according to those rules, equal to $S(S + 1)\hbar^2$, or $2\hbar^2$. You may verify these statements by applying the spin operators (see Exercise 1).

This division of states into a triplet with $S = 1$ and a singlet with $S = 0$ is characteristic of states of any two particles of spin 1/2, whether or not the particles are identical. This fact has a statistical consequence[2] that is well verified experimentally: If two particles of spin 1/2 come together at random, their spins are "parallel" ($S = 1$) three quarters of the time,

[2] This is logical on the assumption that each quantum state is equally likely (as it should be if there is no energy difference between the states). This is the simplest possible assumption, and it yields consequences that agree with observation.

and they are "antiparallel" one quarter of the time. This means that each of the four states of Eq. (11.15) is equally likely to occur. For example, in a random collection of hydrogen molecules, three quarters are *ortho-hydrogen* whose protons have total spin number $S_p = 1$, and the other one quarter are *para-hydrogen* with zero total proton spin.

11.2.2 Exchange Energy

The requirement that the total wave function, involving all coordinate variables, be antisymmetric leads to consequences that resemble the effects of a new force that is unknown in classical physics. This "force" is not a force in the classical sense. However, the effect of this requirement is that the electrons' motions are correlated in a way that suggests the presence of another force in addition to the Coulomb force. (Although we cannot follow the trajectories of the electrons, we deduce from the observed energy levels that this correlation is present.)

The effect may be made plausible from the following considerations: When the space part of the wave function is antisymmetric, the combined wave function must be zero when the space coordinates of the two are equal (as shown for the x coordinates in Figure 11.1). This is not true for the symmetric function. Thus we expect that the electrons tend to be closer together, on the average, when the space function is symmetric, and it can be shown that, in general, $\langle \mathbf{r}_2 - \mathbf{r}_1 \rangle^2$ is smaller when the space function is symmetric. Consequently, *for a given pair of two single-particle wave functions* u_a *and* u_b, the Coulomb energy of two electrons in the symmetric space state is higher than that of two electrons in the antisymmetric space state. This energy difference between the two types of state is called the *exchange energy*.

The symmetric space function can occur only with the antisymmetric spin function (when $S = 0$), and the antisymmetric space function can occur only with the symmetric spin functions (when $S = 1$). Thus the exchange energy appears to be spin dependent; the $S = 1$ states of two electrons have lower energy than the $S = 0$ state. (This effect is not magnetic; the exchange energy is much larger than the energy difference produced by the magnetic dipole interaction between the two electrons.)

11.3 LOCAL REALISM AND THE EINSTEIN–PODOLSKY–ROSEN PARADOX

Consider two identical spin-1/2 particles (X1 and X2) in the singlet ($S = 0$) state. Along any given axis A, one of them must have a spin component of $+\hbar/2$, and the other must therefore have a spin component of $-\hbar/2$ along the same axis. The characteristic of the $S = 0$ state is that the spins must be opposite in direction, although each one's individual spin direction is unknown until a measurement has been made. In other words, after the spin of X1 has been measured along any axis, the spin of X2 *along that axis* is known before we measure it directly.

These facts raise a question similar to those discussed in Section 2.6. Does the spin direction of a particle have any reality, independent of its measurement? Does a direction for this spin exist before it is measured? And if it does not, how does a measurement on one particle *create* a reality for the spin direction of the other particle, even though the particles are widely separated (for example, if they are in an unbound state and moving away from each other)? Quantum mechanics does not address such questions; it simply predicts, with astonishing accuracy, the probabilities of the various results that you will get if you make a measurement.

Questions of this sort are related to what is called *local realism*, or the belief that a measurement of one particle can have no effect on another particle at a distance. If local realism is correct and the particles are far apart, there should be no way for a measurement of one particle to influence the other. Yet when we do the measurements, we find that the results are correlated even if a different axis is measured for each one. To ensure that proton X2 will always give a result that agrees with the measurement of proton X1, when no information about the measurement of proton X1 can reach proton X2, local realism seems to require that each proton carries with it a set of "instructions" on how to react to a spin measurement along any given axis.

11.3.1 The Bell Inequality

It is tempting to believe, as Einstein did, that quantum mechanics is an incomplete theory and that values of unmeasured variables (sets of instructions) do exist in cases where quantum mechanics gives only probabilities of various results. Surprisingly, it is possible to test this belief by means of various experiments, using *the Bell inequality*, initially proposed by John S. Bell in 1964. The inequality is strictly a mathematical consequence of the assumption that each particle with spin 1/2, such as a proton, has its own definite spin component in any given direction, even though we do not know the value of that component until we measure it. In other words, each proton carries instructions (sometimes called hidden variables) that determine how the proton will respond to a spin measurement along any given axis.

One such measurement is done by preparing a large batch of proton pairs in a singlet state. The spin of each proton is then measured along one of three axes A, B, or C. For each pair that is measured, we can write the result of the two measurements by a symbol that shows the following

- The measurement axis for particle X1 and the sign of that particle's spin along that axis
- The measurement axis for particle X2 and the sign of the spin that particle X1 *would have had* if it had been measured along that same axis

This pair of symbols tells us what the instructions were for particle X1, if such a set exists. For example, suppose that we find that X1 had positive spin along axis A, and X2 had positive spin along axis B. Then, since the total spin must always be zero if both spin components are measured along the same axis, we know that the spin component of X1 would have been *negative* if it had been measured along axis B instead of axis A. Thus the instructions for particle X1 must have been A^+B^-.

Considering the axes A, B, and C, eight different instructions are possible for particle X1: $A^+B^+C^+$, $A^+B^+C^-$, $A^+B^-C^+$, $A^+B^-C^-$, $A^-B^+C^+$, $A^-B^+C^-$, $A^-B^-C^+$, and $A^-B^-C^-$. Let us now define $N(A^+B^-C^+)$ as the number of protons that have the instructions $A^+B^-C^+$, $N(A^+B^-C^-)$ as the number with the instructions $A^+B^-C^-$, and $N(A^+B^-)$ as the total number with the instructions A^+B^-, regardless of the instruction for the C axis.

Clearly, since each proton must have either C^+ or C^-, we must have

$$N(A^+B^-) = N(A^+B^-C^-) + N(A^+B^-C^+) \qquad (11.16)$$

if local realism is correct. We also must have

$$N(A^+B^-C^-) \le N(A^+C^-) \quad \text{and} \quad N(A^+B^-C^+) \le N(B^-C^+) \quad (11.17)$$

because $N(A^+B^-C^-)$ is a subset of $N(A^+C^-)$, and $N(A^+B^-C^+)$ is a subset of $N(B^-C^+)$. Therefore Eq. (11.16) can be replaced by the inequality

$$N(A^+B^-) \le N(A^+C^-) + N(B^-C^+) \qquad (11.18)$$

To test this inequality, and thereby confirm or deny the basic premise that an instruction set exists, we can measure the sign of each spin in a large collection of proton singlet pairs. We measure along axes A and B for some pairs, along A and C for an equal number of pairs, and along B and

C for the same number. If (11.18) is satisfied, then we must find that

$$n(A^+B^+) \le n(A^+C^+) + n(B^+C^+) \tag{11.19}$$

where $n(A^+B^+)$ denotes the number of pairs for which both protons have positive spin in the indicated direction, etc. Because a positive spin along the B axis for particle X2 indicates a negative spin for X1 along that same axis, we see that each term in (11.19) is the same multiple of the corresponding term in (11.18) and thus (11.18) implies the experimental result given in (11.19).

Experiments of this type have been done, and (11.19) has been violated. Instead, the results confirm the predictions of quantum mechanics. Therefore, at least in this situation, there are no hidden variables, no instruction sets, and no local realism. Measuring the spin component of particle X1 along axis A tells us the spin component of particle X2 along axis A, just as if we had measured particle X2 itself. Before the measurement of X1 along axis A, there was a 50% probability that X2 would have a positive spin along axis B. After the measurement of X1, the probability that X2 has a positive spin along axis B can be computed by quantum theory, and the result is dependent on the angle between axis A and axis B, as shown in the following example.

Example Problem 11.1 Show that quantum mechanics predicts a violation of the Bell inequality if A is the $+z$ axis, B is the $+x$ axis, and the positive C direction lies at a 45° angle to axis A and to axis B.

Solution. From Section 10.1 we know that the relevant operators are

$$S_z = \frac{\hbar}{2} \begin{pmatrix} 1 & 0 \\ 0 & -1 \end{pmatrix} \quad \text{and} \quad S_x = \frac{\hbar}{2} \begin{pmatrix} 0 & 1 \\ 1 & 0 \end{pmatrix}$$

From these we can deduce that the operator for spin along the intermediate C axis must be

$$S_c = S_x \cos 45° + S_z \cos 45° = \frac{\hbar}{2\sqrt{2}} \begin{pmatrix} 1 & 1 \\ 1 & -1 \end{pmatrix}$$

We can find the eigenstate of S_c by applying S_c to the general vector $\begin{pmatrix} a \\ b \end{pmatrix}$. For the positive spin direction we have

$$\frac{\hbar}{2\sqrt{2}} \begin{pmatrix} 1 & 1 \\ 1 & -1 \end{pmatrix} \begin{pmatrix} a \\ b \end{pmatrix} = \frac{\hbar}{2\sqrt{2}} \begin{pmatrix} a+b \\ a-b \end{pmatrix}$$

which also must equal $\dfrac{\hbar}{2}\begin{pmatrix} a \\ b \end{pmatrix}$ if we have an eigenstate. Therefore $a + b = a\sqrt{2}$. Normalization requires that $a^2 + b^2 = 1$.

Solving these yields $a^2 = 1/(4 - 2\sqrt{2}) = 0.854$ for the probability that the spin along the C axis will be positive, given that the spin along the A axis was positive. Thus 0.146 is the probability of A^+C^- as defined in (11.18). The probability of B^-C^+ must be the same, making the right-hand side of (11.18) proportional to the total probability of 0.292 while the left-hand side is based on a total probability of 0.5, because A is perpendicular to B. Inserting these probabilities into (11.18) gives the numbers

$$0.500 N_T \leq 0.146 N_T + 0.146 N_T \qquad (11.20)$$

where N_T is the total number of pairs whose spins were measured along each set of axes. Thus the inequality is inconsistent with the quantum mechanical calculation. It is also inconsistent with the experimental results, as described next.

11.3.2 Experimental Tests of the Bell Inequality

Many tests of the Bell inequality have been reported, most notably by Alan Aspect and his collaborators.[3] Their results with spin correlations in gamma ray emission showed a clear violation of the Bell inequality and agreement with the predictions of quantum mechanics within the experimental uncertainty.

They also dealt with a possible explanation that would preserve local realism. It had been proposed that the source and the gamma-ray detectors could be in "communication" with each other via some sort of wave, so that each photon's spin direction ("instruction set") could be affected at its creation by the direction that the detectors were set up to measure. Thus the spins could be in some manner be correlated without "spooky action at a distance."

Aspect *et al.* eliminated this possibility by *changing the measured axis* after the photons had been emitted but while the photons were still in flight. This was done by using two polarization detectors for each photon and switching rapidly between them. Each photon was in flight for 50 ns, and the detectors were switched every 10 ns, so several switches occurred during the flight of each photon. Again, the results violated the Bell inequality and agreed with quantum theory.

[3]A. Aspect, P. Grangier, G. Roger, *Phys. Rev. Lett.* **47**, 460 (1981); **49**, 91 (1982). A. Aspect, J. Dalibard, G. Roger, *Phys. Rev. Lett.* **49**, 1804 (1982).

If you are bothered, as Einstein was, by this demise of "local realism," you are in good company. In the words of Richard Feynman,[4]

> *I cannot define the real problem, therefore I suspect there's no real problem, but I'm not sure there's no real problem.*

SUMMARY

Indistinguishable Particles: No physical result can depend on the labels "1" and "2" for two identical particles. Thus it is necessary to have a label-independent way to write a wave function for two identical particles.

Symmetric and Antisymmetric Wave Functions: If one particle is in the state $|a\rangle$ and another particle, identical to the first, is in the state $|b\rangle$, it is no good to write the two-particle state as $|a\rangle_1|b\rangle_2$, because that implies a difference between particle 1 and particle 2. However, we can write the state (unnormalized) as

$$|a\rangle_1|b\rangle_2 + |a\rangle_2|b\rangle_1 \quad \text{(the symmetric state)}$$

because in this form there is no suggestion that a particular one is in a particular state. Similarly, we could write the state as

$$|a\rangle_1|b\rangle_2 - |a\rangle_2|b\rangle_1 \quad \text{(the antisymmetric state)}$$

because the minus sign does not change the fact that either particle could be in either state with equal probability.

Pauli Exclusion Principle: For particles with half-integer spin, only the antisymmetric wave function is permitted. The antisymmetric wave function is zero when state $|b\rangle$ is identical to state $|a\rangle$. Therefore,

for spin-1/2 particles, only one particle can be in any given state.

Normalization: If we observe only one of the two particles we can call that particle number 1. Then there is, logically, a 50% probability that it is in state $|a\rangle$. By our previous rules this probability must be the square of the coefficient of the term $|a\rangle_1$ in the state function. Thus this coefficient (as well as the coefficient of the term containing $|a\rangle_2$) must be

---

[4]R. P. Feynman, *Int. J. Theor. Phys.* **21**, 471 (1982).

$1/\sqrt{2}$, and we write

$$[|a\rangle_1|b\rangle_2 + |a\rangle_2|b\rangle_1]/\sqrt{2} \quad \text{and} \quad [|a\rangle_1|b\rangle_2 - |a\rangle_2|b\rangle_1]/\sqrt{2}$$

for the symmetric and antisymmetric states, respectively.

Exchange Energy: The complete state of a particle may be separated into two factors, a spin part and a space part. When the space part is symmetric, then the spin part must be antisymmetric, and vice versa, so that the entire state is antisymmetric. There are four independent completely antisymmetric wave functions for two electrons in states a and b, as shown in Eqs. (11.15). When the space part is symmetric, the particles are closer together, on the average, and this increases the potential energy of two electrons because of Coulomb repulsion. This has great consequences for atomic spectra, as we shall see in Section 13.2.

ADDITIONAL READING

M. Kafatos (ed.), *Bell's Theorem, Quantum Theory, and Conceptions of the Universe*, Kluwer Academic, Dordrecht (1989).

N. D. Mermin, Bringing Home the Quantum World: Quantum Mysteries for Anybody, *Am. J. Phys.* **49**, 940–943 (1981); Is the Moon There When Nobody Looks? Reality and the Quantum Theory, *Phys. Today* **38**(4), 38–47 (1985).

EXERCISES

1. By direct application of the six spin operators (three operators for each electron) to expressions (11.15), verify that these functions have the respective quantum numbers $S = 0$ and $S = 1$ and the appropriate values for S_z.

2. The four spin states of two electrons [expressions (11.15)] may be combined with the two spin states of a third electron in eight independent ways. Four combinations yield a total spin number $S = 3/2$, and are symmetric in the interchange of *any two* of the three electrons. Write expressions for each of these four states, identifying the value of S_z for each.

3. Following the lines of Example Problem 11.1, show that quantum mechanics predicts a violation of the Bell inequality if A is the $+z$ axis, the positive B axis makes a $60°$ angle with the x axis, and the positive C direction makes $120°$ angle with the x axis.

4. A possible spin state for a system of three electrons is

$$|+ + +\rangle - |- - -\rangle.$$

(a) Show that this state is an eigenstate of the operator $\sigma_{1x}\sigma_{2x}\sigma_{3x}$, and find the eigenvalue. From this result, deduce the *product* of the results of measuring the x component of the spin of each electron.

(b) Show that this state is also an eigenstate of the operator $\sigma_{1x}\sigma_{2y}\sigma_{3y}$, and find the eigenvalue. Also show that the three operators $\sigma_{1x}\sigma_{2y}\sigma_{3y}$, $\sigma_{1y}\sigma_{2x}\sigma_{3y}$, and $\sigma_{1y}\sigma_{2y}\sigma_{3x}$ commute with one another and that they all have the same eigenvalue for the above spin state.

(c) From the result of part (b), determine the eigenvalue of the product $\sigma_{1x}\sigma_{2y}\sigma_{3y}\sigma_{1y}\sigma_{2x}\sigma_{3y}\sigma_{1y}\sigma_{2y}\sigma_{3x}$.

(d) The answer to (c) is not equal to the answer to (a), because of the operations on the y components of σ. Why do those operations make a difference? *Hint:* Use the anticommutation relations for the σ operators (for example, $\sigma_{1x}\sigma_{1y} + \sigma_{1y}\sigma_{1x} = 0$) to reduce the operator of part (c) to a constant multiplied by $\sigma_{1x}\sigma_{2x}\sigma_{3x}\sigma_{1y}^2\sigma_{2y}^2\sigma_{3y}^2$.

CHAPTER 12

Approximate Solutions

In the overwhelming majority of situations of practical interest, it is impossible to find an exact solution of the Schrödinger equation as we have done up to now. Even the hydrogen atom requires an approximation if the effect of spin is to be considered. Fortunately, we have approximation methods that permit us to calculate the energy levels of a wide variety of systems. These calculations have been confirmed by experiment to an astonishingly high degree of accuracy in three-body systems such as the helium atom.

12.1 TIME-INDEPENDENT PERTURBATION THEORY

We are often interested in systems for which we could solve the Schrödinger equation if the potential energy were slightly[1] different. Consider a one-dimensional example for which we can write the actual potential energy as

$$V_{\text{actual}}(x) = V(x) + v(x) \tag{12.1}$$

where $v(x)$ is a small[1] *perturbation* added to the *unperturbed* potential $V(x)$.

---

[1] The meaning of "slightly" and "small" will be clarified as we proceed.

The Hamiltonian operator of the "unperturbed" (i.e., exactly solvable) system is

$$H_0 = -\frac{\hbar}{2m}\frac{\partial^2}{\partial x^2} + V(x) \qquad (12.2)$$

and the Schrödinger equation of that system therefore is

$$H_0 \psi_l = E_l \psi_l \qquad (12.3)$$

with a known set of eigenvalues E_l and eigenfunctions ψ_l.

The Schrödinger equation of the "perturbed" system contains the additional term $v(x)$ in the Hamiltonian:

$$[H_0 + v(x)]\psi_n' = E_n' \psi_n' \qquad (12.4)$$

which may make it impossible to solve directly for the eigenfunctions ψ_n' and eigenvalues E_n'. However, Postulate 3 tells us that any acceptable wave function may be expanded in a series of eigenfunctions of the unperturbed Hamiltonian. Therefore we may write each function ψ_n' as a series of the functions ψ_l with constant coefficients:

$$\psi_n' = \sum_{l=1}^{\infty} a_{nl} \psi_l \qquad (12.5)$$

(The letter l is simply an index, having nothing to do with the angular momentum quantum number; this is a one-dimensional analysis.) Substitution into Eq. (12.4) gives

$$[H_0 + v(x)] \sum_{l=1}^{\infty} a_{nl} \psi_l = E_n' \sum_{l=1}^{\infty} a_{nl} \psi_l \qquad (12.6)$$

The technique used in finding coefficients in a Fourier series may be employed here. We multiply each side of Eq. (12.6) by ψ_m^*—the complex conjugate of a *particular* unperturbed eigenfunction ψ_m. We then integrate both sides over all x, to obtain

$$\int_{-\infty}^{\infty} \psi_m^* [H_0 + v(x)] \sum_{l=1}^{\infty} a_{nl} \psi_l \, dx = \int_{-\infty}^{\infty} E_n' \psi_m^* \left\{ \sum_{l=1}^{\infty} a_{nl} \psi_l \right\} dx \quad (12.7)$$

Integrating each side of Eq. (12.7) term by term and removing the space-independent factors a_{nl} and E'_n from the integrals, we obtain

$$\sum_{l=1}^{\infty} a_{nl} \int_{-\infty}^{\infty} \psi_m^*[H_0 + v(x)]\psi_l \, dx = \sum_{l=1}^{\infty} a_{nl} E'_n \int_{-\infty}^{\infty} \psi_m^* \psi_l \, dx \quad (12.8)$$

Because the functions ψ are normalized and orthogonal, the *right-hand* side of Eq. (12.8) reduces to the single term $a_{nm} E'_n$ for which $l = m$, and we have

$$\sum_{l=1}^{\infty} a_{nl} \int_{-\infty}^{\infty} \psi_m^*[H_0 + v(x)]\psi_l \, dx = a_{nm} E'_n \quad (12.8a)$$

We can rewrite the *left-hand* side of Eq. (12.8) by using the fact that $H_0 \psi_l = E_l \psi_l$ [Eq. (12.3)]; this equation then becomes

$$\sum_{l=1}^{\infty} a_{nl} \int_{-\infty}^{\infty} \psi_m^*[E_l + v(x)]\psi_l \, dx = a_{nm} E'_n$$

or

$$\sum_{l=1}^{\infty} a_{nl} \int_{-\infty}^{\infty} \psi_m^* E_l \psi_l^* \, dx + \sum_{l=1}^{\infty} a_{nl} \int_{-\infty}^{\infty} \psi_m^*[v(x)]\psi_l \, dx = a_{nm} E'_n \quad (12.9)$$

Again, because the functions ψ are normalized and orthogonal, the first term on the left reduces to $a_{nm} E_m$. The second term can be written in abbreviated form as $\sum_{l=1}^{\infty} a_{nl} v_{ml}$, where v_{ml} is an abbreviation for the integral $\int_{-\infty}^{\infty} \psi_m^* v(x) \psi_l \, dx$, called the *matrix element* of the perturbing potential $v(x)$ between the states m and l. (This term can also be written in Dirac notation as $\langle m | v(x) | l \rangle$.) Substitution into Eq. (12.9) now yields

$$a_{nm} E_m + \sum_{l=1}^{\infty} a_{nl} v_{ml} = a_{nm} E'_n$$

and after rearranging terms,

$$\sum_{l=1}^{\infty} a_{nl} v_{ml} = a_{nm}(E'_n - E_m) \quad (12.10)$$

Equation (12.10) is exact. No approximations have been used in deriving it. However, it contains too many unknown quantities to permit an exact solution in most cases.

The most important of the unknown quantities is the perturbed energy of the nth level, or E_n'. To find a first approximation to this energy, we make the arbitrary assumption that each eigenfunction of the perturbed system is identical to one of the eigenfunctions of the unperturbed system. If this were true, it would require that $a_{nl} = 1$ if $l = n$ and $a_{nl} = 0$ if $l \neq n$. In that case, the *left-hand* side of Eq. (12.10) collapses to a single term: v_{mn}. The above assumption also tells us that the *right-hand* side is equal to zero unless $m = n$. Thus Eq. (12.10) is reduced to the *approximation*

$$E_n' - E_n = v_{nn} = \int_{-\infty}^{\infty} \psi_n^* v(x) \psi_n \, dx \qquad (12.11)$$

This is a first-order approximation to the perturbed energy E_n'. You can recognize this integral from the formula for the expectation value of an operator. This formula is not exact because the integral contains the wave function of the unperturbed system rather than the actual wave function.

12.1.1 Second-Order Perturbation Calculation

A more accurate value for the energy can be achieved by using a more accurate wave function, for which we again resort to an approximation. We find this approximation in the values of the coefficients a_{nl}. We now assume that each coefficient a_{nl} is quite small (although not zero) if $l \neq n$ and that $a_{nn} \simeq 1$. On these basis we can solve Eq. (12.10) for each *particular* coefficient a_{nm} in turn, by assuming that each *other* coefficient, except a_{nn}, may be neglected. With this assumption, Eq. (12.10) becomes

$$v_{mn} + a_{nm}v_{mm} = (E_n' - E_m)a_{nm} \qquad (12.12)$$

or

$$a_{nm} = \frac{v_{mn}}{E_n' - E_m - v_{mm}} \qquad (12.13)$$

From Eq. (12.11) we deduce that $v_{mm} = E_m' - E_m$; substitution into (12.13) yields

$$a_{nm} = \frac{v_{mn}}{E_n' - E_m'}, \qquad (n \neq m) \qquad (12.14)$$

Having found the values of the coefficients a_{nm}, we can substitute these values into Eq. (12.5) to find a better approximation to the perturbed eigenfunctions ψ_n'. We can then insert these eigenfunctions (instead of the eigenfunctions of the unperturbed system) into Eq. (12.11) to find a second-order approximation to the energy. This second-order value, E_n'', is

$$E_n'' = E_n' + \sum_{s \neq n} \frac{|v_{sn}|^2}{E_n - E_s} \tag{12.15}$$

This second-order result is particularly important in cases where the matrix elements v_{nn} are all zero, making the first-order perturbation in the energy levels also zero. The following example illustrates this point.

Example Problem 12.1   A one-dimensional harmonic oscillator of charge q is perturbed by the application of an electric field $\mathscr{E}$ in the positive x direction, making the potential energy $V(x) = m\omega^2 x^2/2 - q\mathscr{E}x$. Show that the first-order perturbation in each energy level is *zero* and that the second-order perturbation in the energy of the lowest $(n = 0)$ level is $-q^2\mathscr{E}^2/2m\omega^2$.

Solution. We insert the perturbing potential $v(x) = -e\mathscr{E}x$ into Eq. (12.11). Inserting the harmonic oscillator wave functions [Eqs. (4.29)], we obtain

$$E_n' - E_n = v_{nn} = \int_{-\infty}^{\infty} \psi_n^*(-q\mathscr{E}x)\psi_n \, dx \tag{12.16}$$

which is proportional to

$$\int_{-\infty}^{\infty} \left[(d/dx - ax)^n e^{-ax^2/2}\right](-q\mathscr{E}x)(d/dx - ax)^n e^{-ax^2/2} \, dx$$

The integral reduces to

$$\int_{-\infty}^{\infty} \left[(d/dx - ax)^n e^{-ax^2/2}\right]^2 (-q\mathscr{E}x) \, dx \tag{12.17}$$

Regardless of the value of n, the integrand is an odd function of x. Therefore the integral must be zero, and $E_n' = E_n$ for all values of n.

The second-order result for level 1 is found from Eq. (12.15) with $n = 0$:

$$E_0'' = E_0' + \sum_{s \neq 0} \frac{|v_{s0}|^2}{E_0 - E_s} \tag{12.18}$$

The first term in the series is

$$\frac{|v_{10}|^2}{E_0 - E_1} = -\frac{1}{\hbar\omega}\left|\int \psi_1^*(-q\mathscr{E}x)\psi_0\,dx\right|^2 \tag{12.19}$$

which may be written from Eqs. (4.30b),

$$-\frac{q^2\mathscr{E}^2}{2a\hbar\omega}\left|\int \psi_1^*\psi_1\,dx\right|^2 \tag{12.20}$$

Because ψ is normalized, this becomes $-e^2\mathscr{E}^2/2a\hbar\omega$, or $-e^2\mathscr{E}^2/2m\omega^2$ (since $a = m\omega/\hbar$). Each term in the series (12.18) contains the factor $|\int(\psi_s^*\psi_1\,dx|^2$ [analogous to that in Eq. (12.20)]. When $s \neq 1$, this factor is zero because of the orthogonality of the wave functions). Thus the total energy shift is simply $E_0'' - E_0' = -e^2\mathscr{E}^2/2m\omega^2$.

Example Problem 12.2 From first-order perturbation theory, find the lowest energy level, and the wave function for this level, in the "stepped" potential well given by (see Figure 12.1):

$$V(x) = +\infty \qquad (|x| > d)$$
$$V(x) = 0 \qquad (d/3 < |x| < d)$$
$$V(x) = \delta \qquad (|x| < d/3)$$

where δ is small relative to the energy of the lowest level.

Solution. The unperturbed wave functions are simply those of the infinitely deep square well (Section 3.2), which are (omitting the time factor), for $|x| \leq d$,

$$\psi(x) = N\cos(n\pi x/2d) \quad \text{(for odd } n\text{);}$$
$$\psi(x) = N\sin(n\pi x/2d) \quad \text{(for even } n\text{)}$$

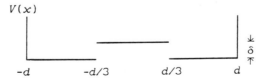

FIGURE 12.1 "Stepped" potential well.

where $N = 1/\sqrt{d}$. The perturbed energy levels are found from Eq. (12.11). For $n = 1$ we have

$$E_1' - E_1 = \int \psi_1^* v(x) \psi_1 \, dx = \delta \int_{-d/3}^{d/3} \psi_1^* \psi_1 \, dx$$

because $v(x)$ equals δ only for $-d/3 < x < +d/3$, and is zero everywhere else.

Not surprisingly, this result is simply the product of δ and the probability of finding the particle in the region where $v(x) = \delta$. The numerical result is

$$E_1' - E_1 = (\delta/d) \int_{-d/3}^{d/3} \cos^2(\pi x/2d) \, dx$$

$$= \delta \sin(\pi/3)[(1/\pi) + (1/3)] = 0.61\delta$$

In other words, because the probability of finding the particle between $x = +d/3$ and $x = +d/3$ is 0.61, the energy is increased by 0.61 times the amount of the additional potential energy in that region.

To find the perturbed wave function ψ_1', we use the series of Eq. (12.5):

$$\psi_1' = \psi_1 + a_{12} \psi_2 + a_{13} \psi_3 + \cdots$$

and we apply Eq. (12.14) to find the coefficients:

$$a_{12} = \frac{\int_{-\infty}^{\infty} \psi_2^* [v(x)] \psi_1 \, dx}{E_1' - E_2} = 0 \quad \text{(because the integrand is an odd function)}$$

$$a_{13} = \frac{\int_{-\infty}^{\infty} \psi_3^* [v(x)] \psi_1 \, dx}{E_1' - E_3'} = \frac{N^2 \delta \int_{-d/3}^{+d/3} \cos \dfrac{3\pi x}{2d} \cos \dfrac{\pi x}{2d} \, dx}{E_1' - E_3'} \quad (12.21)$$

With $N^2 = 1/d$, $E_1' \cong E_1 = h^2/32md^2$ and $E_3' \cong E_3 = (h^2/32md^2)/9$ (Section 3.2), and integration, Eq. (12.21) becomes

$$a_{13} = -\frac{\delta m a^2}{h^2 \pi} \sqrt{3} \quad (12.22)$$

Calculation of the other nonzero coefficients, a_{15}, a_{17}, etc., is straightforward. They become progressively smaller, because of the increasing energy

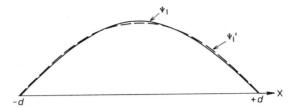

FIGURE 12.2 Wave function for lowest energy in the well of Figure 12.1. Solid line shows the unperturbed wave function for the infinitely deep well.

difference in the denominator. Therefore the perturbed (unnormalized) wave function is given by

$$\psi_1' = \psi_1 - \frac{\delta m d^2 \sqrt{3}}{h^2 \pi} \psi_3 + \cdots$$

$$= \cos \frac{\pi x}{2d} - \frac{\delta m d^2 \sqrt{3}}{h^2 \pi} \cos \frac{3\pi x}{2d} + \cdots$$

Figure 12.2 compares ψ_1' and ψ_1. Notice several features of this figure:

- When $x = 0$, $\psi_1 = 1$, and $\psi_1' < 1$ because of the negative coefficient a_{13}.
- Near $x = \pm d$, $\psi_1' > \psi_1$, and the second derivative of ψ_1' has a greater magnitude than the second derivative of ψ_1, because of the greater value of the perturbed kinetic energy in this region: $E' = E + 0.61\delta$, and $V(x) = 0$ in both cases.
- Near the center of the well, the second derivative of ψ_1' is smaller than the second derivative of ψ_1, because the perturbed kinetic energy $E' - \delta$ is smaller than the unperturbed kinetic energy: $E' - \delta = E - 0.39\delta$.

12.1.2 Perturbation of Degenerate Eigenfunctions

The preceding discussion covers only the special situations in which no eigenstate is degenerate and the perturbation energy is much larger than the separation between eigenvalues. In such situations it is reasonable to assume that for each perturbed eigenfunction a single coefficient in the expansion (12.5) is nearly equal to 1, and other coefficients are quite small.

But in many cases, the unperturbed levels are degenerate. In that case we can only assume that any given perturbed eigenfunction must, in the first approximation, be a linear combination of the eigenfunctions for one of the unperturbed levels. Therefore we must allow as many nonzero coefficients a_{mn} as there are independent eigenfunctions for that unperturbed level. Then we need to use Eq. (12.10) to set up as many *simultaneous* equations as are necessary to find these coefficients and to find the perturbed energy levels.

In some cases we can find these coefficients without much difficulty. For example, consider the functions that are simultaneous eigenfunctions of J^2 and J_z for the hydrogen atom (Section 10.1). A typical pair of orthogonal degenerate eigenfunctions could be written as $R_{nl}(r)|l, l\rangle| - \rangle_s$ and $R_{nl}(r)|l, l - 1\rangle| + \rangle_s$, respectively. If there were no spin–orbit energy, these eigenfunctions would correspond to the same energy level; that is, the level would have a degeneracy of 2. The perturbation that removes this degeneracy is the spin–orbit energy, which is proportional to $\mathbf{L} \cdot \mathbf{S}$, as we saw in Section 10.1 [Eq. (10.11)]. We also saw there that

$$\mathbf{S} \cdot \mathbf{L} = (J^2 - S^2 - L^2)/2 \tag{10.17}$$

and we saw in Example Problem 10.2 how to construct possible linear combinations of eigenfunctions for the perturbed potential in the case $l = 1$. The perturbation removes the degeneracy.

12.2 THE THREE-BODY SYSTEM: THE HELIUM ATOM

In quantum mechanics as in classical mechanics, there is, in general, no exact solution for the motion of three interacting particles. However, there are perturbation techniques that are extremely successful. For example, the helium atom can be solved with great accuracy by taking the "unperturbed" state to be one in which each electron interacts with the proton but not with the other electron. In that case each electron state is a state of the helium *ion*, in which there is no "other electron" with which to interact. The state of the system then must be a superposition like one of the eigenfunctions of Eqs. (11.4), for example,

$$\psi = \frac{1}{\sqrt{2}}[|a\rangle_1|b\rangle_2 + |a\rangle_2|b\rangle_1]\frac{1}{\sqrt{2}}[| + \rangle_1| - \rangle_2 - | + \rangle_2| - \rangle_1]$$

where $|a\rangle$ and $|b\rangle$, respectively, are eigenstates of the one-electron system, the helium *ion* (Table 9.1, with $Z = 2$).

For the lowest energy level, each of these states has $n = 1$; that is, $|a\rangle = |b\rangle = |1, 0, 0\rangle$, and if there were no electron–electron interaction the state would have a total energy of -108.8 eV (or -54.4 eV for each electron).

Obviously, it is a gross oversimplification to overlook the electron–electron interaction energy. The measured ground-state energy of the helium atom is -79.0 eV, or 29.8 eV higher than it would be without the electron–electron interaction. [The value of 79.0 eV is the sum of the first and second ionization energies of the helium atom (Table 12.1).] Can such a large effect be calculated by perturbation methods?

Let us compute the first-order approximation by use of Eq. (12.11). The "perturbing" potential v is the electrostatic potential of two electrons at positions $\mathbf{r}_1$ and $\mathbf{r}_2$, or

$$v = \frac{e^2}{4\pi\varepsilon_0 |\mathbf{r}_1 - \mathbf{r}_2|} \tag{12.23}$$

The energy shift produced by this perturbation is therefore, to first order,

$$E_1' - E_1 = \int\int |u_{100}(1)|^2 |u_{100}(2)|^2 \frac{e^2}{4\pi\varepsilon_0 |\mathbf{r}_1 - \mathbf{r}_2|} \, d\tau_1 \, d\tau_2 \tag{12.24}$$

where $u_{100}(1)$ and $u_{100}(2)$ are simply the hydrogen atom wave functions of Table 9.1 written in terms of the coordinates of particle 1 and particle 2, respectively, and $d\tau_1$ and $d\tau_2$ are volume elements of the respective coordinates. Written in terms of the six spherical coordinates r_1, ρ_1, θ_1, r_2,

TABLE 12.1 Ionization Energies of Light Elements, in Electron Volts

Z	Element	I	II	III	IV	V	VI
1	Hydrogen	13.595					
2	Helium	24.580	54.400				
3	Lithium	5.390	75.619	122.42			
4	Beryllium	9.320	18.206	153.85	217.657		
5	Boron	8.296	25.149	37.920	259.298	340.127	
6	Carbon	11.264	24.376	47.864	64.476	391.986	489.84

From E. U. Condon and H. Odishaw (eds.), *Handbook of Physics*, McGraw-Hill, New York (1958).

ρ_2, and θ_2 for electrons 1 and 2, respectively, I becomes

$$I = \frac{Ze^2}{\pi^2 a_0'} \int \int \int \int \int \int \frac{e^{-2\rho_1}e^{-2\rho_2}}{4\pi\varepsilon_0\,\rho_{12}} \rho_1^2 \sin\,\theta_1\,d\rho_1\,d\theta_1\,d\phi_1\,\rho_2^2\,\sin\,\theta_2\,d\rho_2\,d\theta_2\,d\phi_2$$

(12.25)

Except for the factor $Ze^2/\pi^2 a_0'$, I is identical to the mutual interaction energy of two spherically symmetric charge distributions of charge density $e^{-2\rho_1}$ and $e^{-2\rho_2}$, respectively.

The evaluation of this energy is a standard exercise in electrostatic theory.[2] The result, with $Z = 2$, is

$$I = (5Z/8)m_r c^2\alpha^2 = (5/2) \times 13.60\,\text{eV} = 34.0\,\text{eV} \qquad (12.26)$$

Recall that the energy of the helium ion is $-Z^2(m_r c^2\alpha^2/2)$, and the unperturbed energy E_1 is just twice that, or

$$E_1 = -2Z^2(m_r c^2\alpha^2/2) \qquad (12.27)$$

Therefore the total perturbed energy is the sum

$$E_1' = E_1 + I = -2Z^2(m_r c^2\alpha^2/2) + (5Z/8)m_r c^2\alpha^2$$

$$= (-Z^2 + 5Z/8)(m_r c^2\alpha^2) \qquad (12.28)$$

It should be mentioned that the value of m_r depends on the nuclear mass, but m_r varies only slightly, from $0.9995m_e$ for H^1 to $0.99996m_e$ for C^{12}. The value for $Z = 2$ is -74.8 eV, which differs by only about 6% from the experimental value of -79.0 obtained from Table 12.1. This is a remarkably good result for a perturbation that could hardly be called small, but we can do much better, as shown in the following.

12.2.1 Atoms with Larger Values of Z

Just as the results for the hydrogen atom can be applied to any *one-electron ion* (see text above Table 9.1), Eq. (12.28) can be applied to any *two*-electron ion to yield the approximate total energy that is needed to remove the two electrons. This means that Eq. (12.28) yields the sum of

---

[2] The trick is in integrating energy elements in the most efficient order. It is worked out in *Introduction to Quantum Mechanics*, L. Pauling and E. B. Wilson, Appendix V, McGraw-Hill, New York (1935).

the second and third ionization energies of lithium ($Z = 3$), the sum of the third and fourth ionization energies of beryllium ($Z = 4$), etc. The accuracy of the calculation improves as Z increases; for $Z = 4$, Eq. (12.28) yields $E_1' = (-16 + 2.5)(27.2 \text{ eV}) = -367.2 \text{ eV}$, which is only about 1% different in magnitude from the sum of the third and fourth ionization energies (shown in Table 12.1 to be 371.5). The reason for this improvement is simply that the perturbing potential energy becomes a smaller part of the total energy as Z increases. See also Exercise 7 in this chapter.

12.2.2 The Variational Technique

To obtain greater accuracy, we must vary the wave function. Rather than use the method leading to Eq. (12.15), let us try a different method, called the *variational* technique, to alter the wave function.

The basis of this technique is that, because the allowed wave functions form a complete, orthogonal set of functions (Postulate 3), any well-behaved function ψ' can be expanded in a series of functions belonging to this set. Therefore, if we calculate E' by using the arbitrary function ψ' (which is presumably not one of the eigenfunctions of the perturbed system), the resulting value of E' will be a weighted sum of all of the energy eigenvalues of the perturbed system. We set

$$\psi' = \sum a_n \psi_n \quad \text{and} \quad H\psi_n' = E_n'\psi_n' \tag{12.29}$$

where the ψ_n are precisely the normalized perturbed eigenfunctions, the E_n' are the corresponding energy eigenvalues, and H is the Hamiltonian operator for the complete system, including perturbation(s). The expectation value of the energy is now

$$E' = \int \psi'^* H\psi' \, d\tau \tag{12.30}$$

This becomes, using Eq. (12.29),

$$E' = \int \left\{ \sum a_n^* \psi_n'^* \right\} H \left\{ \sum a_n \psi_n' \right\} d\tau \tag{12.31}$$

Substituting from Eq. (12.29) into Eq. (12.31), expanding both series, and then integrating term-by-term and using the orthogonality of the ψ_n, we finally have

$$E' = \sum |a_n|^2 E_n \tag{12.32}$$

Equation (12.32) shows that the value of E', which is computed from Eq. (12.30), must be greater than or equal to the ground-state energy E_1 for the perturbed system. If $E' = E_1$, the "arbitrary" function ψ' is actually the ground-state eigenfunction.

[If this is not clear, substitute $E_2 = n_2 E_1$, $E_3 = n_3 E_1$, etc. into Eq. (12.32), where n_2, n_3, etc. are numbers greater than 1, because $E_n > E_1$ for $n > 1$. Then

$$E' = |a_1|^2 E_1 + |a_2|^2 n_2 E_1 + |a_3|^2 n_3 E_1 \cdots$$

where each term after the first is of the form $|a_n|^2$ multiplied by a number greater than E_1. Therefore the entire series must have a value greater than or equal to $E_1 \Sigma |a_n|^2$, and since we know that $\Sigma |a_n|^2 = 1$, we see that $E' \geq E_1$. (They are equal only if $a_n = 0$ for $n > 1$, or if there is no energy level higher than E_1.)]

These facts show us how to improve an approximation even when we don't know the actual ground-state energy. If we compute the values of E' for a number of trial wave functions, the *smallest value of E' must be the one that is closest to the actual ground-state energy of the perturbed system.* Therefore, a good strategy for computing the ground state energy of a "perturbed" system is to write ψ' in terms of a parameter that can be varied systematically to find the minimum value of E'.

For example, we can use the eigenfunctions of a two-electron ion as in the first-order calculation above [Eq. (12.25)], except that we substitute a variable Z' in place of the actual atomic number Z in the wave functions (but not in the potential V). The total energy is then

$$E' = \int \psi'^* H \psi' \, d\tau \tag{12.33}$$

where the H operator is given by

$$H = -\frac{\hbar^2}{2m}(\nabla_1^2 + \nabla_2^2) - \left\{ \frac{Ze^2}{4\pi\varepsilon_0 r_1} + \frac{Ze^2}{4\pi\varepsilon_0 r_1} \right\} + \frac{e^2}{4\pi\varepsilon_0 |\mathbf{r}_1 - \mathbf{r}_2|}$$

$$= E_k + V_p + V_e \tag{12.34}$$

In Eq. (12.34), $E_k = -(\hbar^2/2m)(\nabla_1^2 + \nabla_2^2)$ is the operator for the total kinetic energy, $V_p = -(Ze^2/4\pi\varepsilon_0 r_1) - (Ze^2/4\pi\varepsilon_0 r_2)$ is the total potential energy of the interaction between each electron and the nucleus, and

$V_e = e^2/(4\pi\varepsilon_0|\mathbf{r}_1 - \mathbf{r}_2|)$ is the potential energy of the interaction between the two electrons. The first two terms are evaluated by comparison with the corresponding values for hydrogen, as follows:

- Each of the ∇^2 operators yields a factor of Z'^2 in the kinetic energy, which makes E_k the sum of two kinetic energies of $Z'^2 \times 13.60$ eV, because the electron in the hydrogen atom has a kinetic energy of 13.60 eV. Thus $E_k = 2Z'^2 \times 13.6$ eV.
- The expectation value of V_p must be Z' times twice the potential energy of an electron in a one-electron system of nuclear charge Ze, because the exponential factor $e^{-2\rho_1}$ (where $\rho_1 = Zr_1/a_0'$) is now $e^{-2\rho_1'}$, where Z has been replaced by Z'. Thus the combined potential energy of the two electrons in the field of the *nucleus* is $V_p = -2ZZ' \times 27.2$ eV. (One factor of Z remains, because this term is proportional to the nuclear charge Ze.)

The third term, V_e, is computed as in Eq. (12.25), where the result was an energy of $5Z/4 \times 13.60$ eV. When Z is replaced in the wave function by Z', the resulting energy is $5Z'/4 \times 13.60$ eV. The three terms yield a total energy of

$$E' = E_k + V_p + V_e = 2Z'^2 \times 13.6 \text{ eV} - 2ZZ' \times 27.2 - 5Z'/4 \times 13.60 \tag{12.35}$$

or

$$E' = (2Z'^2 - 4ZZ' + 5Z'/4) \times 13.60 \text{ eV} \tag{12.36}$$

For helium, with $Z = 2$, we have

$$E' = (2Z'^2 - 27Z'/4) \times 13.60 \text{ eV} \tag{12.37}$$

The minimum value of E' is -77.46 eV, obtained when $Z' = 27/16$. This is considerably lower than the -74.8 eV given by Eq. (12.28), and it is less than 2% above the experimental value of -78.98 eV. Improvements in the trial function have reduced the result to within a few parts per million of the experimentally observed value.[3]

---

[3]L. Pauling and E. B. Wilson, *Introduction to Quantum Mechanics*, McGraw-Hill, New York (1935), gives details on trial functions that have been used.

12.2.3 Higher Energy Levels

Although the variational technique computes only the lowest energy level of a system, the general perturbation method can be used for any level. Consider the 1s2s and 1s2p states of a two-electron atom or ion. Without the electron–electron interaction, the energies of these states would differ only because of the tiny spin–orbit energy (Section 10.2). The electron–electron interaction not only raises the energy of both of these states, it also greatly increases the splitting between them. The reason for this is seen in the radial probability amplitudes for hydrogen (Figure 9.2b).

The 1s probability amplitude is confined to very small values of r, relative to the 2s or 2p amplitudes, and the 2s amplitude extends to much larger values of r than the 2p amplitude. (The 2s amplitude, with zero angular momentum, can be visualized as representing a more eccentric "orbit" than the 2p amplitude.) Consequently, the mean distance between a 1s electron and a 2s electron is greater than the average distance between a 1s electron and a 2p electron, and therefore the Coulomb repulsion energy is smaller for a 1s2s state than for a 1s2p state, and the energy difference can be computed by the perturbation approach, starting with Eq. (12.24) and using the relevant functions [u_{100}, u_{210}, and u_{200}].

Recall that the spin–orbit energy is proportional to $\mathbf{L} \cdot \mathbf{S}$. Thus the 1s2p, $S = 1$ states (^3P, Figure 12.3)[4] become three energy levels (one for each possible orientation of $\mathbf{S}$ relative to $\mathbf{L}$), spaced about 10^{-4} eV apart as in the spin–orbit splitting of hydrogen. The 1s2s level is unsplit, because $L = 0$ for this state. The other 1s2p states, having $S = 0$, are not split. Thus, when the electron states are $n = 1$ and $n = 2$, there are six distinct energy levels.

12.3 RELATIVISTIC CORRECTION, HYDROGEN ENERGY LEVELS

In Section 1.5, fine structure in the hydrogen was "explained" by the Bohr–Sommerfeld theory as a relativistic effect related to the eccentricity of elliptical electron orbits. Although the Bohr–Sommerfeld theory does not give precisely correct energy levels, there is indeed a relativistic contribution to each energy level, resulting from the fact that the kinetic energy K is not precisely given by the expression $p^2/2m$. This contribu-

---

[4]In this notation, to be discussed in Section 13.2, the superscript 3 indicates a spin triplet, and the letter P indicates that the resultant orbital angular momentum of the two particles has quantum number 1.

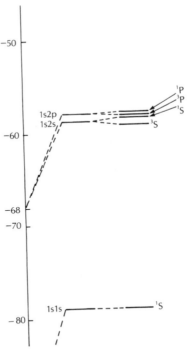

FIGURE 12.3 Energy levels of the helium atom. Levels 1s2s and 1s2p would be −68 eV without the electron−electron interaction. The left-hand column shows the energy that would result if there were no exchange energy. The right-hand column shows experimentally measured energies. The symbol ^{3}P denotes three levels, with spin−orbit splitting too small to see on this scale.

tion may be calculated by an approximation based on expanding K in a power series in p^2.

The accurate formula for K in terms of E, m, and p is given in Appendix C:

$$K = E - mc^2 = (m^2c^4 + p^2c^2)^{1/2} - mc^2$$

$$= mc^2[1 + (p^2/m^2c^2)]^{1/2} - mc^2 \tag{12.38}$$

which can be expanded to

$$K = mc^2[1 + p^2/m^2c^2 - p^4/8m^4c^4 + \cdots]^{1/2} - mc^2$$

or

$$K = \frac{p^2}{2m} - \frac{p^4}{8m^3c^2} + \cdots = \frac{p^2}{2m}\left\{1 - \frac{p^2/2m}{mc^2/2} + \cdots\right\} \quad (12.39)$$

The second term is about 10^{-4}, the same order of magnitude as α^2, so the relativistic effect is comparable to the spin–orbit effect, and we can neglect the third and higher-order terms, which are of order α^4, α^6, etc. We can therefore write an approximate time-independent Schrödinger equation (minus a spin–orbit term) by using the right-hand side of (12.39) in place of the $p^2/2m$ term, obtaining

$$\left\{\frac{p^2}{2m} - \frac{p^4}{8m^3c^2} + V(r)\right\}\psi = E\psi \quad (12.40)$$

where p^2 and p^4 are operators following the general rule $p^2 = -\hbar^2\nabla^2$.

In solving Eq. (12.40) we need only the same degree of accuracy attained in the truncated series (12.39). We begin by rewriting Eq. (12.40) as

$$\frac{p^2}{2m}\psi = \left\{E - V(r) + \frac{p^4}{8m^3c^2}\right\}\psi \quad (12.41)$$

We then apply the operator $p^2/2m$ to each side, obtaining

$$\frac{p^4}{4m^2}\psi = \left\{\frac{p^2}{2m}[E - V(r)] + \frac{p^6}{16m^4c^2}\right\}\psi \cong \left\{\frac{p^2}{2m}[E - V(r)]\right\}\psi \quad (12.42)$$

Using Eq. (12.41), we now replace $p^2/2m$ by $E - V(r) + (p^4/8m^3c^2)$. Dropping the term $p^6/16m^4c^2$, we obtain

$$\frac{p^4}{4m^2}\psi \cong \left\{\left[E - V(r) + \frac{p^4}{8m^3c^2}\right][E - V(r)]\right\}\psi \quad (12.43)$$

Recognizing from Eq. (12.41) that $[E - V(r)]\psi \cong (p^2/2m)\psi$, we see that $(p^4/8m^3c^2)[E - V(r)]$ is of order p^6. Dropping that term in Eq. (12.43) and rearranging, we finally have

$$\frac{p^4}{4m^2}\psi \cong [E - V(r)]^2\psi \quad (12.44)$$

Substitution into Eq. (12.40) then yields (approximately)

$$\frac{p^2}{2m}\psi = \left\{\frac{1}{2mc^2}[E - V(r)]^2 + E - V(r)\right\}\psi$$

whose radial part is

$$-\frac{\hbar^2}{2m_r}\frac{d^2R}{dr^2} = \left\{\frac{1}{2mc^2}[E - V(r)]^2 + E - V(r) - \frac{\hbar^2 l(l + 1)}{2m_r r^2}\right\}R$$

$$(12.45)$$

in terms of the reduced mass m_r and the state function's radial factor $R(r)$. Since $V(r)$ is inversely proportional to r, Eq. (12.45) can be expressed in the form

$$\frac{d^2R}{dr^2} = \left\{A + \frac{B}{r} + \frac{C}{r^2}\right\}R(r) \qquad (12.46)$$

where the constants A, B, and C may be deduced from Eq. (12.45). The problem is now purely mathematical; this equation may be solved by the same power series method used for the uncorrected radial equation (Appendix E), and the energy eigenvalues E_{nl}, in terms of the fine-structure constant α, are

$$E_{nl} = E_n\left\{1 + \frac{\alpha^2 Z^2}{n^2}\left(\frac{3}{4} - \frac{n}{l + 1/2}\right)\right\} \qquad (12.47)$$

where E_n is the eigenvalue for the nonrelativistic equation, without the spin–orbit energy, which is not accounted for in the preceding equations.

Adding the spin–orbit energy W [Eqs. (10.47) and (10.48)] to E_{nl} gives the total theoretical energy levels, as follows

For $j = l + 1/2$:

$$E_{nl} = E_n\left\{1 + \frac{\alpha^2 Z^2}{n^2}\left(\frac{3}{4} - \frac{n}{l + 1/2} + \frac{n}{(2l + 1)(l + 1)}\right)\right\}$$

$$= E_n\left\{1 + \frac{\alpha^2 Z^2}{n^2}\left(\frac{3}{4} - \frac{n}{l + 1}\right)\right\} \qquad (12.48)$$

For $j = l - 1/2$:

$$E_{nl} = E_n\left\{1 + \frac{\alpha^2 Z^2}{n^2}\left(\frac{3}{4} - \frac{n}{l + 1/2} - \frac{n}{l(2l + 1)}\right)\right\}$$

$$= E_n\left\{1 + \frac{\alpha^2 Z^2}{n^2}\left(\frac{3}{4} - \frac{n}{l}\right)\right\} \tag{12.49}$$

Equations (12.48) and (12.49) can be replaced by a single equation written in terms of j:

$$\boxed{E_{nl} = E_n\left\{1 + \frac{\alpha^2 Z^2}{n^2}\left(\frac{3}{4} - \frac{n}{j + 1/2}\right)\right\}} \tag{12.50}$$

Figure 12.4 displays these levels. Notice that there is no splitting when $l = 0$, because in that case the only possible value of j is $+1/2$; j is never negative.

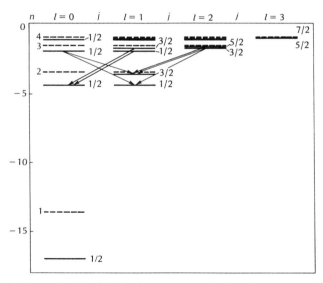

FIGURE 12.4 Energy levels of the hydrogen atom, showing fine structure. Uncorrected levels are shown by dashed lines, corrected levels by solid lines. The energy scale applies to the uncorrected levels only; the displacement of the corrected levels from the uncorrected levels is exaggerated by a factor of α^{-2}, about 19,000.

12.4 TUNNELING AND ALPHA-PARTICLE EMISSION

In Section 5.3 we found the probability that a particle can tunnel through a *square* potential barrier. An important application is analysis of emission of alpha particles (helium nuclei) from heavy elements. For alpha emission the barrier is formed by the Coulomb potential energy; thus we cannot assume that the result for a square barrier is useful. However, a fairly straightforward approximation method gives remarkable agreement with experiment.

Figure 12.5 shows the potential barrier in this case. We assume that there is a preformed alpha particle within the heavy nucleus and that it encounters the potential barrier as it bounces back and forth inside the nucleus. To escape from the nucleus, the alpha particle must traverse the forbidden region extending from $r = R$ to $r = R_1$, where R is the radius of the nucleus and R_1 is the radius at which the potential energy V is equal to the total alpha-particle energy E.

The energy E is equal to the kinetic energy observed when it escapes from the atom (where $V = 0$). The potential energy outside the nucleus is simply the Coulomb potential energy of a charge of $-2e$ and a charge of $+Ze$, separated by a distance r, where $Z + 2$ equals the atomic number of the original atom. It is easy to calculate that

$$V(r) = 2.88Z/r \qquad [V(r) \text{ in Mev, } r \text{ in femtometers (fm)}] \quad (12.51)$$

We expect particles with greater energy E to have a higher probability of penetrating the barrier and escaping from the atom. Thus, all other

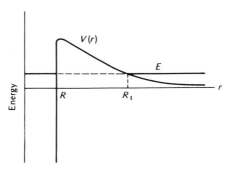

FIGURE 12.5 Sketch of energy level and potential energy $V(r)$ of an α particle in the field of an atomic nucleus of radius $R \approx 9$ fm. For $r > R$, $V(r)$ is the Coulomb potential; for $r < R$ it is the sum of the Coulomb potential and the deep negative potential well resulting from the strong nuclear force.

things being equal, the half-life $t_{1/2}$ of an alpha emitter should decrease with increasing alpha-particle energy, and evidence of this was indeed found in 1911 by Geiger and Nuttal. Let us now try to use the barrier penetration equations to deduce a quantitative relationship between E and $t_{1/2}$.

In Section 5.3 we found the penetration probability T for a *square* barrier:

$$T = \left\{ 1 + \frac{V_0^2 \sinh^2 \alpha a}{4E(V_0 - E)} \right\}^{-1} \quad \text{for } E < V_0 \tag{12.52}$$

where a is the thickness of the barrier, V_0 is its height, and $\alpha = \sqrt{2m(V_0 - E)/\hbar^2}$. For $\alpha a \gg 1$ (which applies for most of the barrier region), Eq. (12.52) can be reduced to

$$T = \frac{16E(V_0 - E)}{V_0^2} e^{-2\alpha a} \tag{12.53}$$

The probability per unit time of alpha emission, denoted by λ, is the product of two factors: the penetration probability T and the rate $\mathcal{R}$ at which the alpha particle strikes the barrier. The rate $\mathcal{R}$ is given by $v/2R$, where v is the speed of the alpha particle inside the nucleus. Since $E = mv^2/2$, where m is the alpha-particle mass, we finally have

$$\lambda = \mathcal{R}T = vT/2R = \sqrt{E/2m} \, \frac{16E(V_0 - E)}{V_0^2 R} e^{-2\alpha a} \tag{12.54}$$

The exponential factor $e^{-2\alpha a}$ is entirely a result of the exponential decay of the wave function inside the barrier, because in that region the equation for the radial probability amplitude $rR(r)$ may be written (Section 8.1) as

$$\frac{d^2}{dr^2} rR(r) = \alpha^2 [rR(r)] \tag{12.55}$$

The solution to Eq. (12.55) is

$$rR(r) = e^{-\alpha r} \quad \text{or} \quad |rR(r)|^2 = |P(r)|^2 = e^{-2\alpha r} \tag{12.56}$$

Thus if the barrier were square, the change in the probability density $|P(r)|^2$ would account for the factor $e^{-2\alpha a}$ in Eq. (12.54).

For the actual barrier, with a slowly varying barrier height [one that varies much more slowly than $P(r)$], we can compute a corrected exponential factor by breaking up the actual barrier into a series of adjoining narrow barriers of width Δr and slowly changing height. We can then replace the factor $e^{-\alpha r}$ by the product of factors $e^{-\alpha_1 \Delta r}, e^{-\alpha_2 \Delta r}, e^{-\alpha_3 \Delta r}, \ldots$, where α_1 is the value of α calculated from the height of the first narrow barrier, α_2 the value for the second barrier, etc. In the limit as $\Delta r \to 0$, the product becomes the exponential $\exp\{-\int_R^{R_1} \alpha(r)\, dr\}$, where $\alpha(r)$ is given by

$$\alpha(r) = \sqrt{2m(V(r) - E)/\hbar^2} \qquad (12.57)$$

in agreement with the definition of α for a square barrier.

The emission probability λ is proportional to the square of this exponential, or to $\exp\{-2\int_R^{R_1} \alpha(r)\, dr\}$. This factor clearly reduces to the exponential factor in Eq. (12.54) with $R_1 - R = a$, if α is constant. As a function of the alpha-particle energy E, the other factors in Eq. (12.54) do not vary nearly as rapidly as this exponential factor varies.

To write our results in terms of the half-life $t_{1/2}$, we use the fact that λ is equal to $0.69/\lambda$ (see Chapter 14) and we have

$$t_{1/2} = 0.69 \left\{ \sqrt{E/2m}\, \frac{16E(V_0 - E)}{V_0^2 R} \right\}^{-1} \exp\left\{ 2\int_R^{R_1} \alpha(r)\, dr \right\} \qquad (12.58)$$

or

$$t_{1/2} = \text{constant} \times E^{-3/2} \times \exp\left\{ 2\int_R^{R_1} \alpha(r)\, dr \right\} \qquad (12.59)$$

Here we have approximated the term $V_0 - E$ as a constant, because $E \ll V_0$ for alpha decays commonly observed. Taking the natural logarithm of each side, we find that

$$\ln t_{1/2} = \text{constant} - 3/2 \ln E + 2\int_R^{R_1} \alpha(r)\, dr$$

$$= \text{constant} - 3/2 \ln E + 2\int_R^{R_1} \sqrt{2m(V(r) - E)/\hbar^2}\, dr \qquad (12.60)$$

Using the fact that $E = V(R_1)$, and knowing that $V(r)$ is proportional to $1/r$, we can rewrite the integral in Eq. (12.60) as

$$2\sqrt{2mE/\hbar^2} \int_R^{R_1} \{(R_1/r) - 1\}^{1/2}\, dr.$$

The integral is now soluble by means of the substitution $r = R_1 \cos^2 \theta$. Using the approximation $\cos^{-1}(R/R_1)^{1/2} \approx \pi/2 - (R/R_1)^{1/2}$, we find that

$$\ln t_{1/2} \approx \text{constant} - 1.5 \ln E + 2(2mE/\hbar^2)^{1/2}\left[(\pi R_1/2) - 2(RR_1)^{1/2}\right]$$
$$(12.61)$$

To show the dependence of the half-life on the energy E, we use Eq. (12.51) to eliminate R_1 via the substitution $E = V(R_1) = 2.88Z/R_1$, with the final result that the decay rate per second is given by

$$\ln t_{1/2} \approx \text{constant} - 1.5 \ln E + 3.92ZE^{-1/2} - 2.94Z^{1/2}R^{1/2} \quad (12.62)$$

or, to base 10

$$\log_{10} t_{1/2} \approx \text{constant} - 1.5 \log_{10} E + 1.702ZE^{-1/2} - 1.277Z^{1/2}R^{1/2}$$
$$(12.63)$$

The constant is dimensionless, but its numerical value does depend on the time unit used in computing the logarithm; the units do not matter as long as the same unit is used consistently for each variable in every case. The constant does depend on the logarithm of the nuclear radius R, but because $\log R$ varies by so little (by less than 0.03 for the known alpha-particle emitters), we can test Eq. (12.63) by using the same constant for each alpha emitter (see Example Problem 12.3). The test shows that there is a remarkable fit of this approximation to observations of alpha-decay rates and alpha energies over a range of 16 orders of magnitude.[5]

Example Problem 12.3 Atoms of ^{238}U (uranium, $Z = 92$) decay with half-life of 4.47×10^9 years (1.41×10^{17} s) to ^{234}Th (thorium, $Z = 90$). The alpha-particle energy E is 4.270 MeV; the radius of the ^{234}Th nucleus is 9.24 fm.

(a) Determine the value of the constant in Eq. (12.63).
(b) Using this value, find the half-life of ^{232}U ($R = 9.16$ fm and $E = 5.414$ MeV).

[5]The recognition that alpha-particle emission was a quantum mechanical barrier penetration phenomenon was a result of the development of this relation between energy and half-life, which was worked out by Gamow and by Condon and Gurney in 1928. For more details see the references in Additional Reading.

Solution. (a)

$$\text{constant} = \log t_{1/2} + 1.5 \log E - 1.702 Z E^{-1/2} + 1.277 Z^{1/2} R^{1/2}$$

$$\begin{aligned}
\text{or constant} &= 17.147 + 1.5 \times 0.630 - 1.702 \times 90/2.067 \\
&\quad + 1.277 \times 9.487 \times 3.040 \\
&= -19.19
\end{aligned}$$

(b) $\log t_{1/2} = \text{constant} - 1.5 \log E + 1.702 Z E^{-1/2} - 1.277 Z^{1/2} R^{1/2},$

$$\begin{aligned}
&= -19.19 - 1.5 \times 0.734 + 1.702 \times 90/2.327 \\
&\quad - 1.277 \times 9.487 \times 3.027 \\
&= 8.87
\end{aligned}$$

Thus $t_{1/2} = 7.41 \times 10^8$ s, or 23.5 years.

The measured value of this half-life is 72 years. This is fairly good agreement, considering that the half-lives of ^{238}U and Th^{234} differ by 4.5 billion years.

ADDITIONAL READING

Robert B. Leighton, *Principles of Modern Physics*, McGraw-Hill, New York (1959), Section 15.7, and John D. McGervey, *Introduction to Modern Physics*, 2nd edition, Academic Press, New York (1983), pp. 566–571 give further details on alpha decay.

L. Pauling and E. B. Wilson, *Introduction to Quantum Mechanics*, McGraw-Hill, New York (1935) gives a good treatment of the variational technique.

EXERCISES

1. For the potential of Example Problem 12.1, show that there is an exact solution for the lowest energy level. *Hint:* The potential may be put into the form $V(x) = \frac{1}{2} m \omega^2 (x - x_0)^2 + V_0$, where V_0 and x_0 are constants. This leads to a Schrödinger equation of the same form as that for the unperturbed simple harmonic oscillator.

2. Use first-order perturbation theory to compute the lowest energy level in the potential well of Exercise 12, Chapter 3. Compare the answer with the exact result. Is the second-order perturbation calculation any better?

3. Use first-order perturbation theory to compute the lowest energy level in a potential well

$$V = kz \quad \text{for } 0 < z < a; \qquad V = \infty \quad \text{for } z < 0 \text{ or } z > a$$

where $k = 0.01$ eV/nm and $a = 3$ nm. Compare the answer with the result obtained by the program QMVGA ($E = 0.05655$ eV).

4. Consider the potential $V = -b\delta(x)$ to be a small perturbation on an infinitely deep square well with $V = 0$ for $|x| \leq L$ and $V = \infty$ for $|x| > L$. The delta function has its usual definition, with $\delta(0) = \infty$.
   (a) Find the ground-state energy by first order perturbation theory.
   (b) Find the coefficients a_{12} and a_{13} in the expansion of the perturbed wave function in a series of unperturbed wave functions:

$$u_1 = u_1 + a_{12}u_2 + a_{13}u_3 + \cdots$$

5. A particle is in a one-dimensional anharmonic oscillator potential given by

$$V(x) = c_1 x^2 + c_2 x^4, \quad \text{where } c_1 = 300 \text{ eV/nm}^2$$

   The particle's ground-state energy would be 6.000 eV if c_2 were zero. Find its ground-state energy to three significant figures if $c_2 = 20$ eV/nm^4.

6. Use the program QMVGA to find the lowest two energy levels for an electron in a one-dimensional square well where

$$V(x) = 0 \quad \text{for } 0.06 \text{ nm} < |x| < 0.30 \text{ nm};$$
$$V(x) = 0.5 \text{ eV} \quad \text{for } |x| < 0.06 \text{ nm}$$

   and $V(x)$ is infinite elsewhere. (In the program, set W = 0.6, WS = 0.12, and V = 0.5 eV.) Compare the result with the levels obtained from first-order perturbation theory, taking the 0.5 eV "step" to be a perturbation. (*Solutions*: $E_1 = 1.23$ eV; $E_2 = 4.20$ eV.)

7. Use Eq. (12.28) to compute the energies of the ground states of Li^+, B^{3+}, and C^{4+}. Compare with the results shown in Table 12.1. Does the accuracy improve as Z increases, as stated in the text?

8. Show that the minimum value for E'_1, using Eq. (12.37), is found for any value of Z when $Z' = Z - 5/16$. Use this value of Z' to compute the energy of the ground state of Li^+ and C^{4+}. How does the accuracy of your answer compare with what you found in Exercise 3?

9. (a) Verify the statement of Section 12.5 that the potential energy between an alpha particle and a nucleus of charge Z is given by $V(r) = 2.88Z/r$, where r is in femtometers and $V(r)$ is in MeV.
   (b) Verify that Eq. (12.53) follows from Eq. (12.52).

10. Use the constant from Example Problem 12.3 to find the half-life of ^{222}Ra ($Z = 88$) with $R = 9.027$ fm and $E = 6.675$ MeV.

CHAPTER 13

Atomic Spectroscopy

Atomic spectroscopy, one of the earliest applications of quantum theory, provides excellent illustrations of approximation methods and angular momentum in quantum mechanics. Although we cannot find an exact solution for any atom except hydrogen, we could, in principle, use perturbation methods to calculate the wave functions of any two-electron system and to determine the energy levels to any arbitrary degree of precision. And if we can do this for a two-electron system, we can do it for larger atoms or even molecules.

Obviously, such calculations would tax the power of the largest and fastest computers. But by knowing the features of the hydrogen atom wave functions and by invoking the Pauli exclusion principle, we can understand some characteristics of atomic and molecular spectra, and we can gain insight into the periodic table of the elements.

13.1 THE PERIODIC TABLE OF THE ELEMENTS

The periodic table (Figure 13.1) was first conceived (by Mendeleev in 1869) from observation of the periodicity in chemical properties of the elements with respect to their atomic masses.[1] A striking illustration of this periodicity is seen by plotting both the first and second ionization energies of the elements as functions of Z, as shown in Figure 13.2.

Notice that the second ionization energy follows the same pattern as the first but is displaced by one unit toward larger Z. The first ionization

---

[1] In Chapter 1 we saw that Moseley found the atomic *number*, rather than atomic *mass*, to be the variable that determined this periodicity.

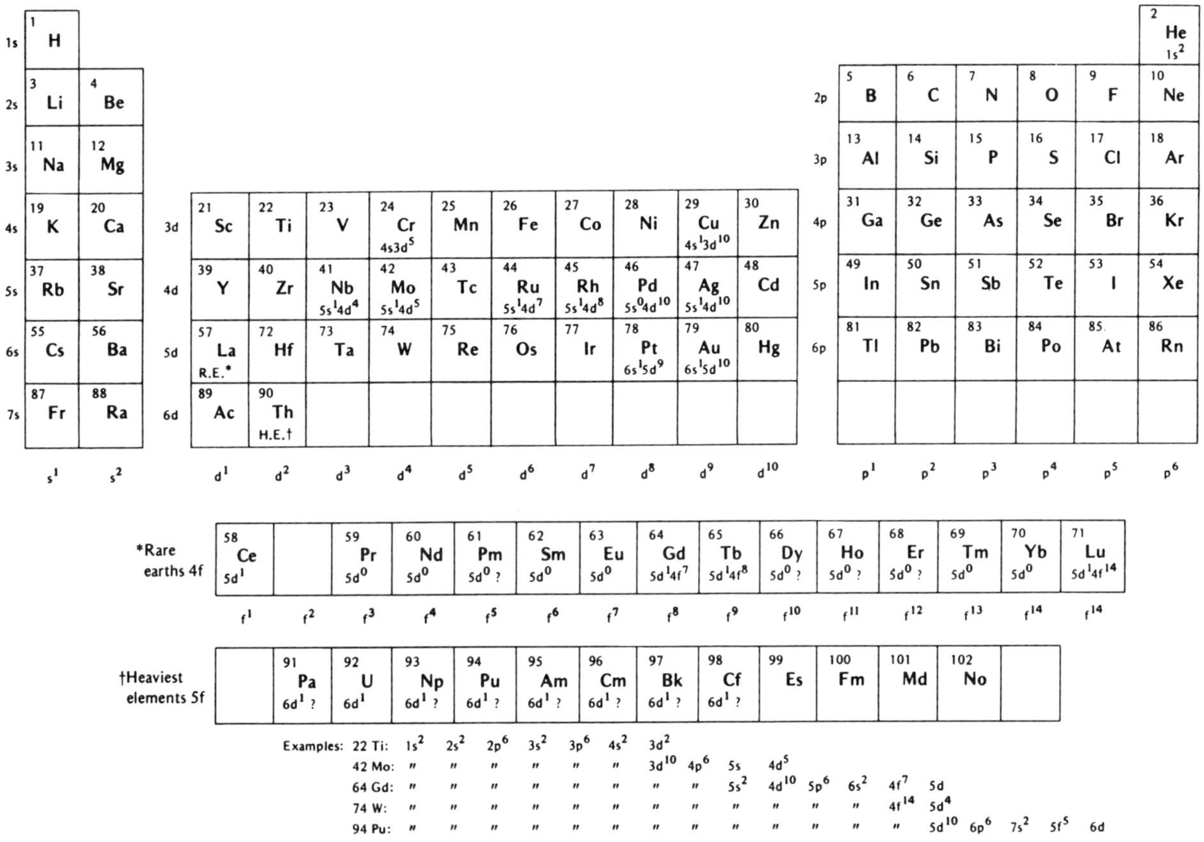

FIGURE 13.1 Electron configurations of the elements. Deviations from a regular filling of shells are indicated in the boxes for the elements involved. (From "Principles of Modern Physics" by R. B. Leighton, McGraw-Hill, New York, 1959. Used by permission.)

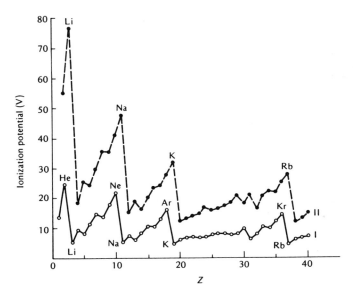

FIGURE 13.2   First (solid line) and second (dashed line) ionization energies of the elements as functions of Z, up to $Z = 40$.

energy increases with increasing Z, except for sharp drops at the alkali metals: lithium (Li, $Z = 3$), sodium (Na, $Z = 11$), potassium (K, $Z = 19$), and rubidium (Rb, $Z = 37$). The second ionization energies have corresponding drops at the values $Z = 4$, 12, 20, and 38.

These numbers are readily explained by the Pauli exclusion principle, on the assumption that each electron occupies a hydrogen-like state. Table 13.1 lists the number of states corresponding to the various quantum numbers. This number is equal to twice the number of possible values of m_l for each value of n, because there are two possible spin states.

Because the energy in a state depends primarily on the principal quantum number n, as it does in hydrogen, the eight states with $n = 2$ have considerably higher energy than the two 1s states. Consequently, the ionization energy for atoms with Z larger than 2 is considerably smaller than it is for $Z = 2$. From $Z = 3$ to $Z = 10$ this energy increases because of the increased charge resulting from the attraction between the additional protons and electrons. It drops again at $Z = 11$, as a 3s state is filled, and at $Z = 19$, as a 4s state is filled. (See Section 13.2 for the evidence that the 19th electron is in a 4s state.)

Why 4s, rather than 3d, which has much lower energy than the 4s state in hydrogen? The answer is that with 19 electrons, interactions between

TABLE 13.1   Electron States

n	l	m_l	Label	Number of states
1	0	0	1s	2
2	0	0	2s	2
2	1	$\pm 1, 0$	2p	6
3	0	0	3s	2
3	1	$\pm 1, 0$	3p	6
3	2	$\pm 2, \pm 1, 0$	3d	10
4	0	0	4s	2
4	1	$\pm 1, 0$	4p	6
4	2	$\pm 2, \pm 1, 0$	4d	10
4	3	$\pm 3, \pm 2, \pm 1, 0$	4f	14

electrons become comparable to the interaction between each electron and the nucleus. However, as a first approximation, we can assume that the 18 lowest-energy electrons form a symmetric distribution and that each set with a given n, l value forms an individual symmetric distribution, or "shell." The 19th electron, therefore, *as long as it is outside these shells*, feels an electric force of a single proton (the net charge of 19 protons and 18 electrons), because of Gauss's law and the symmetry of the situation. But an electron *inside* these shells would feel a stronger attractive force; when an electron is at a distance r from the nucleus, it feels only the effect of the charge distribution lying inside the radius r.

The effect on the 4s and 3d energies can be understood with the aid of Figure 13.3, which shows the radial probability densities for the 2p, 3d, and 4s states of hydrogen. Although these are not the probability densities for such states in other atoms, they have general features that show how it is possible for a 4s state to have lower energy than a 3d state in potassium.

Notice that the 4s electron, unlike an electron in a 3d state, penetrates to very small values of r, because this electron has no angular momentum. Thus in the 4s state the electron has a significant probability of being found in a region inside the shells formed by the other electrons. The result is that an electron in this state is acted upon by an attractive force greater than that produced by a single positive charge. In the case of potassium, the resulting energy is even lower than that of a 3d electron; consequently, in the atom's ground state the 4s state is occupied and the 3d state is not.

13.1.1 Filling of Shells

Each line in Figure 13.1 shows states in a single shell, the top line being the K shell (1s states), the next line the L shell (2s and 2p), and the third

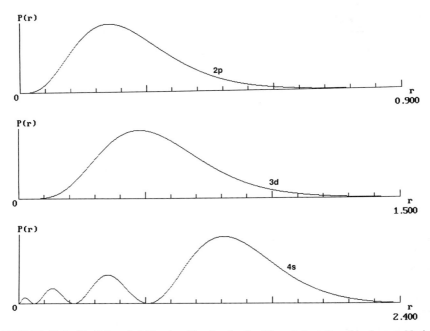

FIGURE 13.3   Radial probability densities for the 2p, 3d, and 4s states of hydrogen. Notice that the 3d state lies outside the 2p state, and the 4s state, whose maximum density is much farther out, nevertheless has a significant fraction lying inside the region occupied by the 3d state. Therefore, in a multi-electron atom, a 4s electron is subject to an attractive force corresponding to a larger effective positive charge.

line the M shell (3s and 3p, but *not* 3d, which has considerably higher energy, with energy comparable to that of a 4s state, for the reason given above).

The fourth line shows the N shell (4s, 3d, and 4p). This is the first "long period," which is followed by the O shell (5s, 4d, and 5p). The 4f electrons do not belong in the O shell, for the same reason that the 3d states do not belong to the M shell; their high angular momentum keeps them outside the smaller shells.

It is instructive to follow each step in filling the N shell as Z increases. Calcium ($Z = 20$) has two 4s electrons, and scandium ($Z = 21$) has two 4s and one 3d electron. You might think that each additional electron would be a 3d electron until Z reaches 30. But chromium ($Z = 24$) upsets the pattern; it has only *one* 4s electron and five 3d electrons. The copper atom also has only one 4s electron, with 10 3d electrons. Thus it is clear that the 3d and 4s levels must be very close in energy.

Confirmation of this small difference between 3d and 4s levels comes from comparison of the scandium *ion* with the calcium *atom*. Each of these contains 20 electrons, but Sc^+ has one 3d and one 4s electron, whereas Ca has two 4s electrons. Even more striking, the titanium atom has two 3d and two 4s electrons, and the vanadium ion V^+ has *four* 3d electrons. The point is that, as Z increases, the effect of the nuclear charge becomes larger, relative to the electron–electron interaction, and the relative energies become more like those in the H atom, where we know that the 3d level is much lower than the 4s level. Thus, as a rule, an increase in Z makes the 3d level drop, relative to the 4s level, and electrons go into the 3d rather than the 4s states.

13.2 ALKALI-METAL SPECTRA

Further principles of quantum mechanics are illustrated by alkali-metal spectra; these spectra are very similar to the spectrum of the hydrogen atom, but their differences are highly instructive. In these spectra, the clearest features involve transitions of the single electron that lies outside a noble-gas core. For example, sodium has a 3s electron with a neon electron core, and the excitation of this electron provides the most prominent features in the sodium spectrum. Figure 13.4 shows the energy levels responsible for this spectrum; notice that the 3d level lies considerably above the 4s level, in agreement with the above discussion.

Notice in Figure 13.4 that transitions occur only between states in adjacent columns. There is a *selection rule* that permits a transition only if the quantum number l changes by exactly 1 unit.[2] That is,

$$\Delta l = \pm 1 \qquad (13.1)$$

As a consequence of this rule, the spectra of sodium and other alkali metals consist of lines belonging to a number of different series. Each series is determined by the initial value of l and the final values of n and l in the transitions of that series. Long before the theory was developed, four series were identified and given the following names:

S̲harp series: Transitions from s̲ states to the 3p state
P̲rincipal series: Transitions from p̲ states to the 3s state
D̲iffuse series: Transitions from d̲ states to the 3p state
F̲undamental series: Transitions from f̲ states to the 3d state

---

[2] So-called "forbidden" transitions with $\Delta l = +2$ or -2 do occur, but far less frequently than the allowed transitions. See Chapter 14 for the theory of these transitions.

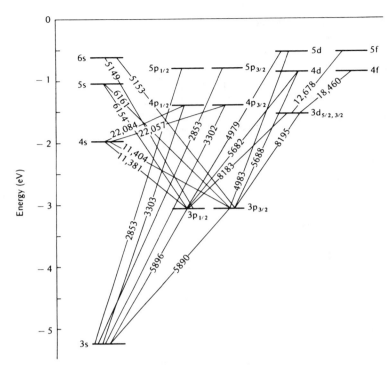

FIGURE 13.4 Energy levels of the sodium atom, labeled according to the hydrogen-like state of the odd electron. Notice the spin–orbit splitting ($3p_{1/2} - 3p_{3/2}$), which can be deduced from the indicated wavelengths.

You see that the transitions in each series obey the selection rule [Eq. (13.1)].[3]

The wavelengths reveal that each line has a *fine structure* that results from the splitting of each level, except the s levels, by the spin–orbit interaction. For example, the transitions $3p_{1/2} - 3s$ and $3p_{3/2} - 3s$ lead to the wavelengths of 589.6 and 589.0 nm, respectively, as a result of an energy difference of about 0.002 eV between the 3p states.

The potassium spectrum is similar, except that the ground state is 4s rather than 3s. In confirmation of the statements in Section 13.1, we now see that the ground state of potassium has to have an electron in a 4s

---

[3]The names of these series were based on the appearance of the spectral lines. For example, for some atoms the lines of the principal series are very intense, and the wavelengths in the fundamental series agree with an empirical equation which has a superficial resemblance to the Rydberg formula for hydrogen.

state, because the series with the shortest wavelength (and therefore the transition of highest energy) consists of doublet lines. This is possible only if the final state is unsplit, which requires that it be an s state.

13.3 SYSTEMS WITH TWO OR MORE ELECTRONS

The spectra of atoms in the second column of the periodic table—the so-called alkaline earths—are clearly more complicated than those of the alkali metals, as you can see by comparing Figure 13.5 with Figure 13.4. The greater complexity can be understood by analyzing the angular momentum states for two electrons.

In addition to the *spin–spin* and *spin–orbit* interactions previously discussed for helium, there is also an *orbit–orbit* interaction that occurs when both electrons have orbital angular momentum. To obtain a coherent picture of the effect of all these interactions, we must examine ways to combine the angular momenta of two electrons to form a resultant angular momentum vector.

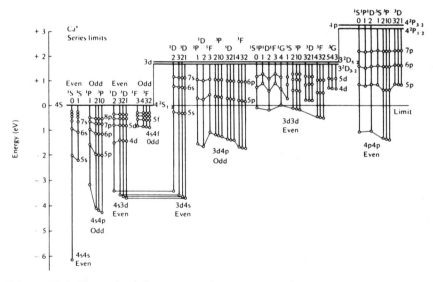

FIGURE 13.5 Energy-level diagram of calcium. States on the same vertical line all have the same value of *j*, shown at the top of the line. (From H. E. White, *Introduction to Atomic Spectra*. Copyright McGraw-Hill, New York, 1934. Used by permission of McGraw-Hill Book Company.)

13.3.1 Angular Momentum in Two-Electron Systems

The resultant angular momentum vector $\mathbf{J}$ may be written as the sum

$$\mathbf{J} = \mathbf{L}_1 + \mathbf{S}_1 + \mathbf{L}_2 + \mathbf{S}_2 \qquad (13.2)$$

Following the results of Section 10.1, we can add these four vectors by considering them to be two pairs, then adding these pairs. This can be done in two different ways:

    1. We write the resultant *orbital* angular momentum: $\mathbf{L} = \mathbf{L}_1 + \mathbf{L}_2$
and the resultant *spin* angular momentum: $\mathbf{S} = \mathbf{S}_1 + \mathbf{S}_2$
then combine these to find the resultant total angular momentum:
$$\mathbf{J} = \mathbf{L} + \mathbf{S}$$
This method is called the *LS coupling* approach; it is useful when the interaction between the two spins is greater than the spin–orbit interactions (as is true for light elements).

    2. We write the resultant angular momentum for one electron:
$$\mathbf{J}_1 = \mathbf{L}_1 + \mathbf{S}_1$$
the resultant angular momentum for the other electron: $\mathbf{J}_2 = \mathbf{L}_2 + \mathbf{S}_2$
then combine these to find the resultant total angular momentum:
$$\mathbf{J} = \mathbf{J}_1 + \mathbf{J}_1$$
This method, called *jj* coupling, is useful when the spin–orbit interactions are much stronger than the orbit–orbit or spin–spin interactions.

    The ideal *LS* (or Russell–Saunders) energy-level diagram is shown in Figure 13.6, where we see each stage of the splitting displayed for a 4p4d state of two electrons. There are three distinct effects:

    • Spin–Spin energy, which cause $S = 0$ states to have higher energy than $S = 1$ states. The reason for this has been discussed in Section 11.2.2.
    • Orbit–orbit energy, which splits states of a given value of S according to their values of L. This energy depends on L because the average distance between electrons with angular momentum vectors $\mathbf{L}_1$ and $\mathbf{L}_2$ is dependent on the value of $\mathbf{L}_1 \cdot \mathbf{L}_2$. This can be understood by a semiclassical picture. If $\mathbf{L}_1 \cdot \mathbf{L}_2$ is positive, the two electrons are revolving around the nucleus in the same direction, and they can stay farther apart and have lower energy than they would have if $\mathbf{L}_1 \cdot \mathbf{L}_2$ were negative. We know that when $\mathbf{L}_1 \cdot \mathbf{L}_2$ is positive, L is larger than it is when $\mathbf{L}_1 \cdot \mathbf{L}_2$ is negative. Thus *larger values of L occur with lower energy levels*, as shown in the third column of Figure 13.6.
    • Spin–orbit energy, magnetic in origin, is the smallest in these atoms. It depends on origin of the resultant $\mathbf{L}$ vector relative to the resultant $\mathbf{S}$ vector, that is, on $\mathbf{L} \cdot \mathbf{S}$. Therefore it depends on the $\mathbf{J}$ vector, making

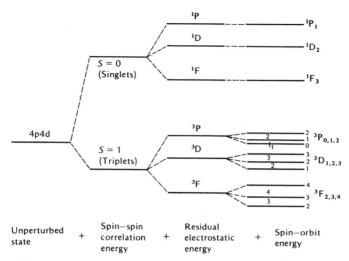

FIGURE 13.6 Fine-structure splitting of a level for a system containing a 4p and a 4d valence electron. (From R. B. Leighton, *Principles of Modern Physics*. McGraw-Hill, New York, 1959. Used by permission.)

states with higher J have higher energy (as they do in hydrogen). This is shown in the fourth column of Figure 13.6.

We therefore find that a given energy level is identified by *seven quantum numbers*: n_1, n_2, l_1, l_2, S, L, and J. In the preceding example, the first four quantum numbers are indicated by the characters 4p4d. The values of S, L, and J are then given in a somewhat indirect way, by a symbol of the form $^{2S+1}L_J$.

The value of L is represented by a capital letter, following the same spectroscopic code used for states of a single electron, the numerical value of $2S + 1$ is written as a superscript, and the value of J is written as a subscript. For example:

$^3P_2$ denotes a state with $S = 1$, $L = 1$, and $J = 2$.
$^1F_3$ denotes a state with $S = 0$, $L = 3$, and $J = 3$.
$^3F_1$ denotes an *impossibility*: a state with $S = 1$, $L = 3$, and $J = 1$. We
    know from the results of Section 10.1 that $|L - S| \leq J \leq |L + S|$

Therefore the complete designation of the lowest-energy state in Figure 13.6 would be $4p4d\,^3F_2$, and for this state $S = 1$, $L = 3$, and $J = 2$.

Example Problem 13.1 Given that the energy difference in spin–orbit splitting is directly proportional to $\mathbf{L} \cdot \mathbf{S}$, show that this energy difference between one level and the next is proportional to the larger of the J values of these two levels. This is called the *Landé interval rule*. For example, for states $^3F_4$ and $^3F_3$, the larger J value is 4; for states $^3F_3$ and $^3F_2$, the larger J value is 3. Therefore the splitting between $^3F_4$ and $^3F_3$ is $4/3$ times the splitting between $^3F_3$ and $^3F_2$.

Solution. The energy is proportional to $\mathbf{L} \cdot \mathbf{S}$, or $(|\mathbf{J}|^2 - |\mathbf{L}|^2 - |\mathbf{S}|^2)/2$, which in terms of the quantum numbers is proportional to $J(J + 1) - L(L + 1) - S(S + 1)$. The difference between levels with quantum numbers $J + 1$ and J, respectively, and equal values of L and S, is thus proportional to $(J + 1)(J + 2) - J(J + 1)$, which reduces to $2(J + 1)$. Q.E.D.

13.3.2 Limitations of the *LS* Coupling Model

The ideal *LS* coupling seen in Figure 13.6 is instructive, but it does not apply to a large number of atoms. The typical case has been shown in Figure 13.5. Compare these two figures by looking at the column marked 3d4p in the middle of the Figure 13.5. The triplets are in roughly the order shown in Figure 13.6, with ^3P lying above ^3D, which in turn is above ^3F, and with the states of higher J having the highest energy in each triplet. But the singlets are not all above the triplets, and ^1F is the highest, not the lowest. In the heaviest elements the spin–spin energy is not the dominant factor that it is in ideal *LS* coupling.

Figure 13.7 shows the gradual transition from *LS* coupling to the extreme case of *jj* coupling. In pure *jj* coupling, each state has well-defined values of j_1 and j_2, making it impossible to associate a definite value of L or S in a given state. This situation was discussed in Section 11.2.2 for single-electron states, for which a state with a known value of j cannot have a known value of l.

13.3.3 States with Positive Energy

Figure 13.5 shows some states that have *positive energy*. That is, the excitation energy is shared by two electrons, so that neither one individually has enough energy to escape from the atom. Nevertheless a calcium atom in this state can emit an electron by a process called autoionization.[4]

In this process, one of the excited electrons drops to a lower energy level and transfers its energy to the other electron, which then escapes

---

[4]Also known as the Auger (pronounced oh-zhay) effect. The spectral lines resulting from this process were called "anomalous" until Bohr and and Wentzel explained that they arose from positive energy states [*Phys. Z.* **24**, 106 (1923)].

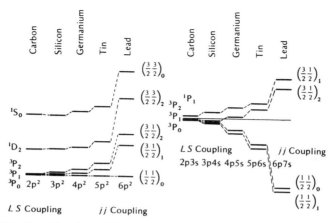

FIGURE 13.7 Transition from LS to jj coupling in the normal and first excited states of the carbon group of elements. [From "Introduction to Atomic Spectra" by H. E. White. Copyright McGraw-Hill, New York, 1934. Used by permission of the McGraw-Hill Book Company.]

from the atom. The energy is transferred directly from one electron to the other, without the creation of a photon.

Example Problem 13.2 Use the information in Figures 13.2 and 13.5 to estimate the energy of a free electron emitted by autoionization from a 4p6p state, if the final state is the free electron plus a Ca^+ ion in its ground state.

Solution. The first ionization energy of Ca is about 6 eV (Figure 13.2). Thus, for the ground state, $Ca^+ + e^-$ has 6 eV more energy than Ca. Figure 13.5 shows that the 4p6p state has 7.5 eV more energy than the ground state of Ca, and therefore 1.5 eV more energy than the lowest energy of $Ca^+ + e^-$. Thus energy conservation requires that $Ca^+ + e^-$ state to have 1.5 eV in kinetic energy.

13.3.4 The Exclusion Principle in States of Two Electrons

In Figure 13.5 you see that certain states do not appear. For example, although there is a $4s5s\,^3S_1$ state, there is no $4s4s\,^3S_1$ state. The reason is very simple; the $4s4s\,^3S_1$ state has the same quantum numbers n, l, and m_l for both electrons, and it is a symmetric spin state. Therefore it is totally symmetric and is not permitted by the Pauli exclusion principle. On the other hand, the $4s5s\,^3S_1$ state, in which the two electrons have different values for quantum number n, can be made antisymmetric in the space coordinates. (See Section 11.1.)

Notice that the combinations $4p4p\,^1P$, $4p4p\,^3S$, and $4p4p\,^3D$ are also missing. This illustrates a general rule for states of two electrons with equal values of l *as well as equal values of* n:

When $l_1 = l_2$, and L is *even* (states labeled S, D, G, etc.), the space part of the wave function is symmetric and the spin part is *anti*symmetric (singlet).

When $l_1 = l_2$, and L is *odd* (states labeled P, F, H, etc.), the space part of the wave function is *anti*symmetric and the spin part is symmetric (triplet).

This general connection between the symmetry of the wave function and the value of L is not immediately obvious, to say the least. Therefore let us see the effect of the exclusion principle on $4p4p$ states in particular. First let us examine the $4p4p\,^3D_3$ states (a set of seven states with $J = 3$). One of these states must have $m_J = 3$; if this state is impossible, then all seven are impossible. But m_J can equal 3 only if both electrons have $m_s = +1/2$, and both have $m_l = 1$. This means that, in violation of the exclusion principle, both electrons have the same set of values for *all* quantum numbers, since $n = 4$ for both of them and $l = 1$ for both.

We can now rule out the $4p4p\,^3D_2$ states and the $4p4p\,^3D_1$ states simply because you can't have $4p4p\,^3D_2$ states and $4p4p\,^3D_1$ states if you don't have $4p4p\,^3D_3$ states. They have to exist as a complete set, because they all have the same values of L and S. They differ only in the orientation of the vectors **L** and **S**. But if a certain **L** and **S** pair are not allowed in one orientation, that pair is not allowed at all, because orientation does not affect the symmetry properties of a state.

The other excluded states in the last column of Figure 13.5 are the $4p4p\,^1P$ and $4p4p\,^3S$ states. To show you why they are excluded, Table 13.2 lists all *allowed* pairs of values of m_l, given that m_l must be $+1$, 0, or -1 in a p state. The states that are symmetric in m_l must be antisymmetric in spin (having $S = 0$), and the states that are antisymmetric in m_l must be

TABLE 13.2 Allowed Pairs of Values of m_l

$S = 1$ (antisymmetric in m_l)	$S = 0$ (symmetric in m_l)
—	$+1, +1$
—	$0, 0$
—	$-1, -1$
$+1, 0$	$+1, 0$
$+1, -1$	$+1, -1$
$-1, 0$	$-1, 0$

symmetric in spin (having $S = 1$), in order to make the entire state antisymmetric. (Remember that these states are all symmetric in n.)

The set with $S = 1$ consists of nine states, or three triplets, which in Dirac notation can be written (omitting normalization) as

$$|+\rangle|+\rangle[|1\rangle|0\rangle - |0\rangle|1\rangle]$$

$$[|+\rangle|-\rangle + |-\rangle|+\rangle][|1\rangle|0\rangle - |0\rangle|1\rangle]$$

$$|-\rangle|-\rangle[|1\rangle|0\rangle - |0\rangle|1\rangle]$$

$$|+\rangle|+\rangle[|1\rangle|-1\rangle - |-1\rangle|1\rangle]$$

$$[|+\rangle|-\rangle + |-\rangle|+\rangle][|1\rangle|-1\rangle - |-1\rangle|1\rangle]$$

$$|-\rangle|-\rangle[|1\rangle|-1\rangle - |-1\rangle|1\rangle]$$

$$|+\rangle|+\rangle[|-1\rangle|0\rangle - |0\rangle|-1\rangle]$$

$$[|+\rangle|-\rangle + |-\rangle|+\rangle][|-1\rangle|0\rangle - |0\rangle|-1\rangle]$$

$$|-\rangle|-\rangle[|-1\rangle|0\rangle - |0\rangle|-1\rangle]$$

The top triplet has $m_L = +1$, the middle triplet has $m_L = 0$, and the bottom triplet has $m_L = -1$ (where m_L is the sum of the m_l values). Thus all of these states have the quantum numbers $L = 1$ and $S = 1$, making them ^3P states, with possible values of 2, 1, or 0 for the J quantum number. In general, it is not possible to identify each of these states with a definite value of J. However, the state $|+\rangle|+\rangle[|1\rangle|0\rangle - |0\rangle|1\rangle]$ is the only state with $m_J = 2$, so it must be identified as one of the $^3P_2$ states, and the state $|-\rangle|-\rangle[|-1\rangle|0\rangle - |-1\rangle|1\rangle]$, with $m_J = -2$, is another $^3P_2$ state. Four of the above states have m_J values of either $+1$ or -1. We can construct two linear combinations of these to form $^3P_2$ states and another two to form $^3P_1$ states. The remaining three states have $m_J = 0$, and can be combined to find the fifth $^3P_2$ state, the third $^3P_1$ state, and the single $^3P_0$ state.

In summary, the nine states with $S = 1$ can be sorted according to their values of J into three groups:

$^3P_2$, a set of five states, with $m_J = +2, +1, 0, -1$, and -2
$^3P_1$, a set of three states, with $m_J = +1, 0$, and -1
$^3P_0$, a single state, with $m_J = 0$

but we cannot, in general, associate one of these states uniquely with a single pair of values of m_S and m_L. With $S = 0$ the table shows six states, all singlets, which are written as

$$[|+\rangle|-\rangle - |-\rangle|+\rangle]|1\rangle|1\rangle$$
$$[|+\rangle|-\rangle - |-\rangle|+\rangle]|0\rangle|0\rangle$$
$$[|+\rangle|-\rangle - |-\rangle|+\rangle]|-1\rangle|-1\rangle$$
$$[|+\rangle|-\rangle - |-\rangle|+\rangle][|1\rangle|0\rangle + |0\rangle|1\rangle]$$
$$[|+\rangle|-\rangle - |-\rangle|+\rangle][|1\rangle|-1\rangle + |-1\rangle|1\rangle]$$
$$[|+\rangle|-\rangle - |-\rangle|+\rangle][|-1\rangle|0\rangle + |0\rangle|-1\rangle]$$

with values of $+2$, 0, -2, $+1$, 0, and -1, respectively, for m_L. Five linearly independent combinations of these states can be found to form the five $^1D_2$ states, and the sixth combination is then the $^1S_0$ state. Thus we observe the levels labeled $^1D_2$, $^1S_0$, $^3P_2$, $^3P_1$, and $^3P_0$ (15 states), and the 21 other possible states are forbidden by the exclusion principle (or to put it another way, by the requirement that the wave function be antisymmetric). The 21 forbidden states, comprising the sets ^3S, ^1P, and ^3D, are of course allowed when the two electrons have different n quantum numbers, as in $4p5p\,^3D$ (Figure 13.5).

The total number of 4p5p states is therefore 36. This is the product $2 \times 3 \times 2 \times 3$, since there are two spin states and three orbital states for each p electron.

Example Problem 13.3 For the $S = 0$ states listed in Table 13.2, find the linear combination that forms the $^1S_0$ state.

Solution. Of the six states shown, the only states that have $m_J = 0$ are the second and fifth ones. Therefore the solution is the combination $[|+\rangle|-\rangle - |-\rangle|+\rangle][a|1,0\rangle|1,0\rangle + b(|1,1\rangle|1,-1\rangle + |1,-1\rangle|1,1\rangle)]$ where, as usual, the numbers in brackets refer to the eigenvalues l and m_l for the respective particles. To find the values of a and b, we use the normalization requirement that $|a|^2 + |b|^2 = 1$, plus the facts that the eigenvalue of the L^2 operator must be zero and that $L^2 = (L_1 + L_2)^2 = L_1^2 + L_2^2 + 2L_1 \cdot L_2$. Thus we have

$$L^2[a|1,0\rangle|1,0\rangle + b(|1,1\rangle|1,-1\rangle + |1,-1\rangle|1,1\rangle)] \equiv L^2F = 0$$

$$\text{or} \quad \left[L_1^2 + L_2^2 + 2L_1 \cdot L_2\right]F = 0 \tag{13.3}$$

Since the eigenvalues of L_1^2 and L_2^2 are each $2\hbar^2$ (because $l = 1$), we know that the eigenvalue of $L_1 \cdot L_2$ must be $-2\hbar^2$ in order to satisfy Eq. (13.3).

Thus in terms of the component operators we have

$$[L_{x_1}L_{x_2} + L_{y_1}L_{y_2} + L_{z_1}L_{z_2}]F = -2\hbar^2 F \qquad (13.4)$$

From the matrices given in Eqs. (7.19) and the column vectors representing the states $|1,0\rangle$, $|1,1\rangle$, and $|1,-1\rangle$, we can now work out the results of these operations to find the ratio of b to a that satisfies Eq. (13.4). (These operations have already been worked out as the solution to Exercise 7, Chapter 7.) The results are:

$$L_{x_1}L_{x_2}F = (a\hbar^2/2)\{|1,1\rangle|1,1\rangle + |1,-1\rangle|1,-1\rangle$$

$$+ |1,1\rangle|1,-1\rangle + |1,-1\rangle|1,1\rangle\} + b\hbar^2|1,0\rangle|1,0\rangle \qquad (13.5a)$$

$$L_{y_1}L_{y_2}F = (a\hbar^2/2)\{-|1,1\rangle|1,1\rangle - |1,-1\rangle|1,-1\rangle$$

$$+ |1,1\rangle|1,-1\rangle + |1,-1\rangle|1,1\rangle\} + b\hbar^2|1,0\rangle|1,0\rangle \qquad (13.5b)$$

$$L_{z_1}L_{z_2}F = -\hbar^2 b\{|1,1\rangle|1,-1\rangle + |1,-1\rangle|1,1\rangle\} \qquad (13.5c)$$

Substituting into Eq. (13.4) and combining terms gives

$$2b\hbar^2|1,0\rangle|1,0\rangle + (a-b)\hbar^2\{|1,1\rangle|1,-1\rangle + |1,-1\rangle|1,1\rangle\} = -2\hbar^2 F$$

$$(13.6)$$

Equation (13.6) is satisfied only if $b = -a$. The normalization equation then gives

$$a^2 + b^2 = a^2 + a^2 = 1 \quad \text{or} \quad \boxed{a = 1/\sqrt{2}, b = -1\sqrt{2}}$$

13.3.5 Systems with Three or More Electrons

Although there are more possible *states* when there are more electrons, no additional *principles* are involved. For example, three-electron states may be constructed from combinations of two-electron states and single-electron states.

For the spin vector, we can combine the four two-electron spin states (three with $S = 1$, one with $S = 0$) with the two one-electron spin states in eight different ways. Four of these states form a "quartet" with $S = 3/2$. Each of these four states is symmetric in the interchange of *any two* electrons. The other four states have $S = 1/2$, and they consist of two spectra, where their splitting shows up in various series of the spectra of three-electron systems.

The orbital angular momentum combines with the spin angular momentum to form $\mathbf{J}$, and the quantum numbers L, S, and J obey the rule $|L - S| \leq J \leq L + S$, as before. The number of possible J values is equal to $2S + 1$ or $2L + 1$, whichever is smaller. For example, for $S = 3/2$ and $L = 3$, there is the quartet $^4F_{9/2}$, $^4F_{7/2}$, $^4F_{5/2}$, and $^4F_{3/2}$, and for $S = 3/2$ and $L = 4$, there is the quartet $^4G_{11/2}$, $^4G_{9/2}$, $^4G_{7/2}$, and $^4G_{5/2}$. These levels are identified by the spectral lines resulting from transitions between a G state and an F state or vice versa. Because of a selection rule requiring that the change in J must be ± 1 or zero, only nine $^4F-^4G$ transitions are allowed.

Because of the exclusion principle, the complexity of the spectra does not continue to increase with the number of electrons in a shell. The number of ways of arranging electrons in any given shell reaches a maximum when the shell is half filled and then decreases. For example, four electrons in a p shell can be combined in exactly the same number of ways as two electrons,[5] because the two *unoccupied* states combine in the same way as two occupied states for two electrons. [Because a filled shell always has zero angular momentum, the angular momentum of the four filled states must always equal (in magnitude) the angular momentum of the two electrons that would be added to fill the shell.]

13.4 THE ZEEMAN EFFECT

The state symbols just presented do not involve the m_J quantum number. In general, in empty space, the value of the z component of angular momentum cannot have any effect on the energy of an atom, because in empty space the direction of the z axis is arbitrary. The energy depends on the *magnitude* of the vector $\mathbf{J}$, not its direction.

However, when a magnetic field is present, the z axis is significant, being defined as the field direction. (See, the discussion of the Stern–Gerlach experiment, Section 6.5.) The interaction energy $\boldsymbol{\mu} \cdot \mathbf{B}$ between an *external* field $\mathbf{B}$ and the atom's magnetic moment $\boldsymbol{\mu}$ is thus proportional to the component μ_z, which in turn depends on J_z. Thus a

---

[5] We identified these 15 states in Section 13.3.4, but we could arrive at the number of states by combinatorial analysis as well. The "first" electron can occupy any of six different states, and the "second" can occupy any of the five remaining states. But interchanging the two electrons does not produce a different state. Therefore we have $6 \times 5/2 = 15$ ways in which we can have two states occupied out of six (just as we have 15 ways to draw two of six different cards from a hat.) See also Chapter 16, Quantum Statistics.

level with quantum number J splits into $2J + 1$ levels (one for each value of m_j) when a magnetic field is applied.

Pieter Zeeman observed as early as 1896 that spectral lines were split into three lines when they were viewed along a line of sight that was perpendicular to an applied magnetic field. He then drilled a hole through one pole piece of the magnet, and he found that the middle line was missing and that the other two lines were circularly polarized in opposite directions. H. A. Lorentz gave a classical explanation of this effect, which was based on the force exerted on an orbiting electron by a magnetic field and the resulting small change in the electron's orbital frequency. This change depends on the direction of motion in the orbit, and that explained why one observed frequency was circularly polarized in the clockwise direction and the other frequency was polarized in the counterclockwise direction.

However, this explanation was inadequate to explain observations of splitting into more than three lines. Thus splitting into more than three lines was dubbed the "anomalous" Zeeman effect, and splitting into three lines was the "normal" Zeeman effect. But both effects have a single quantum-mechanical explanation.

13.4.1 Energy Levels in a Magnetic Field

The Zeeman effect originates in splitting of energy levels, as illustrated in Figure 13.8. With four initial states and two final states, there are $4 \times 2 = 8$

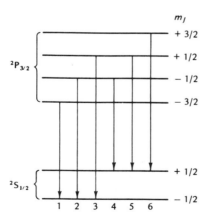

FIGURE 13.8 Splitting of $^2P_{3/2}$ and $^2S_{1/2}$ levels in a magnetic field, showing the values of m_j and the six allowed transitions.

conceivable transitions. However, selection rules require that

$$\Delta m_j = \pm 1 \text{ or } 0, \text{ except that } m_j = 0 \text{ to } m_j = 0 \text{ is forbidden if } \Delta J = 0$$

This rule restricts the number of possible transitions to six in Figure 13.8. The number of distinct *spectral lines* could therefore be six or fewer than six. If the energy splitting between the upper ($^2P_{3/2}$) levels is equal to the splitting between the lower ($^2S_{1/2}$) levels, then transition 1 involves the same energy difference as transition 4, 2 the same as 5, and 3 the same as 6, for *three spectral lines*, and we see the normal Zeeman effect. To compute the magnitude of the splitting we must evaluate $\boldsymbol{\mu} \cdot \mathbf{B}$ for each of the relevant atomic states. The magnetic moment $\boldsymbol{\mu}$ is the vector sum of the moments, which result from the electrons' spin and orbital motions respectively. (See Section 10.2 for $\boldsymbol{\mu}_{spin}$ and Section 6.5 for $\boldsymbol{\mu}_{orbital}$.)

$$\boldsymbol{\mu}_{spin} = \frac{-e\mathbf{S}}{m} \quad \text{and} \quad \boldsymbol{\mu}_{orbital} = \frac{-e\mathbf{L}}{2m} \tag{13.7}$$

Therefore, for the overall magnetic moment $\boldsymbol{\mu}$ we can write

$$-\boldsymbol{\mu} = \frac{e\mathbf{S}}{m} + \frac{e\mathbf{L}}{2m} = \frac{e}{2m}(2\mathbf{S} + \mathbf{L}) = \frac{e}{2m}(\mathbf{S} + \mathbf{J}) \tag{13.8}$$

We see from this equation that $\boldsymbol{\mu}$ *is not parallel to* $\mathbf{J}$ (unless $\mathbf{S} = 0$). This fact is directly responsible for the anomalous Zeeman effect, which classical theory could not explain because it is a consequence of electron spin. To understand this, we make the justifiable assumption that the magnitude of the *external* $\mathbf{B}$ field that produces the Zeeman effect is much smaller than that of the *internal* magnetic field that couples $\mathbf{L}$ and $\mathbf{S}$ and is responsible for spin–orbit energy. As shown in Figure 13.9, the internal field causes $\mathbf{L}$ and $\mathbf{S}$ to precess rapidly around $\mathbf{J}$, while the smaller external field $\mathbf{B}$ causes $\boldsymbol{\mu}$ to precess much more slowly around $\mathbf{B}$.

Because $\boldsymbol{\mu}$ can precess many times around $\mathbf{J}$ in the course of a single revolution of $\mathbf{J}$ around $\mathbf{B}$, we can determine the energy from the *average* value of $\boldsymbol{\mu} \cdot \mathbf{B}$ during one revolution of $\boldsymbol{\mu}$ around $\mathbf{B}$. During one revolution, the component of $\boldsymbol{\mu}$ perpendicular to $\mathbf{J}$ has an average value of zero, and the component of $\boldsymbol{\mu}$ parallel to $\mathbf{J}$ has a constant magnitude equal to $\boldsymbol{\mu} \cdot \mathbf{J}/|\mathbf{J}|$.

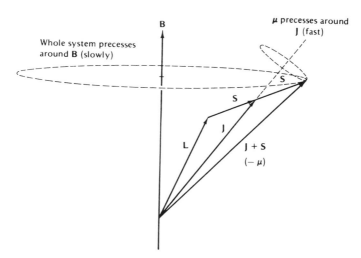

FIGURE 13.9 Relationships among the vectors **L**, **S**, **J**, **B**, and **μ**, and the precessions in a weak external magnetic field **B**.

The average value of the vector **μ**, denoted as **μ**$_{av}$, has magnitude **μ** · **J**/|**J**| and direction parallel to **J**, and thus it can be written as

$$\boldsymbol{\mu}_{av} = \frac{\boldsymbol{\mu} \cdot \mathbf{J}}{|\mathbf{J}|^2}\mathbf{J} \tag{13.9}$$

which you can verify has the correct magnitude and direction. The average value of the interaction energy is then

$$\langle \text{Energy} \rangle = -\boldsymbol{\mu}_{av} \cdot \mathbf{B} = \frac{(\boldsymbol{\mu} \cdot \mathbf{J})(\mathbf{J} \cdot \mathbf{B})}{|\mathbf{J}|^2}\mathbf{J} \tag{13.10}$$

which, using Eq. (13.8), can be written

$$\langle \text{Energy} \rangle = \frac{+e}{2m} \frac{[(\mathbf{J} + \mathbf{S}) \cdot \mathbf{J}][\mathbf{J} \cdot \mathbf{B}]}{|\mathbf{J}|^2} \tag{13.11}$$

or, if **B** is parallel to the z axis,

$$\langle \text{Energy} \rangle = \frac{eBJ_z}{2m} \frac{(|\mathbf{J}|^2 + \mathbf{J} \cdot \mathbf{S})}{|\mathbf{J}|^2} \tag{13.12}$$

We can express $\mathbf{J} \cdot \mathbf{S}$ in terms of $\mathbf{L}$, $\mathbf{S}$, and $\mathbf{J}$ by means of the equation

$$|\mathbf{L}|^2 \equiv (\mathbf{J} - \mathbf{S}) \cdot (\mathbf{J} - \mathbf{S}) = |\mathbf{J}|^2 + |\mathbf{S}|^2 - 2\mathbf{J} \cdot \mathbf{S} \qquad (13.13)$$

or

$$\mathbf{J} \cdot \mathbf{S} = \tfrac{1}{2}(|\mathbf{J}|^2 + |\mathbf{S}|^2 - |\mathbf{L}|^2) \qquad (13.14)$$

and therefore

$$\langle \text{Energy} \rangle = \frac{eBJ_z}{2m} \left\{ 1 + \frac{|\mathbf{J}|^2 + |\mathbf{S}|^2 - |\mathbf{L}|^2}{2|\mathbf{J}|^2} \right\} \qquad (13.15)$$

We now have an expression in terms of quantities with known eigenvalues. Substitution of these into Eq. (13.15) gives

$$\langle \text{Energy} \rangle = \frac{eB\hbar}{2m} m_J g \qquad (13.16)$$

where g is a pure number called the Landé g factor, given by

$$g = 1 + \frac{J(J + 1) + S(S + 1) - L(L + 1)}{2J(J + 1)} \qquad (13.17)$$

The letter g is from the term *gyromagnetic ratio*, a ratio relating magnetic moment to angular momentum. Equations (13.7) could be rewritten in terms of g as

$$\boldsymbol{\mu}_{\text{spin}} = g_e \frac{-e\mathbf{S}}{2m} \quad \text{and} \quad \boldsymbol{\mu}_{\text{orbital}} = g_l \frac{-e\mathbf{L}}{2m} \qquad (13.18)$$

where g_e, the intrinsic gyromagnetic ratio for the electron, equals 2, and g_l, the gyromagnetic ratio for orbital motion, equals 1. Therefore, when $L = 0$, we expect g to equal g_e, and we do find from Eq. (13.17) that $g = 2$ in that case. Similarly, when $S = 0$, then $J = L$ and $g = 1$. In both cases the magnetic moment is parallel to $\mathbf{J}$.

When neither L nor S is zero, g can have a fractional value *which is not necessarily between* 1 and 2. For example, if $L = 2$, $S = 1$, and $J = 1$, then $g = 1/2$. This counterintuitive result occurs because $\mathbf{L}$, $\mathbf{J}$, and $\mathbf{S}$ are vectors, and in this case the $\mathbf{L}$ vector and the $\mathbf{S}$ vector must be opposed, in order to make the resultant $\mathbf{J}$ vector smaller than the $\mathbf{L}$ vector. But since the ratio $\boldsymbol{\mu}_{\text{spin}}/\boldsymbol{\mu}_{\text{orbital}}$ is twice the ratio $|\mathbf{S}|/|\mathbf{L}|$, the magnetic moments are offset more than the angular momenta, making g smaller than 1 (the value that it has when S is zero).

From Eq. (13.16) we see that the energy is linearly dependent on m_J and therefore that the magnetic field B splits each level into $2J + 1$ equally spaced levels. The spacing between levels is proportional to g, so we find that the normal Zeeman effect occurs when the values of g are equal for both states in a transition. This of course is what happens when $S = 0$ for both levels, and that is what made it possible to provide a classical explanation of the effect. Thus the anomalous Zeeman effect depends on the existence of the nonclassical phenomenon of electron spin.

13.4.2 Consequences of Larger Magnetic Fields

If **B** is comparable in magnitude to the internal magnetic field of the atom, then the precession of **μ** around the z axis is too rapid for our averaging procedure to be valid. In that case, the equations following Eq. (13.9) are inapplicable. The splitting observed in such cases is called the *Paschen–Back effect*, in which the split levels are not equally spaced.

This effect can be analyzed by assuming that **L** and **S** precess *independently* around the direction of **B**. The vector **J** is not constant in this case. The transitions are then identified by the values of L_z and S_z rather than J, and the relevant quantum numbers are m_L and m_S rather than m_J. For further details, see the references in the next section.

ADDITIONAL READING

H. E. White, *Introduction to Atomic Spectra*, McGraw-Hill, New York (1934) is the definitive text on this subject, with many good examples.

Robert B. Leighton, *Principles of Modern Physics* (cited previously) also has a number of good examples of atomic spectra and tabulated values of quantum numbers.

EXERCISES

1. As elements of higher Z are produced, at what value of $Z > 100$ would you expect to find another noble gas? What assumptions must you make to arrive at an answer?

2. Deduce, from the wavelengths shown in Figure 13.4, the energy difference between the $3p_{1/2}$ and $3p_{3/2}$ states in sodium. Calculate this difference by using two different pairs of transitions, and show that the two calculations agree.

3. (a) At a temperature of about 1000 K, potassium vapor absorbs light very strongly at wavelengths of 769.9, 766.5, 404.7, 404.4, 344.8, 344.7, 321.8, and 321.7 nm. If all of these lines result from excitation of atoms from their ground state, to which series do these lines belong?

   (b) At higher temperatures another series appears, with wavelengths at 1252.3, 1243.4, 693.9, 691.1, 580.2, and 578.2 nm. From these wavelengths and the ionization energy of potassium (4.34 eV) construct an energy-level diagram for potassium and show that the lowest-energy state is an s state.

4. List the spectroscopic symbols for the allowed 4f4f states and for the 4f5f states (e.g., 4f5f $^3H_6$).

5. For the $S = 0$ states listed in Table 13.2, find the linear combination that forms the 4p4p $^1D_2$ state. *Hint:* This is the second linear combination of the same states that were combined in Example Problem 13.3; it is orthogonal to that combination. Verify that your answer is correct by computing the eigenvalue of L^2 for this state.

6. How many allowed 3d3d3d states are there? How many 3d4d4d states? How many 3d4d5d states?

7. Which of the following are possible states of three electrons in a 3p3p3d configuration? For each one that you do not choose, explain why it is not a possible state. $^4G_{11/2}$, $^2G_4$, $^2G_{9/2}$, $^4F_{7/2}$, $^2F_{7/2}$, $^4S_{3/2}$

8. Identify the nine allowed transitions between ^4G and ^4F states (e.g., $^4G_{9/2} \rightarrow {}^4F_{7/2}$), discussed in Section 13.3. How many transitions are allowed between two ^4F states (e.g., 2p3p5p $^4F \rightarrow$ 2p3p4p ^4F)?

9. Describe the polarization of light emitted in the normal Zeeman effect, if the magnetic field is parallel to the z axis. Consider six possibilities: photons emitted parallel to the field or perpendicular to the field, with frequency shift positive, negative, or zero.

10. (a) Find the energy shifts, in eV, in a magnetic field of 2.0 T, for $^3P_1$ states and for $^3D_1$ states.

    (b) Given a photon wavelength of 310.0000 nm in a transition from $^3P_1$ to $^3D_1$ in zero magnetic field, find the wavelengths of all photons from allowed transitions between $^3P_1$ states and $^3D_1$ states in a **B** field of 2.00 T.

11. Determine the Zeeman splitting of the levels $^4G_{7/2}$ and $^4F_{5/2}$ in a magnetic field of 0.2 T. Use your answer to predict the splitting of the 570-nm line that results from the $^4G_{7/2} \rightarrow {}^4F_{5/2}$ transition in scandium.

CHAPTER 14

Time–Dependent Perturbations and Radiation

In previous chapters you have seen *selection rules* related to transitions between atomic states. These rules are consistent with all observations of atomic spectra; transitions in which these rules would be violated are never (well, hardly ever) seen when we observe atomic spectra. And when "forbidden" transitions (those violating a selection rule) are seen, quantum-mechanical calculations can accurately predict how often they occur, relative to possible allowed transitions from the same initial state.

A rigorous calculation of transition probabilities requires that we go beyond the Schrödinger equation and quantize the classical equations for electric and magnetic fields (Maxwell's equations), so that we can deal with events involving single photons. Such calculations are beyond the scope of this book, so we shall settle for an approximation treatment that gives insight into the reason for selection rules and allows us to gain some understanding of devices such as the laser and the maser.

14.1 TRANSITION RATES FOR INDUCED TRANSITIONS

Transition rates for *induced transitions* can be calculated quite well by means of *time-dependent perturbation theory*. This theory follows the method of Section 12.1, time-independent perturbation theory, in that it assumes that the potential energy of the system contains a small perturbing term and that without this term the Schrödinger equation can be solved exactly. The difference is that the perturbing term, $v(x, y, z, t)$, is assumed to be applied for a limited time, and the result is that the system may make a transition from one unperturbed state to another.

Therefore, we write the time-independent Schrödinger equation (in one dimension for convenience) as

$$[H_0 + v(x,t)]\psi_n' = i\hbar \frac{\partial \psi_n'}{\partial t} \tag{14.1}$$

where, as before, H_0 is the Hamiltonian (or energy) operator for the unperturbed system, and the equation for any eigenfunction ψ_l of the unperturbed system is

$$H_0 \psi_l = i\hbar \frac{\partial \psi_l}{\partial t} \tag{14.2}$$

As in Section 12.2, we now rewrite the perturbed equation [Eq. (14.1)] by expanding the function ψ_n' as a linear combination of the unperturbed eigenfunctions ψ_l, with the important difference that the coefficients in the expansion are, in general, time dependent. Thus

$$\psi_n' = \sum_l a_{nl}(t)\psi_l \tag{14.3}$$

and Eq. (14.1) therefore becomes

$$[H_0 + v(x,t)] \sum_l a_{nl}\psi_l = i\hbar \frac{\partial}{\partial t} \sum_l a_{nl}\psi_l \tag{14.4}$$

After differentiating the right-hand series term by term, we obtain

$$[H_0 + v(x,t)] \sum_l a_{nl}\psi_l = i\hbar \left\{ \sum_l \left\{ \dot{a}_{nl}\psi_l + a_{nl}\frac{\partial \psi_l}{\partial t} \right\} \right\} \tag{14.5}$$

and eliminating the brackets on both sides gives us

$$H_0 \sum_l a_{nl}\psi_l + v(x,t) \sum_l a_{nl}\psi_l = i\hbar \sum_l \dot{a}_{nl}\psi_l + i\hbar \sum_l a_{nl}\frac{\partial \psi_l}{\partial t} \tag{14.6}$$

We now see from Eq. (14.2) that each term in the first series on the left is equal to the corresponding term in the second series on the right-hand side, so we can eliminate both series, reducing Eq. (14.6) to

$$v(x, t) \sum_l a_{nl}\psi_l = i\hbar \sum_l \dot{a}_{nl}\psi_l \tag{14.7}$$

We now proceed as in Section 12.1. We multiply each side of Eq. (14.7) by a particular function ψ_m^* and then integrate over all values of x (or over all space in the three-dimensional world) to obtain

$$\int \sum_l a_{nl}^* \psi_m^* v(x, t)\psi_l \, dx = i\hbar \int \sum_l \dot{a}_{nl} \psi_m^* \psi_l \, dx \tag{14.8}$$

which we integrate term by term as we did with similar expressions in Section 12.1. Because the wave functions ψ_{nl} are normalized and orthogonal to one another, the only nonzero term on the right-hand side is the one for which $l = m$, namely $i\hbar\dot{a}_{nm}$. Using the fact that the time dependence of ψ_m is

$$\psi_m = u_m(x)e^{-iE_m t/\hbar}$$

integrating the left-hand side of Eq. (14.8) term by term, and again using the normalization and orthogonality properties of the wave functions, we finally arrive at an exact equation for the time dependence of the coefficient a_{nm}:

$$\boxed{\dot{a}_{nm} = -\frac{i}{\hbar} \sum_l a_{nl}e^{-i(E_l - E_m)t/\hbar} v_{ml}} \tag{14.9}$$

where, as in Section 12.1, we use an abbreviation: $v_{ml} = \int u_m^* v(x, t)u_l \, dx$.

Equation (14.9) is still exact, but like Eq. (12.10), it contains too many unknown quantities to be useful as it stands. Therefore, we again assume the approximation that the eigenfunctions of the perturbed system differ very slightly from those of the unperturbed system. This permits us to make the approximation that all of the coefficients a_{nl} are very small, except for a_{nn}, which is approximately equal to 1. If we set a_{nn} equal to 1, and all other coefficients a_{nl} equal to zero, Eq. (14.9) becomes[1]

$$\dot{a}_{nm} = -\frac{i}{\hbar}e^{-i(E_n - E_m)t/\hbar} v_{mn} \tag{14.10}$$

[1]Although we set a_{nm} equal to zero for any m not equal to n, at least one of these coefficients must have a nonzero *time derivative* $\dot{a}_{nm}$, or else the perturbation would have no effect at all. The situation is analogous to dropping a ball; at the instant when you let go, the ball's velocity is zero, but its acceleration is not.

If $v(x, t)$ is known for all values of x and t, then it would appear that it is possible to integrate Eq. (14.10) and determine the behavior of the system, with an accuracy that is limited by the size of the neglected coefficients a_{nm}.

14.1.1 Comparison of Time-Dependent and Time-Independent Perturbation Theory

We use time-dependent perturbation theory with a *known set of states*, to calculate probabilities of transitions *between* levels. We know the possible states of the system because the time-dependent perturbation is assumed to continue for a limited time interval, after which the system reverts to one of its *unperturbed* states. Typically, we consider the following sequence of events:

1. At time $t = 0$, the system is in an unperturbed state $|\psi_n\rangle$, an eigenstate of the Schrödinger equation with energy eigenvalue E_n.

2. The perturbing potential is then "turned on." For $t > 0$, the system is then described by the *unperturbed* Schrödinger equation, with a different set of eigenstates $|\psi_n'\rangle$. If the perturbation is small and/or is applied for a very short time, the new state never differs greatly from the state $|\psi_n\rangle$.

3. The perturbing potential is turned off at time $t = t'$, and the system is again described by the *unperturbed* Schrödinger equation. The eigenstate may be the original state $|\psi_n\rangle$, or it may be a different state. In the latter case, we say that the perturbation has *induced a transition* to the new state $|\psi_m\rangle$. The probability that the system will be found in the state $|\psi_m\rangle$ is given by $|a_{nm}|^2$, which is the square of the coefficient of the wave function ψ_m in the expansion of wave function ψ_m in series of eigenfunctions of the original wave function.

Example Problem 14.1   A particle is in its ground state ($n = 1$, kinetic energy E_1, potential energy zero) in an infinitely deep one-dimensional square potential well. A constant perturbing potential $V = \delta$ is turned on at time $t = 0$. Find the probability that the particle will be found in the second excited state ($n = 3$) at time $t = t'$.

Solution. The probability that the particle will be found in the second excited state is $|a_{13}|^2$. The second excited state in this well has kinetic energy $E_3 = 9E_1$. Substitution into Eq. (14.10) gives $\dot{a}_{13} = -(i/\hbar)v_{31}e^{8E_1it/\hbar}$, and since $a_{13} = 0$ at time $t = 0$, we have $a_{13} = \int_0^{t'} \dot{a}_{13}\, dt = -(i/\hbar)v_{31}\int_0^{t'} e^{8E_1it/\hbar}\, dt$. But $v_{31} = 0$, because the functions ψ_1 and ψ_3 are orthogonal. Thus the probability is *zero*. To induce a transition in this potential, the perturbing potential must depend on x.

14.1.2 Dipole Radiation

Let us now apply Eq. (14.10) to atomic radiation, considering an electromagnetic wave as a perturbation that induces a transition between two atomic states. We begin with radiation whose wavelength is much greater than the diameter of the atoms involved, as is true for visible light. In this case, we can make the approximation that at a given time the entire atom feels the same field. That is, the field varies in time but not in space. This is known as the *dipole approximation*, for reasons that will be clear as we develop the equations.

We start with monochromatic (single frequency) radiation, *polarized along the x axis*. Thus the result depends on the x component of the electric[2] field $\mathscr{E}$, or $\mathscr{E}_x = \mathscr{E}_{0x} \cos \omega t$, where $\mathscr{E}_{0x}$ is constant. It is convenient to rewrite this field in complex form as

$$\mathscr{E}_x = \tfrac{1}{2}\mathscr{E}_{0x}(e^{i\omega t} + e^{-i\omega t}) \tag{14.11}$$

The perturbing potential $v(x,t)$ is the potential energy of an electron of charge $-e$ (e is not to be confused with $2.71828\ldots$) in this field, given by

$$v(x,t) = ex\mathscr{E}_x = \tfrac{1}{2}ex\mathscr{E}_{0x}(e^{i\omega t} + e^{-i\omega t}) \tag{14.12}$$

Inserting this expression into Eq. (14.10) gives

$$\dot{a}_{nm} = -\frac{ie\mathscr{E}_{0x}}{2\hbar}\{e^{i(E_m - E_n + \hbar\omega)t/\hbar} + e^{i(E_m - E_n - \hbar\omega)t/\hbar}\}x_{mn} \tag{14.13}$$

where the abbreviation x_{mn} represents the integral $\int_{-\infty}^{\infty} u_m^* x u_n \, dx$.

[We now see the reason for the expression "dipole" radiation. The dipole moment of an electric charge e at a distance x from the origin is given by ex. If the electron were in a stable quantum state $|\psi_n\rangle$, the dipole moment would depend on the probability density for the electron in that state, and thus would be given by the integral $e\int_{-\infty}^{\infty} u_n^* x u_n \, dx$. When there is a transition, the electron (before it is observed) is in a mixed state, with dipole moment given by the integral $e\int_{-\infty}^{\infty} u_m^* x u_n \, dx$. This integral is called the *dipole moment between states n and m*.]

We now assume that the $\mathscr{E}$ field is "turned on" at time $t = 0$ and "turned off" at time $t = t'$. Therefore we must integrate Eq. (14.13) on t between these two limits to find the transition probability from state n to state m, which is given by $|a_{nm}|^2$. From the initial condition $a_{nm}(0) = 0$

---

[2]At this point we neglect the magnetic field, because the magnetic interaction energy for a typical atomic electron is about 1% of the electric interaction energy.

unless $n = m$, we obtain $a_{nm}(t')$:

$$a_{nm}(t') = \frac{ie\mathscr{E}_{0x}}{2} x_{mn} \left\{ \frac{1 - e^{i(E_m - E_n + \hbar\omega)t'/\hbar}}{E_m - E_n + \hbar\omega} + \frac{1 - e^{i(E_m - E_n - \hbar\omega)t'/\hbar}}{E_m - E_n - \hbar\omega} \right\}$$

(14.14)

Rather than attempting to find the complicated general expression for the transition probability $|a_{nm}(t')|^2$, let us examine Eq. (14.14) to gain some insight. The first denominator is zero when $E_m - E_n = -\hbar\omega$; the second is zero when $E_m - E_n = +\hbar\omega$. It is reasonable to suppose that we can neglect the first term for frequencies such that $E_m - E_n \cong +\hbar\omega$. We can then simplify $|a_{nm}(t')|^2$ to:

$$|a_{nm}(t')|^2 = \left| \frac{e\mathscr{E}_{0x}}{2} x_{mn} \frac{1 - e^{i(E_m - E_n - \hbar\omega)t'/\hbar}}{E_m - E_n - \hbar\omega} \right|^2$$

(14.15)

or

$$|a_{nm}(t')|^2 = e^2\mathscr{E}_{0x}^2 |x_{mn}|^2 \frac{\sin^2[(E_m - E_n - \hbar\omega)t'/2\hbar]}{(E_m - E_n - \hbar\omega)^2}$$

(14.16)

If we define the frequency ω_{nm} by

$$\omega_{nm} = (E_m - E_n)/\hbar$$

(14.17)

and the function $f(\omega)$ by

$$f(\omega) = \frac{\sin^2[(\omega_{nm} - \omega)t'/2]}{\hbar^2(\omega_{nm} - \omega)^2}$$

(14.18)

then Eq. (14.16) becomes

$$|a_{nm}(t')|^2 = e^2\mathscr{E}_{0x}^2 |x_{mn}|^2 f(\omega)$$

(14.19)

Figure 14.1 shows a graph of $f(\omega)$ versus $|\omega - \omega_{nm}|$. The maximum value of $f(\omega)$ occurs when $\omega = \omega_{nm}$, the frequency at which the photon energy $\hbar\omega$ is equal to the difference between the energy levels E_n and E_m. This should come as no surprise, but the fact that other frequencies also contribute to transitions appears to violate the law of conservation of energy.

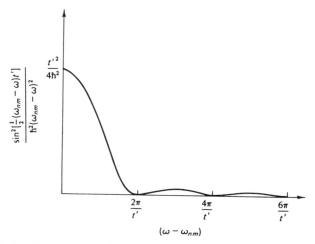

FIGURE 14.1 The function $f(\omega)$ of Eq. (14.18), plotted versus $|\omega - \omega_{nm}|$, in units of $2\pi/t'$; ω is the angular frequency of the perturbing electric potential.

However, when we consider the results of Section 2.4 we find that there is no violation. The fact that the perturbation exists for a limited time t' makes the frequency uncertain, just as confining a particle in a limited space makes its wavelength uncertain. If we let t' approach infinity in Eq. (14.18) we see that $f(\omega)$ approaches a delta function, becoming zero for all frequencies except $\omega = \omega_{nm}$. (For each point on the horizontal axis, the value of $\omega - \omega_{nm}$ is a multiple of $1/t'$; when t' becomes infinite, every point on the horizontal axis represents a value of zero. Thus when t' is infinite the value of $\omega - \omega_{nm}$ is zero over the entire curve.)

14.1.3 Uncertainty Relation for Energy and Time

When t' is finite, a Fourier analysis (Section 2.3) of the light wave would show a sinusoidal distribution of frequencies which is consistent with Eq. (14.18). Therefore Figure 14.1 agrees with the law of conservation of energy and with the condition that a photon of angular frequency ω has energy $\hbar\omega$. This figure shows that, for the overwhelming majority of transitions,

$$\hbar\omega - (E_m - E_n) \leq 2\pi\hbar/t' = h/t' \qquad (14.20)$$

Let us now consider the probable results of a measurement of the energy difference $E_m - E_n$ between two levels in a collection of identical atoms. We might measure this difference by applying a field of angular frequency ω to the atoms for a time t' and measuring the amount of energy that is absorbed. By repeating this procedure at different frequencies, we could plot a graph like Figure 14.1. But Eq. (14.20) shows that any observed photon energy $\hbar\omega$ can differ from the energy difference $E_m - E_n$ by as much as $2\pi\hbar/t'$, or h/t'. Denoting this difference as the uncertainty ΔE in the measurement, we have, for this special case,

$$\Delta E \cong 2\pi\hbar/t' = h/t' \quad \text{or} \quad t'\Delta E \cong h \qquad (14.21)$$

The time interval t' can be thought of as the uncertainty in the time of the measurement of the energy. Thus we have an uncertainty relation involving time and energy, just as we have a relation involving position and momentum. In the general case, the *uncertainty relation for energy and time* is written

$$\boxed{\Delta E \, \Delta t \geq \hbar/2} \qquad (14.22)$$

where ΔE is the uncertainty in a measurement of the energy of a system, and Δt is the time interval over which the measurement is made.

This relation, like the parallel relation $\Delta p_x \, \Delta x \geq \hbar/2$, is based upon the fact that a wave of finite length must consist of a superposition of waves of different frequencies. In the case of the energy measurement, the wave is that of a photon of the radiation field that induces the transition between energy levels, but the mathematics governing this wave is identical to that of a matter wave.

Notice that expression (14.22), like the position–momentum uncertainty relation, is an *inequality*. It becomes an equality only for the case of a Gaussian wave packet in either case. In the case of the energy–time relation, we could in principle turn the radiation field on and off gradually, instead of abruptly as described before, in order to generate a Gaussian shape for the wave packet.

14.1.4 Transition Probability for a Continuous Spectrum of Frequencies

In the general case, Eq. (14.14) cannot give the transition probability directly; it must be modified so that it represents a component in a continuous spectrum of frequencies. When we have a continuous spectrum, there cannot be an amplitude for a single frequency. Instead, there is an *energy density function* $\rho(\omega)$ such that the integral of $\rho(\omega)\,d\omega$ over the range from ω_1 to ω_2 is the energy density of radiation with frequencies between ω_1 to ω_2.

According to classical electromagnetic theory, the quantity $\varepsilon_0 \mathscr{E}_{0x}^2/2$ is the average energy density in the electromagnetic field given by Eq. (14.11).[3] Thus for radiation in a narrow range of frequencies $d\omega$, we have

$$\varepsilon_0 \mathscr{E}_{0x}^2/2 = \rho(\omega)\, d\omega \tag{14.23}$$

and $\mathscr{E}_{0x}^2$ can be replaced in Eq. (14.11) by $2\rho(\omega)\, d\omega/\varepsilon_0$. Then to find the total transition probability T_{nm} resulting from the entire spectrum of radiation, we integrate the resulting expression for $|a_{nm}(t')|^2$ over all frequencies, obtaining

$$T_{nm} = 2e^2 |x_{mn}|^2 \int_0^\infty \frac{\sin^2[(\omega_{nm} - \omega)t'/2]}{\varepsilon_0 \hbar^2 (\omega_{nm} - \omega)^2} \rho(\omega)\, d\omega \tag{14.24}$$

We can simplify Eq. (14.24) by assuming that $\rho(\omega)$ varies much more slowly than $f(\omega)$, and since $f(\omega)$ is symmetric, with a maximum at $\omega = \omega_{nm}$, we can replace $\rho(\omega)$ by the constant value $\rho(\omega_{nm})$, with little loss of accuracy. If we remove $\rho(\omega_{nm})$ from the integral, and we define $\alpha = (\omega_{nm} - \omega)t'/2$, Eq. (14.24) becomes

$$T_{nm} = -\frac{e^2 \rho(\omega_{nm})|x_{mn}|^2 t'}{\hbar^2 \varepsilon_0} \int_0^\infty \frac{\sin^2 \alpha}{\alpha^2}\, d\alpha \tag{14.25}$$

[The reader should verify that Eqs. (14.24) and (14.25) are equivalent, given the substitutions that were made.] The integral is standard, being equal to $\pi/2$, so the transition probability, *for radiation that is polarized in the x direction*, is

$$T_{nm} = -\frac{\pi e^2 \rho(\omega_{nm})|x_{mn}|^2 t'}{2\hbar^2 \varepsilon_0} \tag{14.26}$$

In the general case, when the radiation is randomly polarized, T_{nm} must include equal contributions from $|x_{nm}|^2$, $|y_{nm}|^2$, and $|z_{nm}|^2$, and we

---

[3]This includes the energy of the magnetic field which always accompanies an oscillating electric field.

have

$$T_{nm} = -\frac{\pi e^2 \rho(\omega_{nm})t'}{6\hbar^2 \varepsilon_0} \left\{ |x_{nm}|^2 + |y_{nm}|^2 + |z_{nm}|^2 \right\} \qquad (14.27)$$

where we have divided by 3 because the intensity is equally distributed among the three polarization directions.

The factor t' in Eq. (14.27) requires more scrutiny. It is logical that the probability of a transition should increase with time, but it cannot increase indefinitely, because a probability can never be greater than 1. Obviously the approximation breaks down at times t' such that T_{nm} is no longer small relative to 1.

If the radiation is *coherent* (for example, produced by a laser; see Section 14.4.2), then the perturbation is maintained for times t' that are quite long relative to incoherent radiation, such as that emitted by the sun. Therefore, atoms that are bathed in laser light can be perturbed for such a long time that Eq. (14.27) is no longer valid. (Analysis of such situations falls into the realm of *nonlinear optics*.)

On the other hand, incoherent radiation consists of brief pulses emitted by individual atoms at random; for example, the 2p level of the hydrogen atom survives for about 10^{-8} second. In such cases, the emitted pulse (one photon) can perturb another hydrogen atom for a time t' of the same order of magnitude. This time interval is sufficiently small to satisfy the condition $T_{nm} \ll 1$, and in those cases Eq. (14.27) is quite accurate. After the time t', the perturbation ends, and the hydrogen atom is in its original 1s state or is in the 2p state. The probability that it is in the 2p state is given by Eq. (14.27). This probability can be tested by simply observing that the second atom emits a photon in returning to the 1s state.[4]

14.2 SPONTANEOUS TRANSITIONS

In the previous section we found the probability that a system in one quantum state n will be induced to change to another state m, if it is acted upon by a perturbation such as radiation at the resonant frequency $v_{nm} = |E_n - E_m|/h$. But we still need a way to compute the probability of a *spontaneous* transition—a transition that occurs in the absence of a perturbation.

[4]The photon is emitted by the second atom in a random direction; thus it is easily distinguished from the original radiation.

Fortunately, there is a simple way to attack this problem. Even before quantum mechanics was developed, Einstein was able to derive the rate of spontaneous transitions from basic thermodynamics, given only the induced transition rate. He used the following argument.

14.2.1 Einstein's Derivation

Consider a collection of identical atoms which can exchange energy only by means of radiation. The collection is in thermal equilibrium inside a cavity whose walls are kept at a constant temperature. Because the system is in thermal equilibrium, each atom must be emitting and absorbing radiation at the same average rate, if one averages over a sufficiently long time (such as one second).

Define P_{nm} as the probability of an *induced* transition of a given atom from the state n to state m in a short time interval dt. This probability must be proportional to the probability p_n that the atom is initially in state n, multiplied by the transition probability T_{nm} for an atom in that state, which for unpolarized dipole radiation is given by Eq. (14.27). Thus

$$P_{nm} = T_{nm} p_n \qquad (14.28)$$

Guided by Eq. (14.27), we can now write a general equation for P_{nm} as

$$P_{nm} = A_{nm} \rho(\omega_{nm}) p_n \, dt \qquad (14.29)$$

which expresses the fact that T_{nm} is proportional to the radiation density $\rho(\omega_{nm})$, to the time interval dt (denoted by t' in Eq. (14.27)), and to other factors, incorporated into A_{nm}, which depend on matrix elements.

Equation (14.27) can be applied equally to an induced transition from state n to state m, or from state m to state n. From the symmetry of the equations, we know that $A_{nm} = A_{mn}$ and $\omega_{nm} = \omega_{mn}$. Therefore,

$$P_{mn} = A_{nm} \rho(\omega_{nm}) p_m \, dt \qquad (14.30)$$

Equation (14.30) gives the probability of an induced transition from state m to state n, while Eq. (14.29) gives the probability of an induced transition in the other direction, from state n to state m. The only difference between these probabilities is in the *occupation* probabilities p_n and p_m. These are not equal, because the probability that a state is occupied depends on its energy.

Let us say that state n has the lower energy; that is, $E_n < E_m$. Then $p_n > p_m$, and therefore $P_{nm} > P_{mn}$. There are more induced transitions from n to m than there are from m to n, simply because there are more atoms in state n to begin with. But *the atoms are in thermal equilibrium.* Therefore there must be other transitions, *spontaneous* ones, from m to n, to make the total probability of a transition from m to n equal to the probability of a transition from n to m. This means that

$$P_{nm} = P_{mn} + S_{mn} \tag{14.31}$$

where S_{mn} is the spontaneous transition probability, which may be written

$$S_{mn} = B_{mn} P_m \, dt \tag{14.32}$$

Notice that, unlike P_{mn} or P_{nm}, S_{mn} does not contain the factor $\rho(\omega_{nm})$, because a spontaneous transition, by definition, does not depend on external fields.

Substituting from Eqs. (14.29), (14.30), and (14.32) into (14.31), we have

$$A_{nm} \, \rho(\omega_{nm}) P_n = A_{nm} \, \rho(\omega_{nm}) P_m + B_{mn} P_m \tag{14.33}$$

or

$$B_{mn} = A_{nm} \, \rho(\omega_{nm}) \left\{ \frac{P_n}{P_m} - 1 \right\} \tag{14.34}$$

Remember that B_{mn} is associated with a spontaneous transition, so it does not really depend on the energy density of the electric field. However, we have derived this equation by relating B_{mn} to induced transitions in a cavity; therefore the energy density in the cavity has appeared in the result. We can eliminate $\rho(\omega_{nm})$ from the result by using the formula for the energy density in a cavity [Eq. (16.43), derived by Max Planck and derived in another way in Chapter 16]:

$$\rho(\omega) = \frac{\hbar \omega^3}{\pi^2 c^3 (e^{\hbar \omega / kT} - 1)} \tag{14.35}$$

You can verify that $\rho(\omega)$ has the correct dimensions (energy per unit frequency per unit volume). Inserting this expression into Eq. (14.34) yields

$$B_{mn} = A_{nm} \frac{\hbar \omega_{nm}^3}{\pi^2 c^3 (e^{\hbar \omega_{nm} / kT} - 1)} \left\{ \frac{P_n}{P_m} - 1 \right\} \tag{14.36}$$

To complete the derivation of B_{mn} we need the ratio of the occupation probabilities, p_n/p_m. This ratio is known from Boltzmann statistics (Appendix B and Section 16.1) to be given by

$$p_n/p_m = e^{\hbar \omega_{nm}/kT} \tag{14.37}$$

and therefore Eq. (14.36) becomes simply

$$B_{mn} = A_{nm} \frac{\hbar \omega_{nm}^3}{\pi^2 c^3} \tag{14.38}$$

Using Eqs. (14.27)–(14.29) to find A_{nm}, we find that the spontaneous transition probability in a short time interval dt, from state m to state n, is equal to $\lambda\, dt$, where λ, the probability per unit time for a transition to occur (also called the *decay constant*), is given by

$$\lambda = \frac{e^2 \omega_{nm}^3}{3\pi\varepsilon_0 c^3 \hbar} \left\{ |x_{nm}|^2 + |y_{nm}|^2 + |z_{nm}|^2 \right\} \tag{14.39}$$

A "short" time interval dt is one for which $\lambda\, dt \ll 1$. You should verify that λ has the proper dimensions (reciprocal time, to make $\lambda\, dt$ dimensionless).

When we speak of decay rates, we must remember that the transition is *observed* as a *discontinuous process*; a photon interacts with a measuring instrument as a discrete unit of energy. Here is the same wave–particle duality that has been discussed before.

The term "measuring instrument" has a very broad meaning; it is not necessarily an artifact of our own making. For example, thousands of years ago in Africa a nuclear chain reaction began spontaneously. No measuring instrument could count the decays, but the evidence remains at the site for all to see. (If a tree falls where nobody can hear it, does it make a sound? Of course it does; many animals can hear it.)

Now consider the time at which each atom makes a transition. This is determined by the interaction of a photon with the measuring instrument, which could be any kind of matter on which the photon could leave a lasting imprint. Thus nature makes the measurement without our intervention.

We can make an analogy to alpha-particle emission by a radioactive nucleus. (See Section 12.4.) The alpha particle in a uranium nucleus travels back and forth and has 10^{20} or more opportunities to escape during each second. If it does not escape, the atom is unchanged; a billion-year-old

^{235}U atom is identical to a ^{235}U atom that was just formed by any means whatsoever (perhaps by alpha decay of a ^{239}Pu atom). In a similar way, the oscillating dipole moment of a hydrogen atom in a mixed state, like that of Eq. (14.13), creates an electromagnetic field that, sooner or later, will transfer energy to another hydrogen atom. But the energy can only be transferred by a photon; as long as no transfer has taken place, the original hydrogen atom is unchanged, and thus the probability of decay in the next picosecond is not changed.

14.2.2 Energy Dependence of Transition Rates and Decay of Subatomic Particles

The factor ω_{nm} in Eq. (14.39) tells us that the decay constant is proportional to the cube of the energy difference between states n and m. This is true for any transition that is governed by the electromagnetic force, (where photons are involved). A striking example of this is given by comparing the mean lifetimes of two subatomic systems: the neutral pi meson (pion, π^0) and positronium (Ps), which is a bound state of a positron and an electron. In both cases the entire mass of the system disappears, and two photons (gamma rays) are emitted.[5] The total energy of these photons is equal to $\hbar\omega$, which in this case is just the original rest energy. The rest energy of Ps is twice the electron rest energy, or 1.02 MeV; the rest energy of the pion is 135 MeV. Therefore the value of ω_{nm} for the pion is about 130 times its value for Ps. Since the value of λ is proportional to ω^3, we would expect the ratio of the mean lifetime of Ps to the mean lifetime of the pion to be, neglecting other factors, to be about 130^3, or about 2×10^9. The lifetime of Ps is 1.24×10^{-9} s; that of the π^0 is 0.83×10^{-16} s. The ratio is about 1.5×10^9.

14.2.3 Exponential Decay Law

Given a collection of N_0 identical atoms in the first excited state at time $t = 0$, we expect to find that N of these atoms will remain unchanged when they are observed at time $t > 0$. Given the value of the decay constant λ, let us predict the value of N.

---

[5] A single photon normally cannot be emitted, because it carries momentum as well as energy. In the rest frame of the original particle, the initial momentum is zero. Therefore two photons of equal energy, traveling in opposite directions in this frame, are needed in order to conserve momentum. (Emission of only one photon is possible if a nearby atom participates in the process; the recoil of this atom can then satisfy momentum conservation. Emission of three photons is also possible. Further details are given in J. D. McGervey, *Introduction to Modern Physics*, Section 13.3.)

In any time interval dt, the probability of decay to the ground state will be $\lambda\,dt$ for each atom, so for N the number of decays will be $N\lambda\,dt$. Thus during any time interval dt the change in N will be

$$dN = -N\lambda\,dt \tag{14.40}$$

We can integrate this equation by separating the variables as follows:

$$dN/N = -\lambda\,dt$$

or

$$\ln N = -\lambda t + \text{constant of integration}$$

with the final result that

$$N = N_0 e^{-\lambda t} = N_0 e^{-t/\tau} \tag{14.41}$$

where N_0 is the number of excited atoms at time $t = 0$, and $\tau = 1/\lambda$ is the *mean lifetime* in the excited state. (The total lifetime spent by all the atoms in the excited state is given by $\int_0^\infty N_0 e^{-\lambda t}\,dt = N_0/\lambda$, or $N\tau$, as τ is defined here. Thus τ fits the definition of mean lifetime.)

The time t at which $e^{-\lambda t} = 1/2$ is called the *half-life*, written $t_{1/2}$. Thus, by definition,

$$e^{-\lambda t_{1/2}} = 1/2 \quad \text{or} \quad t_{1/2} = (\ln 2)/\lambda = 0.693/\lambda \tag{14.42}$$

The half-life is independent of the time t. No matter how long the atoms have been in the excited state, one can arbitrarily set t equal to zero and the number at that time equal to N_0, and Eq. (14.41) will hold, with $\lambda = 0.693/t_{1/2}$.

Figure 14.2 is a simulation that illustrates this point. Notice the random fluctuations in the number of atoms decaying in each time interval. The numbers are governed by the laws of Poisson statistics (Appendix A). The simulation was done by using a random-number generator to determine the time at which each atom decays (on the basis of a given half-life), then plotting the results.

14.2.4 Width of an Energy Level

Because an atom spends a limited amount of time in an excited state, the uncertainty relation for energy and time imposes a basic limitation on the accuracy with which the energy of a state can be determined. Therefore

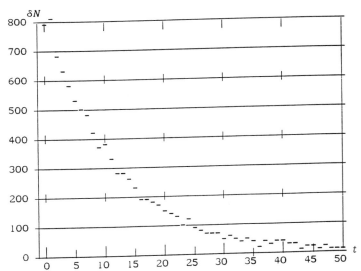

FIGURE 14.2 Computer simulation of the change δN in the number of identical atoms in a state whose half-life is 8 ns. The change in the first time interval is equal to $\lambda N_0 \, \delta t$; the total number in this state at $t = 0$ is $N_0 = 10,000$, and $\delta t = 1$ ns in this case. At $t = 8$ ns, the number emitted in time interval δt is approximately $\lambda N_0 \delta t / 2$. Observe the numbers at $t = 0, 8, 16,$ and 24 ns.

the atom, in making transitions between any two specific states, can emit or absorb photons that have a range of energies. The range of energies is inversely proportional to the mean lifetime of the excited state (just as the scale in Figure 14.1 is inversely proportional to the time t' during which the perturbing field is applied).

Each energy level in a given atom is defined only to the accuracy permitted by the uncertainty relation for energy and time. This uncertainty is called the *natural linewidth* of the state.[6] When the mean lifetime of a state is less than 10^{-17} second, you can be sure that the lifetime was found from the linewidth.[7]

---

[6] This does not mean that energy is not conserved. When a photon is absorbed and a photon is emitted, there is no overall gain or loss of energy, but the energy of *both* photons can vary from one event to another or from one atom to another.

[7] The neutral pion lifetime (8.3×10^{-17} s), found by observing its creation and decay as separate events, was listed in *The Guiness Book of Records* (Guiness Brewing Co.). See *Phys. Rev. D* **45**, #11 (1992) for more information on such measurements.

14.3 DERIVATION OF SELECTION RULES

To illustrate the principles discussed here in the simplest way, we shall derive some selection rules previously stated, without considering spin. We shall also see that there are rules that permit so-called forbidden transitions to occur. Most of these transitions are not strictly forbidden, but their transition rates are much slower than the rates of transitions that are allowed for dipole radiation.

14.3.1 Selection Rules Involving the Magnetic Quantum Number m

The simplest selection rule to derive is the rule for the magnetic quantum number m (Section 13.4):[8]

$$\Delta m = \pm 1 \text{ or } 0 \qquad (14.43)$$

This rule can be deduced by evaluating the dipole matrix elements x_{mn}, y_{mn}, and z_{mn}. When all of these matrix elements are zero for a particular pair of states, the transition is forbidden for dipole radiation. The key to the derivation is the wave function's dependence on ϕ—the factor $e^{im\phi}$. Given that the initial state function is proportional to $e^{im\phi}$ and the final state function is proportional to $e^{im'\phi}$, the matrix element z_{mn} is proportional to

$$\int_0^{2\pi} e^{-im'\phi} z e^{im\phi} \, d\phi = z \int_0^{2\pi} e^{i(m-m')\phi} \, d\phi \qquad (14.44)$$

because z, being equal to $r \cos \theta$, is independent of ϕ. The integral therefore vanishes as long as $m' \neq m$, because the upper limit gives the same result as the lower limit: $e^{2\pi i(m-m')} = e^0 = 1$. However, if $m' = m$, then the integrand becomes simply $d\phi$, and the matrix element is not zero. Thus, for dipole radiation:

> Transitions involving light polarized along the z axis are forbidden unless $\Delta m = 0$

Conversely, as we shall now prove, when $\Delta m = 0$, the emitted light must be polarized along the z axis. This is why the middle spectral line is missing when the light is viewed through a hole in the pole piece (Zeeman's

[8]This m is not to be confused with the letter m used to identify a quantum state!

experiment, Section 13.4); light is a transverse wave, and light that is polarized in the z direction cannot travel in that direction.

The matrix element x_{mn} is proportional to $\int_0^{2\pi} e^{-im'\phi} x e^{im\phi} \, d\phi$. With $x = r \sin \theta \cos \phi$, this integral reduces to $\int_0^{2\pi} e^{-im'\phi} \cos \phi \, e^{im\phi} \, d\phi$. This is evaluated easily by means of the substitution $\cos \phi = (e^{i\phi} + e^{-i\phi})/2$, making the integral proportional to the sum

$$\int_0^{2\pi} e^{i(m-m'+1)\phi} \, d\phi + \int_0^{2\pi} e^{i(m-m'-1)\phi} \, d\phi \qquad (14.45)$$

The first integral is zero unless $m - m' = -1$. The second integral is zero unless $m - m' = +1$. Thus the sum is zero unless $m - m' = \pm 1$.

In the same way, we can show that the matrix element y_{mn} is also zero unless $m - m' = \pm 1$. Therefore we conclude that, for dipole radiation:

> Transitions involving light polarized in
> the xy plane are forbidden unless $\Delta m = \pm 1$

Again, we see the evidence for this rule in the normal Zeeman effect. The shifted lines emitted in the y direction are polarized in the x direction; those emitted in the x direction are polarized in the y direction; and those emitted in the z direction are polarized in the xy plane.

Example Problem 14.2 Show that in the electric dipole approximation, probabilities of transitions induced by *circularly polarized* light are determined by matrix elements of the form $(x + iy)_{mn}$.

Solution. Circularly polarized light can be considered to be a superposition of two equal-amplitude plane waves, polarized at right angles and out of phase by 90°. For the xy plane we can write the two electric fields as

$$\mathscr{E}_x = \mathscr{E}_0 \cos \omega t \quad \text{and} \quad \mathscr{E}_y = \mathscr{E}_0 \sin \omega t$$

or, in exponential notation as

$$\mathscr{E}_x = \mathscr{E}_0 (e^{i\omega t} + e^{-i\omega t})/2 \quad \text{and} \quad \mathscr{E}_y = \mathscr{E}_0 (e^{i\omega t} - e^{-i\omega t})/2i \quad (14.46)$$

The potential energy is then

$$v(x, y, t) = -q \int \mathscr{E} \cdot ds = -(q\mathscr{E}_0/2)\{x(e^{i\omega t} + e^{-i\omega t}) - iy(e^{i\omega t} - e^{-i\omega t})\}$$

$$= -(q\mathscr{E}_0/2)\{(x - iy)e^{i\omega t} + (x + iy)e^{-i\omega t}\} \qquad \text{Q.E.D.}$$

14.3.2 Selection Rule for the Quantum Number l

This rule is derived from the dependence of the state function on the variable θ. This dependence, introduced in Eq. (6.52), is denoted by $P_l^{|m|}(\cos \theta)$ and called the *associated Legendre function*. If the initial state function contains the factor $P_l^{|m|}(\cos \theta)$, and the final state function has the factor $P_{l'}^{|m|}(\cos \theta)$, then it can be shown that in each matrix element the integral on θ vanishes unless $l' - l = \pm 1$. Thus we have the selection rule for dipole radiation:

Electric dipole transitions are forbidden unless $\Delta l = \pm 1$

This rule can be proved by using two formulas involving the *associated Legendre functions*:[9]

$$\cos \theta \, P_l^{|m|} = \frac{(l - m + 1) P_{l+1}^{|m+1|} + (l + m) P_{l-1}^{|m+1|}}{2l + 1} \qquad (14.47)$$

and

$$\sin \theta \, P_l^{|m|} = \frac{P_{l+1}^{|m+1|} - P_{l-1}^{|m+1|}}{2l + 1} \qquad (14.48)$$

Given these formulas, derivation of the selection rule is straightforward. (See Problem 6.)

14.3.3 Parity

There is a general rule for electric dipole transitions that does not depend on any specific eigenfunctions:

The parity of the eigenfunction must change in a electric dipole transition.

The reason is not hard to find. The matrix element is of the general form $\int_{-\infty}^{\infty} \psi_n^* x \psi_{n'} \, dx$, where ψ_n^* and $\psi_{n'}$ are the eigenfunctions for the two states involved in the transition. If these two states are both even or both odd, then the factor x makes the integrand an odd function, and the integral must be zero.

---

[9]The derivation of these formulas may be found in L. Schiff, *Quantum Mechanics*, pp. 72 and 258, McGraw-Hill, New York (1949).

14.3.4 Occurrence of Forbidden Transitions

All of the foregoing rules were derived for dipole radiation; it was assumed that the electric field was uniform over the dimensions of the atom. But if this assumption gives a zero transition probability, then we must go a step further and consider the possibility that a nonuniform field could induce a transition. We do this by writing the electric field as we did in Section 14.1, but including the space dependence as well as the time dependence. Let us consider transitions involving radiation polarized along the z axis, traveling in the x direction, so that the z component of $\mathscr{E}$ is given by:

$$\mathscr{E}_z = \mathscr{E}_{0z} \cos(kx - \omega t) = \mathscr{E}_{0z} \operatorname{Re}[e^{i(kx - \omega t)}]$$
$$= \mathscr{E}_{0z} \operatorname{Re}[e^{ikx}e^{-i\omega t}] \tag{14.49}$$

where Re[ ] denotes the real part of the quantity in brackets. We can expand the space-dependent factor in a power series to obtain

$$e^{ikx} = 1 + ikx + (ikx)^2/2! + (ikx)^3/3! + \cdots \tag{14.50}$$

14.3.5 Electric Quadrupole Transition Rate

The first (dipole) approximation was to cut off series (14.50) at the first term. The quadrupole approximation includes the second term, replacing e^{ikx} by $1 + ikx$. To compute the transition probability we must integrate $\vec{\mathscr{E}} \cdot \vec{ds}$ as before to obtain the potential energy. In this particular case, with $\vec{\mathscr{E}}$ lying along the z axis (the polarization direction), $\vec{\mathscr{E}} \cdot \vec{ds}$ is simply $\mathscr{E}\, dz$. The potential energy then is proportional to $(1 + kx)z$. For light, $k \leq 0.03$ nm^{-1}; for atoms, $x \leq 0.1$ nm. Thus kx is less than 10^{-2}, much smaller than the first term in Eq. (14.50). Nevertheless, we cannot neglect this quadrupole term, because when dipole transitions are forbidden, the quadrupole term is not forbidden, and it determines the entire transition probability.

In the present case, if dipole radiation is forbidden, the transition probability T_{mn} is proportional to the quadrupole matrix element given by $\int u_m^* xz u_n\, d(\text{volume})$, rather than the dipole factor $\int u_m^* x u_n\, dx$ derived earlier for dipole radiation. The quadrupole element differs from the dipole element primarily in the presence of the factor $kx < 10^{-2}$. This factor is squared in the transition rate; thus the typical dipole transition rate is more than 10^4 times that of typical quadrupole transition. Furthermore, because the matrix element is xz instead of z, there are different selection rules for quadrupole radiation. For example, $\Delta l = 2$ is allowed in this case. (See Problem 9.)

These equations hold for nuclear gamma radiation as well as atomic radiation. In this case $x \le 1$ fm, and the value of k for a 1-MeV gamma ray is about 0.01 fm^{-1}, making $kx \approx 10^{-2}$ as before. In a typical nucleus the mean lifetime for emission of a 1-MeV gamma ray is about 10^{-15} second for dipole radiation and about 10^{-11} second for quadrupole radiation.[10] The ratio is 10^4.

14.3.6 Magnetic Dipole Transitions

As you know, an electromagnetic wave has a magnetic field, and this field can also induce transitions. If the electric field has amplitude $\mathscr{E}_0$, the magnetic field amplitude $\mathscr{B}_0 = \mathscr{E}_0/c$. If **B** is parallel to the y axis and we neglect spin, an electron in such a field has energy $-\boldsymbol{\mu} \cdot \mathbf{B}$. Substituting the values of μ and B gives us $(eL_y/2m)(\mathscr{E}_0/c)$ for the energy, and the matrix element is

$$\frac{e\mathscr{E}_0}{2m_e c} \int Y_{l',m'} L_y Y_{l,m} \, d(\text{volume}) \equiv \frac{e\mathscr{E}_0}{2m_e c} \langle l', m' | L_y | l, m \rangle \quad (14.51)$$

determines the magnetic dipole transition probability.

Let us compare expression (14.51) with the corresponding expression

$$\frac{e\mathscr{E}_0}{2} \int \psi_\mu^* x \psi_\nu \, d(\text{volume}) \equiv \frac{e\mathscr{E}_0}{2} \langle \mu | x | \nu \rangle \quad (14.52)$$

for the electric dipole transition probability. Dividing expression (14.51) by expression (14.52), we have

$$\frac{\langle l', m' | L_y | l, m \rangle}{m_e c \langle \mu | x | \nu \rangle} \quad (14.53)$$

For an order-of-magnitude estimate of this ratio, we may set the numerator equal to a typical value for orbital angular momentum, i.e., about $\hbar$, and we set the value of $\langle \mu | x | \nu \rangle$ at the value of a typical atomic radius, or about 0.05 nm. The value of $\hbar/mc$ is about 2×10^{-4} nm, so the order of magnitude of the ratio is about 10^{-2}. This is approximately the same as the ratio of electric quadrupole to electric dipole matrix elements. Thus magnetic dipole transition rates are comparable to electric quadrupole rates, and about 10^{-4} times a typical rate for an electric dipole transition.

---

[10] The dipole transition is associated with a change of one unit ($\hbar$) in the angular momentum quantum number of the nucleus; the quadrupole transition with a change of two units ($2\hbar$). The selection rules are the same as the rules for atomic radiation.

Selection rules for magnetic dipole transitions can be found by writing L_y in terms of stepping operators (Section 7.4):

$$L_y = \left[(L_x + iL_y) - (L_x - iL_y)\right]/2i \qquad (14.54)$$

so the matrix element in (14.51) becomes $\langle l', m' |(L_x + iL_y)| l, m \rangle - \langle l', m' |(L_x - iL_y)| l, m \rangle$. Because $L_x + iL_y$ is a raising operator, it changes $|l, m\rangle$ into $|l, m + 1\rangle$. Therefore the first matrix element is zero unless $l' = l$ and $m' = m + 1$. Similarly, the second matrix element is zero unless $l' = l$ and $m' = m - 1$. Thus we have the selection rules

---

$\Delta l = 0$ and $\Delta m = \pm 1$ for magnetic dipole transitions.

14.3.7 Totally Forbidden Transitions

Although we have seen that various types of radiation have different selection rules, there is a general rule that applies to all of these types (if we neglect spin):

All radiative transitions between $l = 0$ and $l = 0$ are forbidden.

Notice that this rule applies only to *radiative* transitions, i.e., those that involve emission or absorption of photons. The rule is related to the fact that electromagnetic radiation always carries angular momentum. Each photon has an angular momentum of $\hbar$, and if one were emitted from a system that had no angular momentum either before or after the event, the law of conservation of angular momentum would be violated.

Notice also that this rule *does* permit transitions with $\Delta l = 0$, as long as $l \neq 0$. In a transition of this type (e.g., from $l = 1$ to $l = 1$) with emission of a photon, because angular momentum is a vector. When the photo carries away its angular momentum, the direction of the angular momentum of the atom can change even if the magnitude of this vector does not change.

14.3.8 Effect of Spin

Remember that the preceding discussion took no account of electron spin. It is possible for a photon to be emitted if its angular momentum is provided by a flip of the spin of an electron. A famous example of this is the emission of 21-cm-wavelength photons from hydrogen atoms. This radiation is associated with the 1s-to-1s transition. (See Section 10.3.)

14.4 EXAMPLES OF INDUCED AND SPONTANEOUS TRANSITIONS

Quantum mechanics often takes us in unexpected directions. It has given us the laser with its numerous applications and many observations that defy common sense. But some measurements confirm that "a watched pot never boils."

14.4.1 Pot Watching

A dramatic example of the effect of observing a quantum system has been reported in the study of a system of 5000 identical atoms, all of which were initially in their lowest energy state.[11] As these atoms are excited from the lowest state (labeled $|1\rangle$) to a higher energy (state $|2\rangle$) by radiation of the appropriate frequency, the state function of each atom becomes a superposition of the two states, given by

$$|u\rangle = a|1\rangle + b|2\rangle \tag{14.55}$$

with the usual normalization requirement that $|a|^2 + |b|^2 = 1$. This requirement is satisfied by writing $a = \cos(\pi t/2\tau)$ and $b = \sin(\pi t/2\tau)$, where τ is the time required for all of the atoms to be excited to the higher level.

We see that at $t = 0$, $|u\rangle = |1\rangle$, and at $t = \tau$, $|u\rangle = |2\rangle$. We also see that, at $t = 2\tau$, $|u\rangle = |1\rangle$ again. Thus the atoms oscillate between the two states, at a rate that depends on the frequency of the applied radiation; the value of τ is inversely proportional to this rate. This oscillation can continue as long as the system is not disturbed, i.e., if the "pot" is not "watched."

Boiling We define "boiling" as the situation in which all of the atoms have been raised to state $|2\rangle$. Thus it takes a time τ for the system to "boil," *if the pot is not watched.*

Watching the Pot Before the pot boils, at times $0 < t < \tau$, each atom is in the mixed state of Eq. (14.55). To observe the atoms at time $t < \tau$ (to watch the pot), the technique is to try to excite them to a third

---

[11] W. Itano, D. Heinzen, J. Bollinger, and D. Wineland, *Phys. Rev. A* **41**, 2295 (1990).

level (level 3) with radiation that could induce the transition from level 1 to level 3, but not from level 1 to level 2. Then the atoms in level 1 can be detected from the radiation emitted when they returned to level 1 after being excited.

For example, suppose that this is done at a time when $|a|^2 = 1/2$. If the radiation is sufficiently intense, it will be found that 50% of the atoms respond to this radiation, are excited to level 3, and return to level 1. For these atoms, the wave function now changes to the function $|1\rangle$ as a result of this observation; for the other 50%, the wave function is now $|2\rangle$, because the state was known to be a 50–50 mixture of the two states at this time. This means that there are two *distinct populations*, each in a distinct state, either $|1\rangle$ or $|2\rangle$ but not mixed, immediately after the observation.

As time goes on, the population that was observed in state $|1\rangle$ will eventually go to state $|2\rangle$; all of them will be in state $|2\rangle$ at time τ *after the observation*. Meanwhile, the entire population observed in state $|2\rangle$ will arrive in state $|1\rangle$ at time τ after the observation. *There will never be a time when all of the atoms will be in level 2; the pot will never boil.* This is true no matter when the observation is made, if it is done at any time t after $t = 0$ and before $t = \tau$. Exercise 9 gives other examples.

14.4.2 The LASER

LASER is an acronym for light amplification by stimulated emission of radiation. However, the significant characteristic of a laser is not that it amplifies but rather that its beam is *highly directed*, being able to travel great distances with little spreading, and it is *coherent*, consisting of a long wave with a constant wavelength. Coherence comes from the fact that the wave is initiated by a single photon that is "cloned" many times (in contrast to ordinary light, which comes from spontaneous emissions at random times).

Three conditions must be met in order for a laser to function. There must be

- In the material of the laser, a pair of energy levels that can provide transitions of the desired frequency
- A way to create a *population inversion* in the laser, so that the higher of the two levels is more heavily populated than the lower level, and consequently stimulated emission will occur more frequently than absorption
- A way for the photons to remain in the laser long enough to stimulate emission of intense light of the same frequency

Population inversion is often achieved by involving *three* energy levels $E_1 < E_2 < E_3$, with (for example) $l = 0$ for level 1, $l = 2$ for level 2, and $l = 1$ for level 3. Population inversion occurs between levels 1 and 2, as follows:

High-intensity radiation, of frequency $\nu_{13} = (E_3 - E_1)/h$, excites atoms rapidly from level 1 to level 3, after which they decay spontaneously to level 2. A further spontaneous decay to level 1 is forbidden by the selection rule on Δl, because $\Delta l = -2$ for this transition. Thus the population in level 2 increases until it is greater than that of level 1, and population inversion is achieved.

Although this $\Delta l = -2$ transition is forbidden, some spontaneous (electric quadrupole) transitions from level 2 to level 1 do occur; when this happens, each emitted proton can cause induced transitions from level 2 to level 1 (or from level 1 to level 2). These induced transitions do result in amplification, because of the population inversion; a photon is more likely to cause an induced transition from level 2 to level 1, with emission of a second identical photon, rather than be absorbed in inducing a transition from level 1 to level 2.

The only remaining requirement is to have a large enough number of such events. This is achieved by placing parallel plane mirrors at each end of the laser. A photon emitted in a spontaneous transition may travel in any direction, but if it happens to travel in a direction nearly perpendicular to the two mirror faces it will be reflected back and forth many times, stimulating the emission of identical photons which travel in the same direction. (See Figure 14.3).

Coherence Length If the initial photon direction were precisely perpendicular to both mirrors, and level 2 continued to be more populated than level 1, then the laser beam would be a single continuous wave, extending over an unlimited distance. This is not possible in practice, because it is not possible to make the two mirrors *exactly* parallel to each other. Even if a photon's direction were exactly perpendicular to one mirror, after reflection its direction would not be exactly perpendicular to the other mirror, and after a number of such reflections the entire wave train would strike the side of the laser. This wave would of course be replaced by the stimulated emissions from another photon. So in any laser there is a limit to the length of a single coherent wave train; this limit is called the *coherence length*.

Divergence of a Laser Beam Another limit on laser light is determined by the wave nature of light. The beam must diverge by an angle θ of approximately λ/D or more, where D is the initial diameter of the beam.

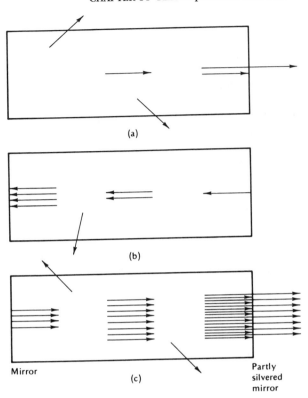

FIGURE 14.3 Buildup of coherent radiation in a laser. (a) Spontaneous emission produces photons traveling in various directions. Some escape; occasionally one travels nearly perpendicular to the mirrors, and it is able to stimulate the emission of a second photon. One of the two escapes; the other is reflected. (b) Reflected photon stimulates emission of more photons. (c) Intensity of photons builds up, while some continue to escape from the partially transmitting end mirror.

Thus a beam with $\lambda \approx 600$ nm, coming from a 1-mm-diameter laser, diverges by an angle of at least 0.6 milliradians. The spreading of the beam can easily be demonstrated in a classroom with a helium–neon laser; the beam spreads to about 6 mm in diameter after it travels 10 meters.

14.4.3 Oscillations of K Mesons and Neutrinos

The neutral K meson leads a double life, as described in Section 7.4, where we saw that it is a superposition of two states, $|K_1\rangle$ and $|K_2\rangle$, with different lifetimes. Each of these states in turn is a superposition of $|K^0\rangle$ and its antiparticle, $|\overline{K}^0\rangle$. Thus we can write, as before,

$$|K^0\rangle = (|K_1\rangle + |K_2\rangle)/\sqrt{2} \tag{7.33}$$

as well as

$$|\overline{K}^0\rangle = (|K_1\rangle - |K_2\rangle)/\sqrt{2}$$

and conversely

$$|K_1\rangle = (|K^0\rangle + |\overline{K}^0\rangle)/\sqrt{2} \tag{7.34a}$$

$$|K_2\rangle = (|K^0\rangle - |\overline{K}^0\rangle)/\sqrt{2} \tag{7.34b}$$

Section 7.4 did not deal explicitly with the time-dependent part of the state function $|\psi\rangle$, which we now address. For a particle that is a K^0 at time $t = 0$, we find that

$$|\psi\rangle = \left[|K_1\rangle(e^{-i\omega_1 t})(e^{-t/2\tau_1}) + |K_2\rangle(e^{-i\omega_2 t})(e^{-t/2\tau_2})\right]/\sqrt{2} \tag{14.56}$$

where $\hbar\omega_1$ is the rest energy of the K_1, $\hbar\omega_2$ is the rest energy of the K_2, τ_1 is the mean lifetime for K_1 decay, and τ_2 is the mean lifetime for K_2 decay. The mean lifetime is multiplied by 2 in the denominator of each exponent, because the *square* of the amplitude determines the probability density, whose time dependence must be $e^{-t/\tau}$.

Because $\tau_2 \gg \tau_1$, we find that the state $|\psi\rangle$, after it is created in a nuclear reaction, eventually turns into state $|K_2\rangle$ (a mixture of $|K^0\rangle$ and $|\overline{K}^0\rangle$). This fact is observed experimentally from the fact that the K_2 meson does not decay into two pions as the K_1 does. (See Figure 7.2 and the discussion in Section 7.4.)

Mass Difference between the K_1 and the K_2   The K^0 and $\overline{K}^0$ must have the same mass, because one is the antiparticle of the other. However, this is not true for the K_1 and the K_2; their mass difference has been measured by observing the oscillation that is implied by the factors $e^{-i\omega_1 t}$ and $e^{-i\omega_2 t}$ in Eq. (14.56). The measurement consists of a careful study of the fraction of $\overline{K}^0$ particles as a function of time. The probability, as a function of time, that a particle which is initially a K^0, with state function $|\psi\rangle$, will react as a $\overline{K}^0$ is found from the absolute square of the scalar product of the state vector $|\psi\rangle$ and the state vector $|\overline{K}^0\rangle$, given by $|\langle\psi|\overline{K}^0\rangle|^2$.

Using the orthogonality of $|K_2\rangle$ and $|K_1\rangle$, we find that

$$|\langle \psi | \overline{K}^0 \rangle|^2 = \frac{1}{4}\left\{ e^{-t/\tau_1} + e^{-t/\tau_2} - 2(e^{-t/2\tau_1})(e^{-t/2\tau_1}) \cos\frac{\Delta mc^2 t}{\hbar} \right\}$$

(14.57)

where Δm is the difference between the masses of the K_1 and K_2 mesons. The measurements have shown that $\Delta mc^2 \cong 4 \times 10^{-6}$ eV, or about 10^{-14} times the rest energy of either particle. Nevertheless, this difference is significant because it is the only measurable mass difference that is known to result from the weak interaction (the interaction responsible for beta decay and neutrino interactions).

Neutrino Oscillations It has always been assumed that neutrinos have no rest energy. Analysis of the energy spectrum of electrons emitted in the beta decay process

$$^3\text{H} \rightarrow {}^3\text{He} + e^- + \bar{\nu}_e$$

(14.58)

puts an upper limit of about 25 eV on the possible rest energy of the electron-type antineutrino $\bar{\nu}_e$, and consequently on that of the neutrino ν_e. Although is difficult to draw a similar conclusion regarding the rest energies of the other members of the neutrino family (ν_μ and ν_τ), theories of elementary particles are much tidier on the assumption that all neutrinos have zero rest energy and consequently travel at the speed of light. But now there are observations that could upset this tidy assumption.

These observations involve neutrinos reaching earth from the sun. Several experiments have found, over a period of decades, that the number of observed neutrinos detected is 30% to 50%, too small to agree with predictions based upon the so-called standard solar model.[12] However, if one assumes that neutrinos have nonzero rest energy, the missing neutrinos can be explained by invoking neutrino oscillations.

The relevant equations are similar to Eqs. (7.34), (14.56), and (14.57). For simplicity we consider just two orthogonal states, $|\nu_1\rangle$ and $|\nu_2\rangle$, and let each of the observed states, $|\nu_e\rangle$ and $|\nu_\mu\rangle$, be superpositions of $|\nu_1\rangle$ and $|\nu_2\rangle$:

$$|\nu_e\rangle = [|\nu_1\rangle + |\nu_2\rangle]/\sqrt{2}$$

(14.59a)

$$|\nu_\mu\rangle = [|\nu_1\rangle - |\nu_2\rangle]/\sqrt{2}$$

(14.59b)

[12]Alfred K. Mann and Raymond Davis, Jr., *Neutrino Observational Astronomy*, *Encyclopedia of Astronomy and Astrophysics* (Steven Shore, ed.), Academic Press, San Diego (1989), 323–331.

The states $|\nu_e\rangle$ and $|\nu_\mu\rangle$ are produced in conjunction with electron emission and muon production, respectively. If each of these is a superposition, as in (14.59), we can write a time-dependent neutrino state as

$$|\psi\rangle = (|\nu_1\rangle e^{-i\omega_1 t} + |\nu_2\rangle e^{-i\omega_2 t})/\sqrt{2} \qquad (14.60)$$

where $\omega_1 = m_1 c^2/\hbar$, $\omega_2 = m_2 c^2/\hbar$, and when $t = 0$ the state is $|\nu_e\rangle$. Let us assume that $m_2 > m_2$, making $\omega_2 > \omega_1$. Then, there will come a time t when $\omega_2 t = \omega_1 t + \pi$. At that time the sign of the amplitude of the coefficient of $|\nu_2\rangle$ will be opposite to the sign of the amplitude of the coefficient of $|\nu_2\rangle$, and the particle will be in the state $|\nu_\mu\rangle$. It will have changed into a muon-type neutrino, and it will be unable to interact with a detector designed to find electron-type neutrinos.

As time goes on, this state will oscillate back and forth between a muon-type and an electron-type neutrino. At any given time, there will be a 50% chance of it detecting it as an electron type. Such oscillation could solve the so-called "missing neutrino" problem. Efforts are now being made to detect muon-type neutrinos coming from the sun, which are not produced in the sun. If this type is found, there will be a strong case for the phenomenon of neutrino oscillations.

ADDITIONAL READING

L. Pauling and E. B. Wilson, *Introduction to Modern Physics*, McGraw-Hill, New York (1935), in Chapter 9, work out transition probabilities by the methods of this chapter, then apply the theory to molecular states. Most recent quantum mechanics textbooks are not so helpful in this area.

EXERCISES

1. A hydrogen atom is in the 1s state at time $t = 0$. At this time an external electric field of magnitude $\mathscr{E}_0 e^{-\tau/t}$ is applied along the z direction. Find the first-order probability that the atom will be in the 2p state (u_{210}) at time $t \gg \tau$, assuming that the spontaneous transition probability for the 2p $\rightarrow$ 1s transition is negligible at that time.

2. (a) Find the Fourier transform $\phi(\omega)$ of the function

$$f(t) = \mathscr{E}\cos\omega t \quad (0 < t < t'); \qquad f(t) = 0 \quad \text{otherwise}$$

(Use the equations of Section 2.3, with k and x replaced by ω and t, respectively.)

(b) Compute $|\phi(\omega)|^2$, and show that the distribution of frequencies in this function is in agreement with Eq. (14.16), as stated in Section 14.1.

3. (a) Use Eq. (14.38) to compute the mean lifetime of the 2p state of hydrogen.
   (b) Find the approximate uncertainty in the energy of this state, and compare it with the fine-structure splitting.

4. In Figure 14.2, the plotted numbers δN_t are, for $t < 17$ (covering two half-lives)

t	0	1	2	3	4	5	6	7	8
δN_t	793	812	679	628	576	534	503	476	424

t	9	10	11	12	13	14	15	16
δN_t	368	375	331	283	277	255	228	188

(a) Show that the expected number of atoms in the excited state at time t is given by the equation $N = 10^4(2^{-t/\ell})$, where $\ell \equiv t_{1/2}$ is the half-life of the state.

(b) Use the result of (a) to find the values of N when $t = n$ ns, for $0 < n < 25$.

(c) The expected number of photon emissions in each picosecond is approximately $\lambda N_n \, \delta t$, where $\lambda = 0.693/8 \text{ ns}^{-1}$ and $\delta \, dt = 1$ ns. Use this relation to find the number δN_t of photons emitted during each nanosecond for $0 \le t \le 24$.

(d) Find the difference between each calculated value of δN_t and the corresponding value shown in the figure. Are these differences within the expected range, on the basis of standard deviations and Poisson statistics (Appendix A)?

5. Show that dipole radiation from a simple one-dimensional harmonic oscillator obeys the selection rule $\Delta n = \pm 1$, by computing the matrix elements X_{mn}. Hint: Use the stepping operators introduced in Section 4.2, where it was shown that

$$\left\{\frac{d}{dx} - ax\right\}u_n = au_{n+1} \quad \text{and} \quad \left\{\frac{d}{dx} + ax\right\}u_n = \beta u_{n-1}$$

The selection rule follows from the orthogonality of the eigenfunctions.

6. Use Eqs. (14.47) and (14.48) to show that $\Delta l = \pm 1$ for electric dipole transitions.

7. Use the results of Example Problem 14.2 to explain the polarization of the shifted lines (for which $\Delta m = \pm 1$) in the normal Zeeman effect.

8. For 0.1-MeV gamma rays and dipole transitions, the mean lifetime is 10^{-12} s.
   (a) What mean lifetime do you expect for 1.0-MeV gamma-ray emission? *Hint:* Remember the ω^3 factor in the transition rate.
   (b) What mean lifetime do you expect for 1.0-MeV quadrupole transitions?

9. *Electric quadrupole selection rule.* For electric quadrupole transitions, the matrix element $(xy)_{mn}$ can be written as a linear combination of spherical harmonics. From this result deduce a selection rule for Δl in electric quadrupole transitions.

10. *Quantum pot watching.* Using Eq. (14.55) and the accompanying discussion, with 5000 atoms in level 1, find the *maximum* number of atoms that will reach level 2 if the system is observed as follows:
    (a) Once, at time $t = 0.5\tau$
    (b) Twice, at $t = 0.25\tau$ and 0.50τ.

11. *Laser weapons.* Much research has been done on the possibility of using orbiting lasers to disable ballistic missiles. Assuming that the target is at a distance of 2000 km and the laser beam is to be no more than 1 meter in diameter when it hits the target (in order to have sufficient intensity), calculate the diameter that would be required for the laser mirrors.

12. *The sudden approximation.* In beta decay, an atom of ^{3}H suddenly changes to singly ionized ^{3}He by emitting a high-energy electron (beta particle) from its nucleus. Assuming that the electron in the original atom was in the 1s state, we wish to find the probability that this electron is in the 1s state of ^{3}He immediately after the beta decay has occurred. To find this probability, proceed as follows: Write the initial and final operator equations as $H_{01}|u_1\rangle = E_1|u_1\rangle$ and $H_{0f}|v_m\rangle = E_n|v_m\rangle$, respectively, where H_{01} is the operator for the initial state, H_{0f} the operator for the final state, and $|u_1\rangle$ and $|v_1\rangle$ are the respective eigenstates of these operators. Then, including time dependence, we can write the initial state as

$$\psi_{\text{initial}} = |u_1\rangle e^{-iE_1 t/\hbar} \quad \text{for } t \leq 0.$$

The final state is a superposition of the eigenstates $|v_m\rangle$ with the appropriate time dependence, or $\psi_{\text{final}} = \sum b_m|v_m\rangle e^{-iE_m t/\hbar}$, for $t > 0$, where $|b_m|^2$ is the probability that the system will be in state $|v_m\rangle$ when it is observed. Because the wave function must be a continuous function of time, we know that at $t = 0$, $\psi_{\text{initial}} = \psi_{\text{final}}$, or $|u_1\rangle = \sum b_m |v_m\rangle$.

You can now find the coefficient b_1 by the methods used previously in finding Fourier coefficients, and determine the probability that the system is in state $|v_1\rangle$ (the 1s state of singly ionized helium-3) at $t = 0 + \varepsilon$. (Use eigenfunctions from Table 9.1.)

13. *Neutrino oscillations.* Experiments on muon decay put an upper limit of about 1 keV on the rest energy of the mu-type neutrino (state $|v_\mu\rangle$).

   (a) If $\hbar\omega_1$ differs from $\hbar\omega_2$ by 1 keV, how much time would it take for the state $|v_e\rangle$ to change into the state $|v_\mu\rangle$ (neglecting the existence of the tau-type neutrino)?

   (b) If you could measure the period of oscillation between the state $|v_e\rangle$ the state $|v_\mu\rangle$, what could you deduce about the masses of these two neutrinos (neglecting the existence of the tau-type neutrino)?

CHAPTER 15

Molecular Structure and Spectra

The structure of molecules can also be understood from basic principles of quantum mechanics, provided that appropriate approximation methods are brought to bear on the problem. The theoretical results are verified by analyzing the molecule's absorption and emission spectra. Here we analyze only the simplest systems, diatomic molecules, to bring out the principles.

15.1 IONIC AND COVALENT BINDING

15.1.1 Ionic Binding

Our first goal is to understand why molecules exist at all. We begin with the alkali halide molecules. We have seen (Section 13.1) that the valence electron in an alkali-metal atom is relatively loosely bound, with a typical ionization energy of about 5 eV. For example, in sodium this energy is 5.14 eV, and in potassium it is 4.34 eV. Thus it takes an input of 4.34 eV to remove an electron from potassium and form a K^+ ion.

Now suppose that this electron attaches itself to a neutral chlorine atom. It can do this because of the polarization of Cl when an electron is nearby. The incomplete M shell (states with $n = 3$) is distorted by the repulsion of this electron, making the atom into a dipole whose positive end is closer to the electron and can hold it. The binding energy of the extra electron in a Cl^- ion is 3.82 eV.

An isolated K^+ ion is spherically symmetric, because its outer shell is filled, and an isolated Cl^- ion is spherically symmetric for the same reason. Because of this symmetry they attract each other as if each were a point charge, with a Coulomb potential energy of the two ions given by $V_c = -1.44$ eV-nm/r, provided that r is sufficiently large.

In general, the total potential energy is $V = V_c + 0.52$ eV, relative to that of a neutral K and a neutral Cl atom. The additional 0.52 eV is the result of the work done in transferring the electron from the potassium atom to the chlorine atom. This potential energy is shown in Figure 15.1. Notice that the equation for the Coulomb potential here is $V_c = -1.44$ eV-nm/r, except when $r \ll 1$ nm. For small r, the force between the ions is not like that of a point charge, because at close distances the ions are polarized and not spherically symmetric.

At $r = 0.28$ nm, the potential energy curve has a minimum which determines the size of the KCl molecule. For $r < 0.28$ nm the potential energy becomes very large because of the exclusion principle; when the 3p electrons of potassium share the same space with the 3d electrons of chlorine, some of them must be excited to higher energy states.

In the ground state of the molecule, the two ions oscillate about the equilibrium separation; they have the "zero-point energy" that is always associated with a harmonic oscillator. They can be excited to vibrational states with higher energies; the resulting spectrum will be discussed shortly.

15.1.2 Covalent Binding: The Hydrogen Molecule

In contrast to ionic binding, which involves a transfer of an electron from one atom to another, in covalent binding a single electron is bound to two

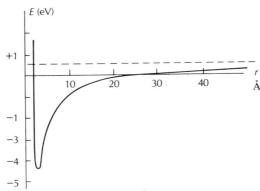

FIGURE 15.1 Potential energy of K^+ and Cl^- as a function of the internuclear separation r. Minimum energy, -4.42 eV, is at $r = 0.279$ nm. As $r \to \infty$, E approaches $0.52 - 14.4/r$.

different atoms, forming the "glue" between them. For example, in an H^+ molecule-ion there is one electron and two protons.

The electron's probability density is concentrated between the two protons, attracting both of them. The total energy, as a function of the proton–proton separation, has a minimum value at a separation of 0.106 nm between the protons.

In the neutral hydrogen molecule *both* electrons are concentrated between the protons. The Schrödinger equation for this system may be written

$$\left\{ \frac{1}{m_p}(\nabla_1^2 + \nabla_2^2) + \frac{1}{m_e}(\nabla_3^2 + \Psi\nabla_4^2) \right\} \psi = -\frac{2}{\hbar^2}(E - V)\Psi \quad (15.1)$$

where ψ is a function of *twelve* space coordinates (three for each particle); ∇_1^2 operates on the coordinates of one proton, ∇_2^2 on the coordinates of the other proton, ∇_3^2 on the coordinates of one electron, and ∇_4^2 on the coordinates of the other electron.

The potential energy V may be written

$$V = \frac{1}{4\pi\varepsilon_0}\left\{ \frac{e^2}{r_{12}} + \frac{e^2}{r_{34}} - \frac{e^2}{r_{13}} - \frac{e^2}{r_{14}} - \frac{e^2}{r_{23}} - \frac{e^2}{r_{24}} \right\} \quad (15.2)$$

where r_{12} is the distance between the two protons, r_{34} is the distance between the two electrons, r_{13} is the distance between one of the protons and one of the electrons, etc.

The operators can be rewritten in terms of the relative coordinates of the protons and the coordinate of the center of mass, as was done already for the hydrogen atom (Section 9.1). As before, this gives us an equation that involves the reduced mass M_r of the two nuclei:

$$M_r = M_1 M_2 / (M_1 + M_2) \quad \text{with } M_1 = M_2 = M_p \text{ in hydrogen} \quad (15.3)$$

The ground-state energy of the molecule is determined by the electron state. As shown in Section 11.2, when there are two identical electrons in one quantum state, the space part of the wave function is always either symmetric or antisymmetric in the exchange of the two electrons, and the average distance between the electrons is smaller when they are in a symmetric space state.

In the case of H_2, the energy difference is so great that the only stable states are antisymmetric. Figure 15.2 shows the potential energy, as a function of the distance r between the protons, for both the symmetric and

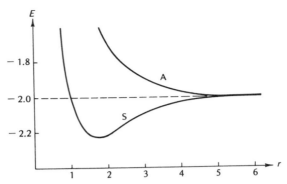

FIGURE 15.2 Calculated potential energy of two hydrogen atoms, in units of the H-atom ionization energy, as a function of the distance between the two protons, in units of the first Bohr radius. Curve S is for the symmetric spatial wave function. (From L. Pauling and E. B. Wilson, *Introduction to Quantum Mechanics*, McGraw-Hill, New York, 1935. Used by permission.)

the antisymmetric case. For the latter, we see that no bound state occurs; the minimum potential energy is at infinite r.

15.1.3 Polyatomic Molecules

For systems with more electrons, Eq. (15.2) becomes increasingly difficult to handle. However we can draw some useful conclusions from general principles. For example, the H_3 molecule is not stable because of the exclusion principle. (A third electron cannot be accommodated in the lowest energy level in the region of the first two electrons.) Other polyatomic molecules are stable whenever the electrons can be transferred (ionic binding) or shared (covalent binding) without being forced into higher energy states by the exclusion principle. For example, since oxygen is missing two electrons in its L shell, it can share an electron from each of two hydrogen atoms to form an H_2O molecule. Further analysis is beyond the scope of this book.

15.2 MOLECULAR SPECTRA: VIBRATIONAL AND ROTATIONAL STATES

15.2.1 Vibrational States of Diatomic Molecules

The minimum in the potential energy curve of Figure 15.1 shows the size of the KCl molecule. But from the uncertainty principle we know that the two nuclei do not remain at a fixed distance apart; there is a zero-point energy.

Near the potential minimum the curve can be approximated as a parabola. For KCl, the best-fitting parabola is given by

$$V = -4.42 + 226(r - 0.279)^2 \tag{15.4}$$

where V is in eV and r is in nanometers.

Recalling that the harmonic oscillator potential is given by $m\omega^2 r^2/2$ [Eq. (6.16)], we see that the potential of Eq. (15.4) is simply the harmonic oscillator potential with a shift of origin. Therefore the energy levels should be those of a simple harmonic oscillator.

Using the reduced mass m_r in place of m, we set $m_r\omega^2/2$ equal to the coefficient of 226 eV/nm^2 in Eq. (15.4). For $^{39}K^{35}Cl$, the reduced mass is

$$m_r = \frac{M_1 M_2}{M_1 + M_2} = \frac{39 \times 35}{39 + 35}u \tag{15.5}$$

where u is the atomic mass unit ($uc^2 = 931.5$ MeV).

We can solve for ω, and then multiply by $\hbar$, with the result that $\hbar\omega = 0.0347$ eV. Thus the energy levels are given by[1]

$$E_v = -4.42\text{ eV} + \hbar\omega(v + \tfrac{1}{2}) = -4.42\text{ eV} + 0.0347(v + \tfrac{1}{2})\text{ eV} \tag{15.6}$$

where v is the vibrational quantum number (equal to a positive integer or zero), and the term -4.42 eV gives the energy at the bottom of the well.

This equation is supported by the observation of an intense KCl spectral line at a wavelength of 3570 nm, which corresponds to a photon energy of 0.0347 eV. This line is intense because of the selection rule for all simple harmonic oscillators, that the quantum number (v in this case) changes by only one unit in an allowed transition. Applied to Eq. (15.6), this rule requires that the change in energy be 0.0347 eV in all allowed transitions.

However, there are fainter lines in the KCl spectrum corresponding to photon energies that are integral multiples of the 0.0347 eV. These frequencies result from transitions in which v changes by more than one unit, and they exist because the potential energy is not precisely represented by the simple harmonic oscillator potential. (See Section 4.2.)

---

[1]The zero-point energy here is that of a one-dimensional oscillator, because the potential is independent of θ and ϕ, leaving only one degree of freedom for *vibration* as opposed to rotation.

Another potential that works better is the *Morse potential*, given by

$$V(r) = D\{1 - e^{-a(r-r_e)}\}^2 \tag{15.7}$$

where D is the dissociation energy, a is an adjustable constant, and r is the equilibrium distance between the two nuclei. [Notice that $V(r) \to D$ as $r \to \infty$, and $V(r) = 0$ for $r = r_e$. Thus energy D is required to separate the atoms completely.]

For $a(r - r_e) \ll 1$, we can approximate the exponential as

$$1 - a(r - r_e),$$

and Eq. (15.7) becomes

$$V(r) = Da^2(r - r_e)^2 \tag{15.8}$$

Comparison of Eqs. (15.8) and (15.4) shows that, for KCl, $Da^2 = 266$ eV/nm². The dissociation energy D is 4.42 eV, so a^2 must equal 60.2 nm⁻².

It is possible to solve the Schrödinger equation exactly for the Morse potential and find the theoretical vibrational energy levels E_v.[2] The result, for $E_v \ll D$, is

$$E_v = \hbar\omega(v + \tfrac{1}{2}) - \frac{\hbar^2\omega^2}{4D}(v + \tfrac{1}{2})^2 \tag{15.9}$$

where v is again the vibrational quantum number (equal to $0, 1, 2, \ldots$), and $\omega^2 = 2Da^2/m_r$. Substituting the numerical values for the KCl molecule, we find that, in electron volts,

$$E_v = 0.037(v + \tfrac{1}{2}) - 0.0000068(v + \tfrac{1}{2})^2 \quad \text{(for KCl)} \tag{15.10}$$

and the ground-state energy is equal to about 18 milli-electron-volts.

These energies are extremely close to the observed energies for KCl, and they show the deviation from simple harmonic oscillator levels. For $v = 5$, the difference is about 1%, and the deviations of course become larger as v increases.

15.2.2 Rotational States of Diatomic Molecules

The vibrational mode accounts for one of the three degrees of freedom of relative motion in a diatomic molecule. If the x axis is along the line joining the two atoms, the other two degrees of freedom are associated

---

[2] L. Pauling and E. B. Wilson, pages 271–274.

with rotation in the yz plane. The energy associated with this rotation is given by

$$E_r = K(K + 1)\hbar^2/2m_r r^2 \qquad (15.11)$$

where K is a positive integer and r the variable separation between the atoms.

Variation in r (vibration) combined with rotation may seem difficult to analyze, but we can simplify matters by first considering the atomic separation to be constant, at the equilibrium value r_e. Eq. (15.11) then becomes

$$E_r = K(K + 1)\hbar^2/2m_r r_e^2 \qquad (15.12)$$

where r_e is constant, and the molecule can radiate energy by making transitions in which K changes by one unit (just as l changes by one unit in allowed atomic transitions). In such a transition, the emitted energy is

$$\Delta E = [K(K + 1) - (K - 1)K]\hbar^2/2m_r r^2 = \hbar^2 K/m_r r^2 \quad (15.13)$$

and the emitted frequency is

$$\nu_K = \hbar K/2\pi m_r r^2 \qquad (15.14)$$

This result is in agreement with the observation of molecular spectra showing a series of lines that are equally spaced in frequency; the difference between one frequency and the next is

$$\nu_{K+1} - \nu_K = \hbar/2\pi m_r r^2 \qquad (15.15)$$

as you can easily verify.

15.2.3 Combined Effects of Vibration and Rotation

When both vibration and rotation are considered, the result is a so-called *vibrational band* in the observed photon spectrum, as shown in Figure 15.3.

Table 15.1 lists the frequencies observed in this band. Each frequency is associated with a change from a $v = 0$ state to a $v = 1$ state while the rotational quantum number changes from K to $K \pm 1$. The lines on the right-hand side of Figure 15.3 are produced by changes from K to $K + 1$; the others are produced by changes from K to $K - 1$. The gap between the two groups results from the fact that transitions with $\Delta K = 0$ are forbidden.

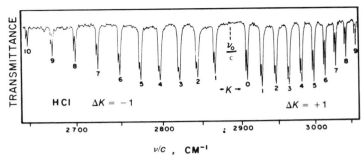

FIGURE 15.3 HCl vibrational–rotational band for transitions from an initial state for which $v = 0$ to a final state for which $v = 1$. Each absorption line is labeled by the value of the quantum number K in the initial state. Lines of higher frequency are those for which $\Delta K = +1$. Lines of lower frequency are those for which $\Delta K = -1$.

TABLE 15.1   Vibrational1Rotational Band of $H^{35}Cl^a$
(ν/c in cm^{-1})

K of initial state ($v = 0$)	ν/c ($\Delta K = +1$)	ν/c ($\Delta K = -1$)
0	2906.24	
1	2925.90	2865.10
2	2944.90	2843.62
3	2963.29	2821.56
4	2981.00	2798.94
5	2998.04	2775.76
6	3014.41	2752.04
7	3030.09	2727.78
8	3045.06	2703.01
9	3059.32	2677.73

[a]Data are from N. J. Colthup, L. H. Daly, and S. E. Wiberley, *Introduction to Infrared and Raman Spectroscopy*, Academic Press, New York (1962). Values shown are $H^{35}Cl$ only. The normal isotopic mixture of $H^{35}Cl$ and $H^{37}Cl$ gives a set of double values, which are shown in the double peaks of Figure 15.3.

Example Problem 15.1   (a) Use the data in Table 15.1 to deduce the value of $\hbar^2/2m_r r^2$ [Eq. (15.11)] for the states listed there.

Solution. We can convert each value of ν/c (Table 15.1) to a photon energy by multiplying by $hc = 1239.85$ eV-nm, or 1.23985×10^{-4} eV-cm. This yields the five photon energies, in eV, for transitions from $v = 0$

states to $v = 1$ states

Change in K:	0 to 1	1 to 2	2 to 3	3 to 4	4 to 5
Photon energy:	0.360330	0.362768	0.365123	0.367404	0.369599

and (also from $v = 0$ states to $v = 1$ states),

Change in K:	1 to 0	2 to 1	3 to 2	4 to 3	5 to 4
Photon energy:	0.355229	0.352566	0.349831	0.347027	0.344153

Let us use the Dirac symbol $|v, K\rangle$ for a state with quantum numbers v and K, respectively. To excite a molecule from the state $|0, 0\rangle$ to the state $|1, 1\rangle$ requires an energy of 0.360330 eV. To excite one from the state $|0, 2\rangle$ to the same final state $|1, 1\rangle$ requires an energy of 0.352566 eV. Therefore the energy difference between $|0, 2\rangle$ and $|0, 0\rangle$ is $(0.360330 - 0.352566)$ eV, or 0.007764 eV.

This difference is purely rotational energy, because the states differ only in the value of K, not v. From Eq. (15.12) we know that the rotational energy of the state $|0, 0\rangle$ is zero, and we can substitute $K = 2$ into Eq. (15.12) to obtain the rotational energy of the state $|0, 2\rangle$:

$$E_r = K(K + 1)\hbar^2/2m_r r_e^2 = 6(\hbar^2/2m_r r_e^2) = 0.007764 \text{ eV} \quad (15.16)$$

Therefore, $\hbar^2/2m_r r_e^2 = 1.294 \times 10^{-3}$ eV.

Let us see if the value of $\hbar^2/2m_r r_e^2$ is the same when other transitions are involved. When we compare the transition $|0, 1\rangle \to |1, 2\rangle$ with the transition $|0, 3\rangle \to |1, 2\rangle$ we find the energy difference between states $|0, 1\rangle$ and $|0, 3\rangle$ to be $(0.362768 - 0.349831)$ eV, or 0.01294 eV. From Eq. (15.16), substituting $K = 3$ and $K = 1$, we then find that this energy difference can be written $[3(4) - 1(2)]\hbar^2/2m_r r_e^2 = 10\hbar^2/2m_r r_e^2 = 0.01294$, which again makes $\hbar^2/2m_r r_e^2$ equal to 1.294×10^{-3}. Thus r_e is unchanged for small K.

15.2.4 Raman Scattering

When light is scattered by a molecule, the scattered light may have a longer wavelength than the incident light, because of excitation of the molecule to higher energy states. The change in wavelength can be used to deduce rotational and or vibrational energies of molecules. The selection rule for transitions involving this Raman effect is $\Delta K = 0, +2,$ or -2.

Why should the rule for this situation differ from the rule $\Delta K = \pm 1$ for emission of absorption of a photon? The answer is quite simple. Scattering is a two-step process consisting of absorption followed by emission. In each step, the rule is $\Delta K = \pm 1$. As a result, the net change in K can be one of three possibilities: $+1 + 1 = +2$, or $+1 - 1 = 0$, or $-1 - 1 = -2$.

ADDITIONAL READING

L. Pauling and E. B. Wilson, *Introduction to Modern Physics*, McGraw-Hill Book Co., (1935), gives a thorough treatment of the theory of molecular structure.

EXERCISES

1. For $^{23}\text{Na}^{35}\text{Cl}$, $r_e = 0.251$ nm, $D = 3.58$ eV, and ω [Eq. (15.9)] is 7.16×10^{13} s^{-1}.
   (a) From Eq. (15.9) determine the value of a that should appear in the Morse potential for this molecule.
   (b) For $v = 8$, compare the vibrational energy given by Eq. (15.6) with that given by Eq. (15.9); compute the percentage difference between the two.
   (c) Find the rotational energy for $K = 2$.

2. Compute the formula for the wavelengths of the absorption lines in the far infrared (pure rotational spectrum) for $^{7}\text{Li}^{1}\text{H}$ (for which $r_e = 0.1596$ nm).

3. From the figures in Table 15.1, compute the first six wavelengths in the pure rotational spectrum of H^{35}Cl.

4. Compute the ratio of the wavelengths in the rotational spectrum of H^{35}Cl to the corresponding wavelengths in the spectrum of $^{2}\text{H}^{37}\text{Cl}$.

5. From Eq. (15.14) compute the lowest photon energy in the rotational spectrum of KCl, and compare with the lowest photon energy in the purely vibrational spectrum of KCl.

6. *Measuring zero-point energy.* The potential energy for H_2, HD, and D_2 (D $= ^{2}\text{H}$) is the same function $V(r)$. However, the dissociation energies D differ, because the zero-point energies depend on m_r. The values of D are, in eV, 4.477, 4.513, and 4.556, respectively, for H_2,

HD, and D_2. Use this information to calculate each zero-point energy in eV. [*Hint:* Work out the *ratios* of these energies by using the values of m_r and Eq. (15.9).]

7. From the data in Table 15.1 and in Example Problem 15.1,
   (a) Determine the average distance r_e between the proton and the ^{35}Cl nucleus in the ground state of H^{35}Cl.
   (b) Find this distance to five significant figures for $K = 8$, with $v = 0$.

8. *Rotational energy:*
   (a) Use Eq. (15.12) with the value of $\hbar^2/2m_r r_e^2$ found in Example Problem 15.1 to find the rotational energy of each of the states with $K = 1$ through 6.
   (b) By comparing two different transitions, as was done in Example Problem 15.1, find the rotational energy of state $|0,5\rangle$, and compare the result with your result of part (a).

9. *Rotation and stretching of a molecule.* Like any macroscopic object, a diatomic molecule stretches as it rotates. This effect can be related to the rotational energy E_r [Eq. (15.11)].
   (a) Write the effective potential V_{eff} in terms of K by adding E_r to the Morse potential, and solve for the equilibrium radius r_e—the value of r that minimizes V_{eff}.
   (b) Use this result to calculate the change in r_e for KCl when the quantum number K increases from 0 to 10.

10. *Raman scattering.* Using the selection rule for Raman scattering, find the frequencies that should appear in the ^{23}Na^{35}Cl spectrum of the scattered light when the incident light has a frequency of 1.50×10^{13} Hz. In this molecule, $r_e = 0.251$ nm.

CHAPTER 16

Quantum Statistics

As you now know, quantum mechanical laws are valid on the macroscopic scale as well as the atomic scale. What you may not know is that these laws are just as essential for the understanding of large systems as they are for atomic phenomena.

Although classical statistical theory describes some properties of gases quite successfully, it fails completely when it is applied to such seemingly simple tasks as computation of the specific heat of a solid. This failure is a consequence of the indistinguishability of elementary particles or identical atoms. In a gas, identical atoms can be distinguished by their positions, but the electrons in a solid or liquid, even those belonging to different atoms, tend to be so close together that their probability distributions overlap, and we cannot distinguish their trajectories.

No matter how many particles are involved, there is a single state function that tells us all that we can possibly know about the state of the system. This state function must be consistent with the exclusion principle for a system containing electrons, protons, and/or neutrons.[1] This requirement has huge effects on the population of the allowed energy levels for systems of electrons, and these effects show up on a large scale in the phenomena that we will address in this chapter.

You may find it hard to believe that the simple act of counting the number of possible states of a system can lead to great insights into previously unexplained phenomena, but this chapter provides many such situations.

---

[1] Any particle with spin $1/2 + n$, where n is an integer, is subject to the exclusion principle; such particles are called *fermions*. Fermions include muons and neutrinos, plus other highly unstable heavy particles. (See Appendix H.)

16.1 THREE KINDS OF STATISTICS

The basic task of a statistical theory in quantum mechanics is as follows: Given a set of allowed energy levels in a system, the degeneracy of each level, the total number of particles in this system, and the temperature of the system, determine the most probable number of particles occupying each energy level.

Given this information, there are in general *three possible results*. These results correspond to three different assumptions about the particles.

Assumption 1 (Boltzmann statistics): Distinguishable particles.

Assumption 2 (Fermi–Dirac statistics): Indistinguishable particles that obey the exclusion principle.

Assumption 3 (Bose–Einstein statistics): Indistinguishable particles that *do not* obey the exclusion principle.

Assumption 1 is classical, but we present it here to show how a system can obey classical mechanics if the particle density is sufficiently low (as in the atmosphere, treated in Appendix B). In such cases the particles are widely separated in space and thus can be distinguished by their locations.

16.1.1 Example: A System of Four Identical Particles with Equally Spaced Energy Levels

The energy levels are ε_0, ε_1, ε_2, etc., with degeneracies g_0, g_1, g_2, etc., as shown in Figure 16.1. The states are depicted as boxes in which the particles reside at any given time. As time passes the particles can exchange energy, but we assume that the total energy remains fixed as the energy is redistributed among the particles. The figure shows a system in which the total energy is 10 eV, at an instant when one particle's energy is 8 eV, another's is 2 eV, and two have zero energy (neglecting zero-point

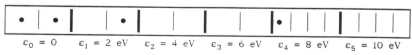

$$\varepsilon_0 = 0 \qquad \varepsilon_1 = 2 \text{ eV} \qquad \varepsilon_2 = 4 \text{ eV} \qquad \varepsilon_3 = 6 \text{ eV} \qquad \varepsilon_4 = 8 \text{ eV} \qquad \varepsilon_5 = 10 \text{ eV}$$

FIGURE 16.1 Graphical representation of energy levels and particle states as "boxes." Heavy lines outline each energy level; lighter lines outline individual states of particles in a given energy level; dots indicate the current states of four particles. In the situation shown, two particles have zero energy but are in different states, another particle has an energy of 2 eV, and a fourth particle has an energy of 8 eV.

energy). The total energy of the particles depends on the temperature of the system.

To proceed further we must distinguish between a *distribution* and an *arrangement*, as follows:

A *distribution* is a set of numbers n_s that describes the number of particles in each energy level ε_s.

An *arrangement* is a "microscopic distribution," which tells us how many particles are in each state within each energy level.

Thus there are usually many possible arrangements that belong to the same distribution. For example, for the distribution shown in Figure 16.1, if the particle with energy 2 eV were in the other state with that energy, the arrangement would be different but the distribution would be unchanged.

Computation of the Average Distribution The average distribution of particles is found by making the reasonable assumption that *each arrangement is equally probable*. Therefore the probability of observing a given *distribution* is directly proportional to the number of different *arrangements* that are possible for that distribution. For example, if the 8-eV level in our example had a degeneracy of only two instead of four, we would be less likely to see a distribution in which one of the particles had an energy of 8 eV.

Table 16.1 shows the possible distributions of the four particles for this example, using each of the three types of statistics. Notice that the number of arrangements for a given distribution depends strongly on the type of statistic that is applied. For example, any arrangement that puts more than one particle in a state is ruled out for Fermi–Dirac statistics, and any two arrangements that differ only in the interchange of two identical particles

TABLE 16.1 Possible Distributions of Four Particles

		Distribution					Arrangements			Probability		
	n_0	n_1	n_2	n_3	n_4	n_5	Boltz.	FD	BE	Boltz.	FD	BE
a:	3	0	0	0	0	1	128	0	16	128/1152	0	4/21
b:	2	1	0	0	1	0	384	8	24	384/1152	8/20	6/21
c:	2	0	1	1	0	0	192	4	12	192/1152	4/20	3/21
d:	1	2	0	1	0	0	192	4	12	192/1152	4/20	3/21
e:	0	3	1	0	0	0	64	0	8	64/1152	0	3/21
f:	1	1	2	0	0	0	192	4	12	192/1152	4/20	3/21
							1152	20	84	1	1	1

are considered to be a single arrangement for either Fermi–Dirac or Bose–Einstein statistics.

Table 16.1 shows all of the distributions that satisfy the given parameters, and it also shows the number of possible arrangements for each distribution. Finally, it shows the probability of observing each *distribution*, on the assumption that each *arrangement* is equally probable. You should verify that for each of these distributions the total energy is 10 eV and that there is no other distribution for which this is true.

Let us now calculate of the number of arrangements for line (b).

Boltzmann: There are four particles to choose for level 4, and the particle chosen can go into any of four possible states in that level, making 16 possibilities for that level. After level four is determined, level one will have one of the three remaining particles, and it can be in either of two states. Thus level one has six possibilities; combining these with the 24 possibilities in all possible ways gives us $6 \times 16 = 96$ arrangements in those two levels. There is no further choice of the particles for level zero, but there are two choices of the state for each of them, making four ways to place them into the two states. Thus we have $4 \times 96 = 384$ possible arrangements.

Fermi–Dirac: Level 4 takes one particle in one of four possible states, and level 1 takes a particle in one of two possible states, making eight possibilities. In level zero there is only one possibility, because there are two particles and two states in that level, and the exclusion principle forbids two particles to be in the same state. Thus there are only eight arrangements.

Bose–Einstein: As in Fermi–Dirac statistics, there are eight possibilities for the two upper levels. But with no exclusion principle in effect, level zero has three possibilities: both particles in one state, both in the other state, or one in each state. Thus there are $3 \times 8 = 24$ arrangements.

This analysis suggests that quantum statistics is simpler than the classical Boltzmann statistics, because it is not necessary to label the particles.

The probability shown in the table for each distribution is easily found from the assumption that each arrangement is equally probable, and this is how the probability column in the table was constructed. The probability for each distribution is the ratio of the number of arrangements for that distribution [24 for Bose–Einstein in line (b)] to the total number of arrangements for all distributions [100 for Bose–Einstein].

Table 16.2 shows the total probability of occupation of each level on

TABLE 16.2   Average Distributions Derived from Table 16.1

Statistics	n_0	n_1	n_2	n_3	n_4	n_5
Boltzmann	1.667	1.00	0.556	0.333	0.333	0.111
Fermi–Dirac	1.60	1.00	0.60	0.40	0.40	zero
Bose–Einstein	1.714	1.00	0.524	0.286	0.286	0.190

the basis of the distributions of Table 16.1. The numbers in Table 16.2 are computed from Table 16.1 as follows: The number n_s in level s for a given distribution is multiplied by the probability of that distribution, and the results are added for each of the six possible distributions to find the average number of particles expected to be found in level s. For example, the number n_2 in level 1 is the sum of contributions from distributions (b), (d), and (f), namely $1 \times 8/20 + 2 \times 4/20 + 1 \times 4/20$, which gives a total of 1. Notice that the sum of all the average numbers for each distribution is 4, equal to the total number of particles.[2]

It is useful to define the *occupation index* as the ratio n_s/g_s, which tells us the average number of particles *per state* in level s. Table 16.3 gives the occupation indices corresponding to the numbers in Table 16.2. In the next section we derive the general formula for n_s/g_s for large numbers, for each type of statistics.

Notice that, for the lowest energy level, the Fermi–Dirac occupation index is smaller than the classical (Boltzmann) index, because the exclusion principle prevents too many particles from being in the lowest level. On the other hand, the greatest number of particles in the lowest level is found with Bose–Einstein statistics. We shall see in the following sections that this comparison is true in general; this so-called *Bose–Einstein condensation* explains many important phenomena.

TABLE 16.3   Occupation Indices Computed from Table 16.2

	n_0	n_1	n_2	n_3	n_4	n_5
Boltzmann	0.833	0.50	0.278	0.167	0.083	0.028
Fermi–Dirac	0.80	0.50	0.30	0.20	0.10	zero
Bose–Einstein	0.857	0.500	0.262	0.143	0.072	0.048

[2] Notice that the numbers decline with increasing s until level 4 is reached. At that point the degeneracy suddenly increases from 2 to 4, giving this level and level 5 a greater occupancy. Notice also that level 5 has a greater average number than level 4, for B-E statistics. This is an anomaly that may occur when the number of particles is so small.

16.2 DERIVATION OF THE GENERAL FORM OF EACH DISTRIBUTION FUNCTION

We now apply the foregoing principles and assumptions to the general case, when the actual number of particles is too large to count. In this case it is not necessary to find the average distribution directly. Instead we find the *most probable* distribution. This distribution, like any other *specific* distribution of an enormous number of particles, has a very small probability of occurrence, but the overwhelming majority of observed distributions differ by a negligible amount from the most probable distribution.

First we must find an expression for P_s, the number of ways in which n_s particles in the sth energy level can be put into the g_s states in that level. Then we find the total number of arrangements for the *whole set* of numbers $n_0, n_1, \ldots, n_s, \ldots$ by taking the product of all the numbers $P_0, P_1, \ldots P_s, \ldots$ This product gives the statistical *weight* $W(n_0, n_1, \ldots n_s, \ldots)$ of the distribution $(n_0, n_1, \ldots n_s, \ldots)$. That is,

$$W(n_0, n_1, \ldots n_s, \ldots) = \prod_{s=0}^{\infty} P_s \qquad (16.1)$$

To find the most probable distribution, we maximize W while imposing two conditions:

1. The total number N of particles must be fixed:

$$\sum_{s=0}^{\infty} n_s = N \qquad (16.2)$$

2. The total energy E must be fixed:

$$\sum_{s=0}^{\infty} n_s \varepsilon_s = E \qquad (16.3)$$

16.2.1 Computation of Statistical Weights

We now find expressions for the most probable P_s and W under the given conditions, for each type of statistics.

Boltzmann: In this case only, the particles are distinguishable, and therefore the interchange of two particles yields a different arrangement. Consider level 0 first. From the N particles present, we can choose n_0

particles for this *level* in $N!/n_0!(N - n_0)!$ different ways.[3] These particles can then be distributed among the g_0 states of level 0 in $g_0^{n_0}$ different ways. Thus the total number of ways to put n_0 particles into level 0 is

$$P_0 = \frac{N!g_0^{n_0}}{n_0!(N - n_0)!} \tag{16.4}$$

Having placed n_0 particles into level 0, we have $N - n_0$ remaining particles. We put n_1 of these into level 1, distributed among g_1 states. We find that the number of ways to put n_1 particles into level 1, when there are P_0 particles in level 0, is

$$P_1 = \frac{(N - n_0)!g_1^{n_1}}{n_1!(N - n_0 - n_1)!} \tag{16.5}$$

By now we can see that, after we put n_0 particles in level 0, n_1 into level 1, n_2 into level 2, and so on, up to n_{s-1} into level $s - 1$, the number of ways to put n_s particles into level s is, in general,

$$P_s = \frac{(N - n_0 - n_1 - n_2 - \cdots - n_{s-1})!g_s^{n_s}}{n_s!(N - n_0 - n_1 - n_2 - \cdots - n_s)!} \tag{16.6}$$

The total number of ways to put the N particles into the levels to produce a given distribution $n_0, n_1, \ldots n_s, \ldots$ must be equal to the product $P_0 P_1 \ldots P_s \ldots$ It is easy to see that in this product there are intermediate factors of the form $(N - n_0)!(N - n_0 - n_1)!$, etc. in both numerator and denominator, which therefore cancel out, making the final result simply

$$W(n_0, n_1, \ldots n_s, \ldots) = \prod_{s=0}^{\infty} P_s = N! \prod_{s=0}^{\infty} \frac{g_s^{n_s}}{n_s!} \quad \text{Boltzmann} \tag{16.7}$$

Fermi–Dirac: When the particles are indistinguishable, the factor $N!/n_0!(N - n_0)!$ does not appear, because that factor is related to a choice of particles for a given level, and all choices are identical. Therefore we need only find the factor analogous to $g_0^{n_1}$—the factor that gives the

---

[3]There are N choices for the first particle, $N - 1$ for the second, etc., so there are $N(N - 1)(N - 2)\ldots(N - n_1 + 1)$ ways to choose the particles in *a given order*. But we have the same n_1 particles regardless of the order of choosing them, so we must divide the above number by the number of different orders in which we could choose them. This number is $n_1!$ Thus the total number of ways of choosing these particles is as stated above.

number of ways in which the n_0 particles can go into the g_0 states of level zero. For Fermi–Dirac we know that there can be only zero or one particle in each state, so we simply divide the states in two groups. In any level s, there are n_s occupied states, and therefore there must be $g_s - n_s$ unoccupied states. The number of arrangements must be simply the number of ways of selecting n_s objects (the occupied states) from a total collection of g_s objects (all of the states). Elementary statistics gives this number as $g_s!/n_s!(g_s - n_s)!$. (Remember that the *states* are distinguishable, even though the *particles* are not.) Therefore, for Fermi–Dirac statistics

$$W(n_0, n_1, \ldots n_s, \ldots) = \prod_{s=0}^{\infty} W(n_0, n_1, \ldots n_s, \ldots) = P_s = \prod_{s=0}^{\infty} \frac{g_s!}{n_s!(g_s - n_s)!}$$

(16.8)

Bose–Einstein: Again, with indistinguishable particles, we do not count choices for a given particle; we look only at the occupation of states. See Figure 16.2, which shows the states in a single level s, with n_s particles in g_s states. We find a new arrangement if we move a number of particles from one state to another, but no new arrangement if we simply exchange two particles.

We can count the arrangements by a simple trick. There are $g_s - 1$ partitions, or dividing lines, between the states, making the total number of particles plus partitions equal to $n_s + g_s - 1$. We obtain new arrangements by permuting particles and partitions; for example, we obtain the arrangement of Figure 16.2b from the arrangement of Figure 16.2a by interchanging the second particle and the first partition (the partition between states 1 and 2).

There are $(n_s + g_s - 1)!$ possible permutations of particles and partitions, but each permutation does not lead to a new arrangement. Permutation of particles among themselves, or of partitions among themselves, changes nothing physically. There are $n_s!$ ways to permute the particles among themselves, and $(g_s - 1)!$ ways to permute the partitions among themselves. This means that there are $n_s!(g_s - 1)!$ permutations of each

a   b

FIGURE 16.2 (a) One arrangement of n_s particles in the g_s states of level s, with $g_s = 5$. (b) Another arrangement of the same n_s particles, obtained from the arrangement of (a) simply by interchanging the second particle and the first partition between states.

arrangement, and the total number of arrangements must be the quotient

$$P_s = \frac{(n_s + g_s - 1)!}{n_s!(g_s - 1)!} = \frac{\text{number of permutations}}{\text{number of permutations per arrangement}} \quad (16.9)$$

Therefore

$$W(n_0, n_1, \ldots n_s, \ldots) = \prod_{s=0}^{\infty} P_s = \prod_{s=0}^{\infty} \frac{(n_s + g_s - 1)!}{n_s!(g_s - 1)!} \quad \text{Bose–Einstein}$$

$$(16.10)$$

Computation of the most probable distribution   For each type of statistics we can now find the most probable set of numbers n_s—the set that maximizes the value of W for the given conditions [Eqs. (16.2) and (16.3)]. The easiest way to do this is to maximize the *logarithm* of W rather than W itself. Thus we set the variation in $\ln W$ equal to zero, just as we find the maximum of a function of one variable by setting the derivative equal to zero, and we write

$$\delta(\ln W) = 0 \quad (16.11)$$

We compute $\delta(\ln W)$ by first computing W for one set of the numbers n_s, varying these numbers by increments δn_s in a way that is consistent with Eqs. (16.2) and (16.3), then recomputing W. The difference between the new value of W and the previous value is $\delta(\ln W)$. To be consistent with Eqs. (16.2) and (16.3), we require that

$$\delta N = \sum_{s=0}^{\infty} \delta n_s = 0 \quad (16.12)$$

$$\delta E = \sum_{s=0}^{\infty} \varepsilon_s \, \delta n_s = 0 \quad (16.13)$$

We can now use the method of *Lagrange multipliers* to maximize W.[4] We multiply both sides of Eq. (16.12) by an arbitrary number α, and we multiply both sides of Eq. (16.13) by an arbitrary number β. Then we

---

[4] This method is often used to maximize a function of several variables while satisfying certain conditions on these variables.

combine Eqs. (16.11), (16.12), and (16.13) to obtain

$$\delta(\ln W) = \alpha \sum_{s=0}^{\infty} \delta n_s + \beta \sum_{s=0}^{\infty} \varepsilon_s \, \delta n_s = 0 \qquad (16.14)$$

We shall see the reason for this step in a moment, as we introduce the expressions for W for each type of statistics.

Boltzmann:

$$\ln W = \ln N! + \sum_{s=0}^{\infty} (n_s \ln g_s - \ln n_s!) \qquad (16.15)$$

We are concerned with cases in which n_s is quite large, for which we can use the Stirling approximation, valid for large n:

$$\ln n! \rightarrow n(\ln n - 1) \quad \text{as } n \rightarrow \infty \qquad (16.16)$$

With this approximation, Eq. (16.15) becomes

$$\ln W = \ln N! + \sum_{s=0}^{\infty} (n_s \ln g_s - n_s \ln n_s + n_s) \qquad (16.17)$$

Applying the standard equation for a total differential to the quantity $\delta(\ln W)$, we next have

$$\delta(\ln W) = \sum_{s=0}^{\infty} \frac{\partial(\ln W)}{\partial n_s} \, \partial n_s \qquad (16.18)$$

$$\delta(\ln W) = \sum_{s=0}^{\infty} (\ln g_s - \ln n_s - 1 + 1) \, \delta n_s \qquad (16.19)$$

Equating the expressions for $\delta(\ln W)$ in Eq. (16.14) and Eq. (16.19), and rearranging terms, gives

$$\boxed{\sum_{s=0}^{\infty} (\ln g_s - \ln n_s - \alpha - \beta \varepsilon_s) \, \delta n_s = 0} \qquad (16.20)$$

Fermi–Dirac:

$$\ln W = \sum_{s=0}^{\infty} [\ln g_s! - \ln n_s! - \ln(g_s - n_s)!] \qquad (16.21)$$

The Stirling approximation gives

$$\delta(\ln W) = \sum_{s=0}^{\infty} [-\ln n_s + \ln(g_s - n_s)] \, \delta n_s \qquad (16.22)$$

and substitution into Eq. (16.14) gives

$$\sum_{s=0}^{\infty} [-\ln n_s + \ln(g_s - n_s) - \alpha - \beta\varepsilon_s] \, \delta n_s = 0 \qquad (16.23)$$

Bose–Einstein:

$$\ln W = \sum_{s=0}^{\infty} [\ln(n_s + g_s - 1)! - \ln n_s! - \ln(g_s - 1)!] \qquad (16.24)$$

Now the Stirling approximation gives

$$\delta(\ln W) = \sum_{s=0}^{\infty} [\ln(n_s + g_s) - \ln n_s] \, \delta n_s \qquad (16.25)$$

and substitution into Eq. (16.14) gives

$$\sum_{s=0}^{\infty} [\ln(n_s + g_s) - \ln n_s - \alpha - \beta\varepsilon_s] \, \delta n_s = 0 \qquad (16.26)$$

Equation (16.20), (16.23), or (16.28) can give us the most probable distributions if we can establish that *for every value of s, the coefficient of* δn_s *is equal to zero*. The following logical steps show that this in fact is true.

1. *If the values of* n_s *could be varied at will*, then the left-hand side of Eq. (16.20), (16.23), or (16.26) would be equal to zero only if all of the coefficients of n_s were zero.

2. Because the distribution must satisfy Eqs. (16.2) and (16.3), we are *not* free to vary the values of n_s at will. However, we *can* vary *all but two* of these values (say all but n_0 and n_1) arbitrarily. Then *we can choose the values of α and β to make the coefficients of δn_0 and δn_1* [in the series of Eq. (16.20), (16.23), or (16.26)] equal to zero.

3. Having done this, we know that the remaining series, running from $s = 2$ to infinity, must equal zero. Since the values of n_s for s in this series can be chosen arbitrarily (step 2), this forces the coefficients of δn_s to be zero for $s > 1$. Thus *all* of the coefficients of δn_s must be zero in each series, and we have (for *all* values of s), the *Boltzmann* distribution:

$$\ln g_s - \ln n_s - \alpha - \beta \varepsilon_s = 0, \quad \text{or}$$

$$\boxed{\frac{n_s}{g_s} = e^{-\alpha - \beta \varepsilon_s}} \tag{16.27}$$

the *Fermi–Dirac* distribution:

$$-\ln n_s + \ln(g_s - n_s) - \alpha - \beta \varepsilon_s = 0, \quad \text{or}$$

$$\boxed{\frac{n_s}{g_s} = \frac{1}{e^{\alpha + \beta \varepsilon_s} + 1}} \tag{16.28}$$

and the *Bose–Einstein* distribution:

$$\ln(n_s + g_s) - \ln n_s - \alpha - \beta \varepsilon_s = 0, \quad \text{or}$$

$$\boxed{\frac{n_s}{g_s} = \frac{1}{e^{\alpha + \beta \varepsilon_s} - 1}} \tag{16.29}$$

Let us now compare Eq. (16.27) with Eq. (B.5) of Appendix B

$$n = n_0 e^{-Mgz/RT} = n_0 e^{-E/kT} \tag{B.5}$$

which also applies to classical particles. By rewriting Eq. (16.27) as

$$n_s = g_s e^{-\alpha} e^{-\beta \varepsilon_s} \tag{16.30}$$

we can identify β with the reciprocal of kT, and we can see a connection between $g_s e^{-\alpha}$ and the number of particles. This is consistent with Eqs.

(16.14), where β and α were introduced in applying the constraints of constant energy and a fixed number of particles, respectively.

By considering a system in which energy is exchanged between particles obeying different types of statistics, we can now show that, if $\beta = 1/kT$ for Boltzmann statistics, then β must be equal to $1/kT$ for Fermi–Dirac and Bose–Einstein as well.

Proof: We consider two sets of particles in thermal contact: set A, obeying Boltzmann statistics, and set B, obeying either Bose–Einstein or Fermi–Dirac statistics. We label the respective energy levels and occupation numbers as ε_s and n_s for set A, and ε_t' and n_t' for set B. The total number of arrangements of all particles combined, for a given overall distribution, is the product $P(n_s, n_t') = P_A(n_s)P_B(n_t')$. To find the most probable overall distribution for the complete set of particles we must maximize $P(n_s, n_t')$ by the method used before, this time with *three* conditions:

$$\delta N_A = \sum_{s=0}^{\infty} \delta n_s = 0, \qquad \delta N_B = \sum_{t=0}^{\infty} \delta n_t = 0,$$

$$\delta E = \sum_{s=0}^{\infty} \varepsilon_s \, \delta n_s + \sum_{t'=0}^{\infty} \varepsilon_{t'} \, \delta n_{t'} = 0$$

Therefore we introduce three constants, α_A, α_B, and β, into the variation equation analogous to Eq. (16.14):

$$\delta(\ln W_A + \ln W_B)$$

$$= \alpha_A \sum_{s=0}^{\infty} \delta n_s + \alpha_B \sum_{t=0}^{\infty} \delta n_t + \beta \sum_{s=0}^{\infty} \varepsilon_s \, \delta n_s + \beta \sum_{t'=0}^{\infty} \varepsilon_{t'} \, \delta n_{t'} = 0$$

$$(16.31)$$

Notice that we need a different α for each type of particle, because the number of each kind of particle is conserved, but we have the same β for both kinds, because only the total energy of both kinds is conserved. Therefore, if both kinds are at the same temperature T, and $\beta = 1/kT$ for the particles in group A, then β must equal $1/kT$ for the other particles as well.

The interpretation of the coefficient α depends on the type of statistics. Eqs. (16.2) and (16.30) (with $\beta = 1/kT$) lead to an expression for α in

Boltzmann statistics as follows:

$$N = \sum_{s=0}^{\infty} n_s = \sum_{s=0}^{\infty} g_s e^{-\alpha} e^{-\varepsilon_s/kT} \tag{16.32}$$

We can divide both sides by the factor $e^{-\alpha}$ to obtain

$$e^{-\alpha} = \frac{N}{\sum_{s=0}^{\infty} g_s e^{-\varepsilon_s/kT}} \tag{16.33}$$

The quantity $\sum_{s=0}^{\infty} g_s e^{-\varepsilon_s/kT}$ is called the *partition function* of the system and is denoted by the letter Z. Thus the occupation indices for the Boltzmann distribution are written concisely as

$$\frac{n_s}{g_s} = \frac{N}{Z} e^{-\varepsilon_s/kT} \tag{16.34}$$

Example Problem 16.1   Consider a system with only three equally spaced energy levels $\varepsilon_1 = 0$, $\varepsilon_2 = kT$, and $\varepsilon_3 = 2kT$, and degeneracies $g_1 = 1$, $g_2 = 2$, and $g_3 = 2$. If $N = 10^5$ and the particles are distinguishable, find the most probable set of values for n_1, n_2, and n_3, and verify the result.

Solution. From the given quantities we find Z to be:

$$Z = \sum_{s=0}^{\infty} g_s e^{-\varepsilon_s/kT} = 1 + 2/e + 2/e^2 = 2.00643$$

Therefore

$$n_1 = g_1 \frac{N}{Z} e^{-\varepsilon_1/kT} = \frac{100,000}{2.00643} e^0 = 49,840$$

$$n_2 = g_2 \frac{N}{Z} e^{-\varepsilon_2/kT} = \frac{100,000}{2.00643} e^{-1} = 36,670$$

$$n_3 = g_3 \frac{N}{Z} e^{-\varepsilon_3/kT} = \frac{100,000}{2.00643} e^{-2} = 13,490$$

for a total of $n_1 + n_2 + n_3 = 100,000$, as required.

To verify that this distribution gives the maximum value of W, we now vary the values of n_1, n_2, and n_3 while keeping E and N constant. Let us compare the values of $W(n_1, n_2, n_3)$, $W(n_1 + 1, n_2 - 2, n_3 + 1)$, and $W(n_1 - 1, n_2 + 2, n_3 - 1)$. From Eq. (16.7) we have

$$W(n_1, n_2, n_3) = N! \prod \frac{g_s^{n_s}}{n_s!} = \frac{100{,}000!}{49840!36670!13490!} 2^{36670} 2^{13490}$$

$$W(n_1 + 1, n_2 - 2, n_3 + 1) = \frac{100{,}000!}{49841!36668!13491!} 2^{36668} 2^{13491}$$

$$W(n_1 - 1, n_2 + 2, n_3 - 1) = \frac{100{,}000!}{49839!36672!13489!} 2^{36672} 2^{13489}$$

Our calculator is not up to the task of computing these factorials as they stand, but we can find their ratios. We find that

$$\frac{W(n_1, n_2, n_3)}{W(n_1 + 1, n_2 - 2, n_3 + 1)} = \frac{49841 \times 13491 \times 2}{36670 \times 36669} \approx 1.0001$$

$$\frac{W(n_1, n_2, n_3)}{W(n_1 - 1, n_2 + 2, n_3 - 1)} = \frac{36672 \times 36671}{49840 \times 13490 \times 2} \approx 1.0001$$

Thus either increasing or decreasing n_2 by the smallest possible amount (consistent with energy conservation and constant N) leads to a smaller value of W, and we are confident that we have found the most probable distribution.

16.3 APPLICATION OF FERMI–DIRAC STATISTICS: FREE ELECTRONS IN METALS

The large electric and thermal conductivity of metals is evidence that many electrons (valence electrons) in a metal are not bound to individual atoms, but can move freely throughout the metal's volume. It is only when an electron reaches the metal surface that it is restricted in its motion, being pulled back into the metal. Thus valence electrons in a metal behave like a gas confined within a box. This view is supported by study of many phenomena, with the aid of Fermi–Dirac statistics.

16.3.1 Electronic Specific Heat

A difficulty in considering metallic electrons to be a gas is that this gas makes a very small contribution to the specific heat C_p of the metal. The electrons, if they were like a monatomic gas, would contribute a constant

$3R/2$ per mole to C_p. They contribute far less, and this contribution is not constant but is directly proportional to the temperature.

All of these facts are readily explained by the application of Fermi–Dirac statistics. We begin by investigating the parameter α. We write $\alpha = \mu/kT$, where μ in general may be a function of temperature. We shall soon see how this substitution simplifies the analysis. We now write the occupation index, with reference to Eq. (16.28) as

$$\frac{n_s}{g_s} = \frac{1}{e^{(\varepsilon_s - \mu)/kT} + 1} \tag{16.35}$$

or for a continuous distribution,

$$\frac{n(\varepsilon)}{g(\varepsilon)} = \frac{1}{e^{(\varepsilon - \mu)/kT} + 1} \tag{16.36}$$

We can see that when $\mu = \varepsilon$ the occupation index is $1/2$, and this fact can serve as a definition of μ. If we wished, we could find the value of μ at any temperature by applying the condition that $N = \int_0^\infty n(\varepsilon)\,d\varepsilon$ and solving the resulting equation for μ.

Fortunately we can solve many problems without knowing the temperature dependence of μ. Consider the situation at $T = 0$ K. By definition, when $T = 0$ K, all particles are in their lowest possible energy levels. Because only one particle can occupy each state, we expect the occupation index to be 1 for the lowest N states and zero for all higher states. Equation (16.35) is consistent with this statement; at $T = 0$, the exponent becomes $+\infty$ for $\varepsilon > \mu$ and $-\infty$ for $\varepsilon < \mu$. Thus $n(\varepsilon) = 0$ for $\varepsilon > \mu$ and $n(\varepsilon) = 1$ for $\varepsilon < \mu$. When $\varepsilon = \mu$, $n(\varepsilon) = 1/2$, and half of the states are occupied at that energy. Thus μ must equal the energy of the most energetic particle. By definition, the energy of the highest occupied state at temperature $T = 0$ is called the Fermi energy, ε_f.

Figure 16.3 shows the effect of raising the temperature. At the point where the occupation index is $1/2$, ε must always be equal to μ, to satisfy Eq. (13.6). For a given value of T, μ is fixed, but μ becomes smaller as T increases. However, for $kT < 0.1$, ε_f, $\varepsilon_f - \mu < 4.5 \times 10^{-6}$ ε_f, so we can replace μ by ε_f with little loss of accuracy in that temperature range.

At temperatures such that $kT \gg \varepsilon_f$, μ becomes a large negative number. In that case, we can neglect the 1 in the denominator of Eq. (16.36), making the distribution identical to a Boltzmann distribution. This is consistent with reality, because metals are vaporized and ionized at such

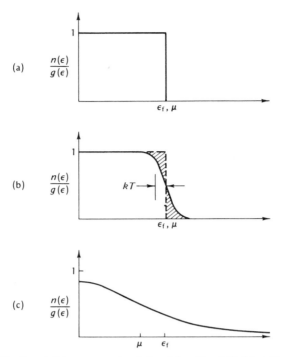

FIGURE 16.3 The Fermi–Dirac occupation index. (a) at $T = 0$ K; (b) at $T > 0$ but $\ll kT$; (c) at $T \gg kT$, showing the change in μ.

temperatures; the valence electrons are then far apart and can be distinguished by their positions.

For any solid metal, $kT \ll \varepsilon_f$. Since the temperature dependence of μ is negligible at those temperatures, we shall simply set μ equal to ε_f in the following discussion.

We are now in a position to explain (qualitatively) the above-mentioned linear temperature dependence of the electronic specific heat (C_p) of metals. We can see that only a small fraction of the electrons gains energy when the temperature is raised from 0 K to the temperature T shown in Figure 16.3b. Electrons are transferred only from the shaded area below ε_f to the shaded area above ε_f. The *average* gain in energy per electron is the change in energy between the centroids of these two areas, or an amount of about kT. The *number* of electrons that are transferred is proportional to the magnitude of the shaded area; that is, it is proportional to kT.

The total energy E gained by these electrons is proportional to the product of the gain in energy per electron and the amount of electrons that gain energy; this product is proportional to $(kT)^2$. Since C_p is proportional to $\partial U/\partial T$, we see that the electronic contribution to C_p is proportional to T.

Figure 16.4 shows that the total specific heat is the sum two parts; one is linear in temperature, and the other is cubic in temperature. The cubic term is the lattice contribution to C_p, which will be discussed in Section 16.4.

16.3.2 Calculation of the Fermi Energy for a Metal

To calculate the Fermi energy we must determine $g(\varepsilon)$; to do this we must know the potential energy function for the valence electrons. A simple assumption is that the potential energy is constant inside the metal. We know that this cannot be strictly true, because the electrons are strongly attracted to the positively charged atomic cores. However, it gives values of ε_f that are in agreement with experiment.

To calculate ε_f, we assume that the metal is a cube of volume a^3. From Eq. (6.14) we find that the energy levels of a particle of mass m in such a cube are given by

$$\varepsilon = \frac{h^2}{8ma^2}\left(n_x^2 + n_y^2 + n_z^2\right) \tag{16.37}$$

where each of the quantum numbers n_x, n_y, and n_z can be any positive integer.

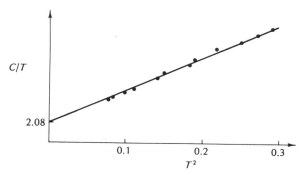

FIGURE 16.4 Specific heat C of potassium at low T, plotted as C/T vs. T^2. The fit to a straight line is $C/T = 2.08 + 2.57\,T^2$, which shows that C is the sum of a linear term and a cubic term [W. Lien and N. E. Phillips, *Phys. Rev.* **133**, A1370 (1964)].

The integers for which $\varepsilon \le \varepsilon_f$ obey the relation

$$n_x^2 + n_y^2 + n_z^2 \le \frac{8ma^2}{h^2} \varepsilon_f \qquad (16.38)$$

All of the states with quantum numbers that satisfy Eq. (16.38) can be plotted as points on a graph whose axes are n_x, n_y, n_z; all of these points lie within a sphere whose radius is the maximum value of $\sqrt{n_x^2 + n_y^2 + n_z^2}$. From Eq. (16.38) we see that the radius of this sphere is equal to $(a/h)\sqrt{8m\varepsilon_f}$, and its volume is $V_s = (4\pi/3)(a^3/h^3)(8m\varepsilon_r)^{3/2}$.

Because the numbers n_x, n_y, and n_z must all be positive, the number of points satisfying Eq. (16.38) lie within an octant of this sphere. Each point within this octant represents two states, because there are two spin states for each electron. The number of *points* in an octant is equal to $V_s/8$, and therefore the number of *states* is $V_s/4$. This number must be equal to the number N of particles, because all of these states are occupied at temperature $T = 0$. Putting all of these factors together, we have

$$N = V_s/4 = (\pi a^3/3h^3)(8m\varepsilon_f)^{3/2} \qquad (16.39)$$

and the Fermi energy is

$$\varepsilon_f = \frac{h^2}{8m}(3N/\pi V)^{2/3} \qquad (16.40)$$

where V, the volume of the *metal*,[5] has been substituted for a^3. We see that ε_f is proportional to the two-thirds power of the *valence* electron density N/V in the metal. Although we derived this result for a cube, there is no reason why it should not be equally valid for any other shape.

Example Problem 16.2   Compute the Fermi energy of aluminum.

Solution. The atomic mass is 26.98, the density is 2.702 gm/cm³, and the valence is 3, so

$$\frac{N}{V} = \frac{3 \times 6.022 \times 10^{23}}{26.98 \text{ g}} \times \frac{2.702 \text{ g}}{\text{cm}^3} = 1.809 \times 10^{23} \text{ cm}^{-3} = 180.9 \text{ nm}^{-3}$$

[5] Not to be confused with V_s, the volume of the imaginary sphere.

TABLE 16.4 Fermi Energies and Equivalent Temperatures for Monovalent Metals

Metal	ε_f (eV)	$\dfrac{\varepsilon_f}{k}$ (K)
Li	4.7	5.5×10^4
Na	3.1	3.7×10^4
K	2.1	2.4×10^4
Rb	1.8	2.1×10^4
Cs	1.5	1.8×10^4
Cu	7.0	8.2×10^4
Ag	5.5	6.4×10^4
Au	5.5	6.4×10^4

Thus

$$\varepsilon_f = \frac{h^2}{8m}(3N/\pi V)^{2/3} = \frac{h^2 c^2}{8mc^2}\left\{\frac{180.9 \text{ nm}^{-3}}{1.047}\right\}^{2/3}$$

$$= \frac{1240 \text{ eV}^2 - \text{nm}^2}{8 \times 5.11 \times 10^5 \text{ eV}} \times 172.7^{2/3} \text{ nm}^{-2} = 11.66 \text{ eV}$$

Table 16.4 lists Fermi energies for some monovalent metals, and also shows the equivalent temperatures (ε_f/k). The ratio of ε_f to kT at room temperature (about 300 K) shows that for copper this ratio is about 270, and its lowest value is about 60 (still far above the value that would cause us to rethink our approximations).

The accuracy of this model and the resulting Fermi energy have been checked by using positrons as probes to measure the momenta, and thus the energies, of valence electrons in metals. The results are in agreement with those listed in Table 16.4.[6]

Example Problem 16.3 A Neutron Star as a "Fermi Gas"

A neutron star is a star that is compressed into such a small volume that charged particles cannot remain inside. It has been speculated that ε_f for the remaining neutrons could be so high that some neutrons could be transformed into lambda (Λ) particles, whose rest energy is 176.0 MeV greater than the rest energy of a neutron. If ε_f were greater than 176.0

[6]John D. McGervey, *Introduction to Modern Physics*, 2d edition, Academic Press, (1983). Pages 490–501 describe the positron method and examples of its applications, including the PET scanner. See also Positron Annihilation, in *The Encyclopedia of Modern Physics*, Academic Press (1989).

MeV, a neutron in one of the highest energy levels could become a lambda. Since it would no longer be excluded from the lower states, it could fall to a low energy level, converting some of its kinetic energy into the required rest energy. Calculate the neutron density (N/V) that would be required for lambdas to be stable in a neutron star.

Solution. We solve Eq. (16.40) for the value of N/V with $\varepsilon_f = 176$ MeV and $m = 939.6$ MeV/c^2 (since the neutron rest energy is about 939.6 MeV). We have

$$\varepsilon_f = \frac{h^2}{8m}\left[\frac{3N}{\pi V}\right]^{2/3}, \quad \text{so that}$$

$$\frac{N}{V} = \frac{\pi}{3}\left\{\frac{8m\varepsilon_f}{\hbar^2}\right\}^{3/2} = \frac{\pi}{3}\left\{\frac{8mc^2\varepsilon_f}{\hbar^2 c^2}\right\}^{3/2}$$

$$= \frac{\pi}{3}\left\{\frac{8 \times 939.6 \times 176 \text{ MeV}^2}{(197.3)^2 \text{ eV}^2\text{-nm}^2}\right\}^{3/2}$$

$$= \frac{1.593 \times 10^{27} \text{ eV}^3}{7.68 \times 10^6 \text{ eV}^3\text{-nm}^3} \approx 2.1 \times 10^{20} \text{ neutrons/nm}^3$$

This means that the density of such a star would be almost 10^{17} times the density of the earth (an "astronomical" number). A piece the size of a grain of sand ($\sim 10^{-12}$ m^3) would have a mass of more than 300 million kg.

16.3.3 Work Function of a Metal

We saw in Section 1.1 that the work function $e\phi$ of a metal is defined as the minimum photon energy required to remove an electron from that metal. If we assume that the valence electrons are the easiest to remove and that the potential energy of these electrons can be approximated as a square well of depth W, then the minimum energy required is the difference between W and ε_f, and thus the work function is

$$e\phi = W - \varepsilon_f \tag{16.41}$$

Figure 16.5 shows the relationship between W and ε_f.

Contact Difference of Electrical Potential   Figure 16.6 illustrates a related phenomenon. When a wire is connected between two different

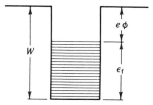

FIGURE 16.5 Relationship between well depth W, Fermi energy ε_f, and work function $e\phi$ for electrons in a metal. Horizontal lines indicate filled energy levels.

metals, electrons usually flow from one metal to the other. This flow continues until the highest occupied states in each metal are at the same energy; this energy is called the *Fermi level*.[7] The reason for the flow is clear from the figure, as follows:

Figure 16.6a shows the situation when the metals are separate and uncharged; they have different well depths, different Fermi energies, and different work functions. The net effect is that there are many valence electrons in metal A with higher *total* energy than any valence electron in metal B. When metals are connected (Figure 16.6b), these highest-energy electrons flow to B in order to occupy the unoccupied states of lower energy there. Metal B becomes negatively charged, A becomes positively charged, and the potential energy well for electrons in B rises, relative to the well in A, until the energy at the top of the filled levels in B equals that in A.

When metals A and B are connected, the number of electrons transferred is minuscule[8] in comparison to the total number of valence electrons in either metal, so *the values of W and ε_f are not changed in either metal*. As Figure 16.6 makes clear, the difference in electrostatic potential between A and B becomes $e(\phi_b - \phi_a)$. This difference is known as the *contact potential* difference. It is also clear from Figure 16.6 that, when A and B are connected and then irradiated, the photon energy required to free a valence electron from either metal is equal to the *larger* of the work functions of the metals (or $e\phi_b$ in the figure).

Measurement of Contact Potential   If you try to measure a contact potential with a conventional voltmeter, you are doomed to disappoint-

---

[7]Be aware of the distinction between Fermi *level* and Fermi *energy*. The Fermi *level* is the *total* energy of the highest-energy electrons with respect to an outside reference energy. The Fermi *energy* is the *kinetic* energy of these electrons. This perhaps subtle distinction is clarified in Figure 16.5.

[8]See Exercise 5 for verification of this statement.

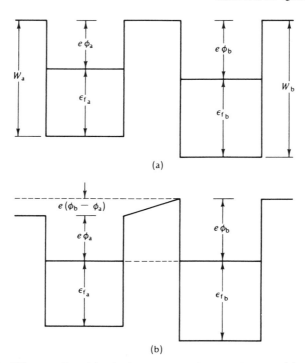

(a)

(b)

FIGURE 16.6 Effect on Fermi level when two metals, A and B, are (a) separated, (b) connected. Change in Fermi level results from change in the overall level of the potential well, with negligible change in Fermi energy of either metal.

ment, because the two metals are normally in equilibrium and no current will flow through your meter. But you can measure this potential difference if you upset this equilibrium by varying the distance between the two metals while measuring the current that flows in a wire connecting these metals. This current flows because the two metals form a "capacitor," whose capacitance changes when the distance changes, while the potential difference does not change (being a property of the metals). Consequently the charge on this capacitor must change, and a current must flow, when the capacitance is changed.

The magnitude of the contact potential difference can be found by inserting a known additional potential difference into the circuit and varying this additional potential until no current flows when the distance between the metals is changed. When this situation is reached, the added potential must be just enough to cancel the contact potential.

16.3.4 Emission of Electrons from Metals

Thermal Emission   In the preceding discussion we ignored the electrons whose kinetic energy exceeds ε_f. As the temperature of a metal increases, many electrons acquire enough thermal energy to escape from the metal, thereby making cathode-ray tubes (and television) possible.

The current density of escaping electrons can be related to the temperature of the metal by means of the Fermi–Dirac distribution, which shows that this density is proportional to $T^2 e^{-W/kT}$. (According to Boltzmann statistics, it should be proportional to $T^{1/2} e^{-W/kT}$. Details are given in Exercise 5.)

Field Emission   Application of an electric field will change the potential energy of an electron near the surface of a metal, as shown in Figure 16.7. Even if there is no externally applied field, the potential energy well for an electron cannot be perfectly "square" at the surface of the metal. Standard electrostatic theory tells us that an electron *outside* a metal at a distance x from the surface is attracted to the metal by a force equal to that of an "image" charge. This image charge is a fictitious charge of $+e$, located inside the metal at a distance x from the surface, so that the charge and its image are separated by a distance $2x$. The force on the electron is $e^2/4\pi\varepsilon_0(4x)^2$, and the potential energy is therefore

$$V = -\frac{e^2}{16\pi\varepsilon_0 x} \tag{16.42}$$

A modified potential curve is shown in Figure 16.7a, where the value of V outside the metal is given by Eq. (16.42) except at values of x that would make V less than $-W$. For those values, V is set equal to $-W$, as shown. This potential, although still crude, can be used for approximate calculations of the effect of an externally applied electric field in the $-x$ direction. The potential energy, for an electron, is then approximately as in Figure 16.7b.

Because the maximum value of the potential energy is negative rather than zero as in the field-free situation, the energy required to liberate an electron is smaller than W. Thus the *effective* work function is smaller in the presence of an external electric field. Furthermore, the escape of electrons from the metal is enhanced by the possibility of tunneling through the barrier, made possible by the fact that the external potential energy continues to decline until it becomes smaller than $-W$. Calculation

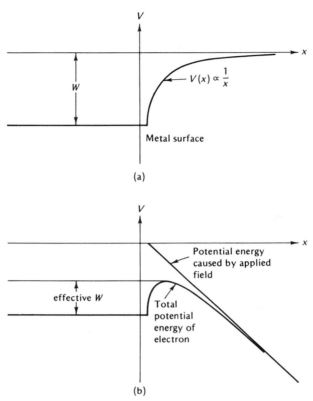

FIGURE 16.7 (a) Potential energy of an electron near the surface of a metal, including the effect of the image charge but ignoring the effect of individual atoms. (b) Potential energy of an electron after a uniform electric field is applied to the surface; the linear potential of the applied field is added to the potential shown in (a).

of the current of emitted electrons, taking these two effects into account, gives results in reasonable agreement with experiment.

16.4 APPLICATIONS OF BOSE–EINSTEIN STATISTICS

16.4.1 A Photon Gas: Radiation from a Blackbody

The fact that bodies of different temperatures radiate light of different colors was known for a long time, but the actual dependence of color on temperature was a mystery before Max Planck, by inventing the concept of quantization, derived the blackbody radiation law that now carries his name:

$$u(\nu) = \frac{8\pi\nu^2}{c^3} \left\{ \frac{h\nu}{e^{h\nu/kT} - 1} \right\} \tag{16.43}$$

In this formula, $u(\nu)$ is the energy density of radiation of frequency ν in a hollow cavity. That is, $\int_{\nu_1}^{\nu_2} u(\nu)\, d\nu$ equals the energy per unit volume in the cavity, in the frequency range between ν_1 and ν_2. The function $u(\nu)$ can be measured by observing the radiation through a small hole in the cavity, which emits a random sample of the radiation within the cavity. When the radiation density in the hole is low, the hole is observed to be black, because the hole does not reflect any incident light. Thus we call it blackbody radiation.

Equation (16.43) can be derived without recourse to Planck's derivation[9] by applying Bose–Einstein statistics to the photon "gas," using the same approach that we have used for the electron gas in a metal. We approximate the cavity as a cube, of side a, in which the radiation forms standing waves. The momentum in each wave is determined by the boundary condition that the wave must be zero at any wall of the cube. Therefore the three components of momentum of a photon must be

$$p_x = \frac{h}{\lambda_x} = \frac{hl_x}{2a}$$

$$p_y = \frac{h}{\lambda_y} = \frac{hl_y}{2a} \tag{16.44}$$

$$p_z = \frac{h}{\lambda_z} = \frac{hl_z}{2a}$$

where l_x, l_y, and l_z are positive integers. The photon energy is then given by

$$\varepsilon^2 = c^2 p^2 = c^2 \left(p_x^2 + p_y^2 + p_z^2 \right)$$

$$= \frac{c^2 h^2}{4a^2} \left(l_x^2 + l_y^2 + l_z^2 \right) \tag{16.45}$$

---

[9] Planck's derivation is given in many texts, including *Introduction to Modern Physics* by John D. McGervey (Academic Press, New York, 1983).

Following the reasoning of Section 16.3, we find that all sets of integers for which the photon energy is ε or less obey the relation

$$l_x^2 + l_y^2 + l_z^2 \leq \frac{4a^2\varepsilon^2}{h^2c^2} \tag{16.46}$$

Thus, as in Sec. 16.3, the sets of positive integers that satisfy Eq. (16.46), when plotted in three dimensions, lie within an octant of a sphere of radius $2a\varepsilon/hc$. There are two modes of oscillation (two polarization directions) for each set of integers; the final result is that the number G of states of energy ε or less must be

$$G = \frac{\pi}{3}\left(\frac{2a\varepsilon}{hc}\right)^3 \tag{16.47}$$

The density of states is then

$$g(\varepsilon) = \frac{dG}{d\varepsilon} = \frac{8\pi V\varepsilon^2}{h^3c^3} \qquad (V \equiv a^3) \tag{16.48}$$

and, from Eq. (16.29), with $\alpha = 0$ and $\beta = KT$,[10] we find that

$$n(\varepsilon) = \frac{g(\varepsilon)}{e^{\varepsilon/kT} - 1} = \frac{8\pi V\varepsilon^2}{h^3c^3(e^{\varepsilon/kT} - 1)} \tag{16.49}$$

The energy density dU of the radiation of frequency between ν and $\nu + d\nu$ is found from Eq. (16.49) by multiplying the energy $h\nu$ per photon by the number of photons per unit volume, $n(\varepsilon)\,d\varepsilon/V$. We find that

$$
\begin{aligned}
dU &= \frac{8\pi\varepsilon^3\,d\varepsilon}{h^3c^3(e^{\varepsilon/kT} - 1)} = \frac{8\pi h\nu^3\,d\nu}{c^3(e^{h\nu/kT} - 1)} \\
&\equiv \frac{8\pi\nu^2}{c^3}\left\{\frac{h\nu}{e^{h\nu/kT} - 1}\right\}d\nu
\end{aligned}
\tag{16.50}
$$

Equation (16.50) is exactly the same equation that Planck derived in a completely different way, before Einstein had suggested the photon concept. It has a number of applications, such as the observation of colors to

---

[10] We found in Section 16.3 that $\beta = kT$. We can set α equal to zero because α was introduced to meet the requirement that the total number of particles is constant, and there is no such requirement for photons.

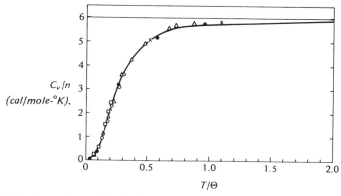

FIGURE 16.8 Comparison of the Debye theoretical curve with the observed specific heats of some simple substances. ($\bullet$) Ag; ($\triangle$) Al; ($\square$) C (graphite); ($\bigcirc$) Al_2O_3; ($\times$) KCl. (From *The Modern Theory of Solids*, by F. Seitz, McGraw-Hill, New York, 1940. Used by permission.)

determine the temperature of stars and furnaces. Another very important application is to the specific heats of solids.

16.4.2 The Specific Heat of a Solid

According to the classical specific heat "law" (law of Dulong and Petit), solids have a temperature-independent specific heat of $3R$ per mole ($3k$ per atom, on the assumption that the solid has an internal energy of kT per degree of freedom of oscillation of an atom). However, P. Debye developed a theory that fitted all of the available data, as shown in Figure 16.8.[11] He assumed that the problem is similar to that of cavity radiation, except that it involves *sound* waves rather than electromagnetic waves.

We can understand the Debye theory by introducing a "particle" called a *phonon* that plays a role similar to that of the *photon* in a cavity. Each phonon, like a photon, has a characteristic frequency ν and an energy of $h\nu$. The difference is that a limited number of frequencies are available in a solid, because there is only one mode of oscillation per degree of freedom, or $3N_A$ modes for one mole (or N_A atoms).

The number of modes of oscillation per unit frequency range, $dN/d\nu$, is proportional to ν^2. [This is the same frequency (or energy) dependence that we saw in Eq. (16.48) for the number of modes in a cavity per unit frequency range.]. We can thus write the integral for the total number of

[11] P. Debye, *Ann. Phys.* (*Leipzig*) **39**, 789 (1912).

modes in terms of the maximum frequency ν_m as

$$3N_A = \int_{\nu=0}^{\nu=\nu_m} dN = \int_0^{\nu_m} A\nu^2 \, d\nu = A\nu_m^3/3 \qquad (16.51)$$

where the constant A is therefore equal to $9N_A/\nu_m^3$.

We can now write the integral for the total energy as $E = \int_{\nu=0}^{\nu=\nu_m} \langle \varepsilon \rangle \, dN$, where $\langle \varepsilon \rangle$, the average energy of a mode whose frequency is ν, is $\langle \varepsilon \rangle = h\nu/(e^{h\nu/kT} - 1)$ [as in Eq. (16.50)]. Therefore, with $dN = A\nu^2 \, d\nu$ and $A = 9N_A/\nu_m^3$, we have

$$E = \int_0^{\nu_m} \frac{h\nu}{e^{h\nu/kT} - 1} \frac{9N_A}{\nu_m^3} \nu^2 \, d\nu \qquad (16.52)$$

We can show the essential features of Eq. (16.52) by replacing $h\nu/kT$ by the dimensionless variable x and introducing the *Debye temperature* $\Theta_D = h\nu_m/k$. The equation then becomes, after changing the variable from ν to x,

$$E = 9N_A kT \left\{ \frac{T}{\Theta_D} \right\}^3 \int_0^{\Theta_D/T} \frac{x^3}{e^x - 1} \, dx \qquad (16.53)$$

When $\Theta_D/T \ll 1$, we can approximate e^x as $1 + x$; the integrand is then roughly equal to x^2 over the entire range of integration. The energy E then becomes equal to $3N_A kT$, which gives a specific heat of $3R$ per mole —just the classical value. Clearly, those solids obeying the Dulong–Petit "law" are just those for which $\Theta_D \ll T$ over the range of temperatures that had been measured.

At the other extreme, when $\Theta_D \gg T$, the integral in Eq. (16.53) approaches a constant limit, independent of temperature, of $\pi^4/15$.[12] In this case E is proportional to T^4 and the specific heat is proportional to T^3. We have seen this temperature dependence in Figure 16.4. Although the Debye theory is only a first approximation (based as it is on the assumption of a ν^2 dependence of dN), it is remarkable that it fits the data so well with the introduction of the single parameter Θ_D.

16.4.3 Bose–Einstein Condensation

Many peculiar properties of liquid ^{4}He can be explained if it is a Bose–Einstein "gas," which undergoes a strange condensation into a

---

[12] For $x \gg 1$, the integrand is approximately $x^3 e^{-x}$, and the upper limit is infinite.

"superfluid" at a temperature of 2.16 K. In this condensation a large fraction of a macroscopic amount of liquid helium occupies the lowest-energy state, in which it behaves like a fluid with zero viscosity. In this type of fluid, quantum mechanical effects are visible on a macroscopic scale, in apparent defiance of the correspondence principle.[13]

To understand the role of Bose–Einstein statistics in this unique condensation, let us compare Bose–Einstein with Boltzmann statistics for a system containing N particles having a set of nondegenerate energy levels. For large N the number of possible distributions is enormous, but we gain an insight by comparing just two possibilities with the same total energy:

(a) One particle occupies each of the lowest N levels.

(b) $N - 1$ particles are in the lowest level, and one particle is in a higher level, with all of the remaining energy.

You can verify that both distributions are equally likely in Bose–Einstein statistics; each consists of a single arrangement. However, in Boltzmann statistics (a) has $N!$ arrangements while (b) has only N arrangements; (a) occurs $(N - 1)!$ times as often as (b). Thus a distribution like (b) is so improbable that it is never seen except in a system obeying Bose–Einstein statistics.

How is it seen in a Bose–Einstein system? Atoms of ^{4}He, having zero total spin, obey Bose–Einstein statistics. At atmospheric pressure they undergo a normal condensation into a liquid at a temperature of 4.2 K. They can be cooled below that point by reducing the pressure, causing the liquid to boil until the temperature reaches 2.16 K. At this *critical temperature*, T_c, called the lambda point,[14] the liquid suddenly stops *boiling*, although it continues to *evaporate*. The cessation of boiling is an indication of a sharp increase in thermal conductivity, which appears to become infinite, for all practical purposes. Thus the liquid stays at a perfectly uniform temperature throughout, without the "hot" spots that are responsible for bubble formation (boiling).

Another peculiarity of liquid ^{4}He is that it can flow through the tiniest capillary, as if its viscosity were zero, and yet when a plate is dragged through it the viscosity is *not* zero.

The explanation of these apparently contradictory facts is that liquid ^{4}He below the λ point (known as liquid He II) is a mixture of *two* fluids: a

---

[13] The defiance is not real, because the principle applies to large *quantum* numbers, not to large numbers of *particles*. The particles displaying the effect have the smallest quantum number of all.

[14] Called λ because the curve of specific heat vs. T resembles the letter λ at this point.

normal fluid and a "superfluid."[15] The superfluid consists of atoms in the ground state; the normal fluid is the rest. The normal fluid causes the drag on a plate; the superfluid flows through capillaries.

Being a single quantum state, the superfluid cannot be localized. When the liquid is heated, superfluid is destroyed uniformly throughout the liquid. Thus the temperature rises everywhere simultaneously and the liquid appears to have infinite thermal conductivity. Let us see the theory of this two-fluid system.

Theory of an Ideal Bose–Einstein Gas   Equation (16.29) gives the Bose–Einstein distribution for discrete states:

$$\frac{n_s}{g_s} = \frac{1}{e^\alpha e^{\beta \varepsilon_s} - 1}$$

As we did for the Fermi–Dirac distribution, we can modify this equation to

$$\frac{n(\varepsilon)}{g(\varepsilon)} = \frac{1}{e^\alpha e^{\varepsilon/kT} - 1} \tag{16.54}$$

Notice the similarities and differences between Eq. (16.54) and the corresponding equation for Fermi–Dirac statistics [Eq. (16.26)].

Next we must find an expression for the density of states, $g(\varepsilon)$. Here the problem is almost identical to the problem for Fermi–Dirac statistics, a problem that we already solved. From Eq. (16.39) we can deduce that, in a volume V containing particles of mass M, the *number G* of states is

$$G = (\pi V / 3h^3)(8M\varepsilon)^{3/2} \tag{16.55}$$

[Notice the difference between this G and the G in Eq. (16.47); this reflects the fact that Eq. (16.47) refers to massless photons.]

The *density* of states, $g(\varepsilon)$, is found as before by differentiating G with respect to ε, with the result that

$$g(\varepsilon) = dG/d\varepsilon = 2\pi V (2M/h^2)^{3/2} \varepsilon^{1/2} \tag{16.56}$$

We have not yet dealt with the factor e^α, which disappeared in the photon gas because the number of photons is not fixed. In this case the number of helium atoms is fixed at N, which we relate to α by integrating $n(\varepsilon)$ from Eq. (16.54) over all energies (assuming for convenience that the

---

[15] This explanation was published by Fritz London in *Phys. Rev.* **54**, 947 (1938).

lowest level has $\varepsilon = 0$):

$$N = \int_0^\infty n(\varepsilon) \, d\varepsilon = \int_0^\infty \frac{g(\varepsilon) \, d\varepsilon}{e^\alpha e^{\varepsilon/kT} - 1} \tag{16.57}$$

$$= 2\pi V \left(\frac{2M}{h^2}\right)^{3/2} \int_0^\infty \frac{\varepsilon^{1/2} \, d\varepsilon}{e^\alpha e^{\varepsilon/kT} - 1} \tag{16.58}$$

which may be rewritten as

$$N = AT^{3/2} \int_0^\infty \frac{x^{1/2}}{e^{\alpha+x}} (1 - e^{-\alpha-x})^{-1} \, dx \tag{16.59}$$

where $A \equiv 2\pi V(2M/h^2)^{3/2}$ and $x \equiv \varepsilon/kT$. With the aid of the gamma function, defined as $\Gamma(n) = \int_0^\infty x^{n-1} e^{-x} \, dx$, we finally obtain

$$N = AT^{3/2}\Gamma(3/2)f(\alpha), \quad \text{where } f(\alpha) \equiv \sum_{p=1}^\infty \frac{e^{-p\alpha}}{p^{3/2}} \tag{16.60}$$

Solving for $f(\alpha)$ in terms of T yields

$$f(\alpha) = \frac{N}{AT^{3/2}\Gamma(3/2)} \tag{16.61}$$

The preceding two expressions for $f(\alpha)$ are mutually inconsistent! The maximum value of $f(\alpha)$, as defined in Eq. (16.60), occurs when $\alpha = 0$ $[f(0) = 2.612$, as you may verify].[16] But putting T equal to zero in Eq. (16.61) makes $f(\alpha)$ go to infinity. How can we reconcile these two results?

The problem arose when we went from Eq. (16.29) to Eq. (16.54) to obtain a continuous density of states $g(\varepsilon)$ instead of the discrete distribution g_s. As a result, in Eq. (16.58) $n(\varepsilon)$ is proportional to $\varepsilon^{1/2}$, which makes $n(\varepsilon)$ go to zero as $\varepsilon \to 0$. Thus *we have not counted the particles in the lowest-energy state*, which we defined to have $\varepsilon = 0$ in Eq. (16.57). This state is precisely the state into which the particles "condense" at low temperatures!

---

[16] By definition, α can never be negative.

We can rectify this situation by simply adding n_0 to the expression for N in Eq. (16.58). Writing n_0 as $1/(e^\alpha - 1)$, we obtain

$$N = \frac{1}{e^\alpha - 1} + AT^{3/2}\Gamma(3/2)f(\alpha) \qquad (16.62)$$

We can now draw two conclusions:

- As $T \to 0$, the second term in Eq. (16.62) goes to zero, and $n_0 \to N$.
- Since $e^\alpha - 1 = 1/n_0$, we have $e^\alpha = 1 + 1/n_0$, and therefore for large values of n_0 (from the approximation $e^x \approx 1 + x$ for $x \ll 1$), we have

$$\alpha = 1/n_0 \quad \text{(for } n_0 \gg 1) \qquad (16.63)$$

Calculation of the Critical Temperature T_c   We define the critical temperature as the temperature at which Eq. (16.60) can no longer be satisfied by any value of α, that is, the temperature at which n_0 must be accounted for separately, as in Eq. (16.62). At this temperature we know that $f(\alpha)$ has reached its maximum value of 2.612. Setting T equal to T_c and $f(\alpha)$ equal to 2.612 in Eq. (16.61), we obtain (after some algebra)

$$T_c = \left\{ \frac{N}{2.612\,A\Gamma(3/2)} \right\}^{2/3} \qquad (16.64)$$

Calculation of n_0 versus Temperature   We now have several clues to the temperature dependence of n_0. We know that $n_0 = N$ at $T = 0$ K, n_0 is negligible at $T = T_c$, and $f(\alpha) = 2.616$ for $0 \le T \le T_c$. Writing Eq. (16.62) as

$$N = n_0 + 2.612\,AT^{3/2}\Gamma(3/2) \qquad (16.65)$$

Substituting this expression for N into Eq. (16.63) gives, after some manipulation,

$$n_0 = N\left[1 - (T/T_c)^{3/2}\right] \quad (T < T_c) \qquad (16.66)$$

Be aware that n_0 need not be zero at $T = T_c$; it could be a "small" number like 10^9, which is so tiny (of order $10^{-15}\,N$ in one mole) that it does not show up on a plot of n_0 versus T (Figure 16.9).

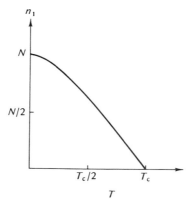

FIGURE 16.9 Temperature dependence of n_0 for an ideal Bose–Einstein gas. When $T > T_c$, n_0 is so small (compared to N) that it does not show up on this graph.

Calculation of Critical Temperature for Liquid ^{4}He   Substitution of the density and particle mass of liquid ^{4}He into Eq. (16.61) yields $T_c = 3.13$ K instead of the actual value of 2.16 K. This should be regarded as confirmation of the basic concept. We do not expect our analysis to be numerically correct for a real situation, because it neglects interaction between the particles, which certainly has an effect on the energy levels and thus on $g(\varepsilon)$.

There have been attempts to deny that this phenomenon involves Bose–Einstein statistics. But it is significant that liquid ^{3}He is a superfluid only at temperatures below 0.0025 K; ^{3}He atoms have spin 1/2, so they obey Fermi–Dirac statistics rather than Bose–Einstein statistics. The only reason for the actual condensation in ^{3}He is that at very low temperatures the atoms form *pairs* whose *total* spin is 1.[17]

Although it appears that condensation into the ground state is a necessary condition for superfluid behavior, it is not a *sufficient* condition. An *ideal* Bose–Einstein gas (in which there are no interactions between the particles) does *not* become a superfluid when it condenses into the ground state. Superfluid behavior also requires that there be no low-energy states that could be easily excited by friction with the walls of a thin tube. If there is a sufficiently large energy gap, and the speed of the fluid is sufficiently low, then the atoms cannot acquire enough energy to cross the gap, and superfluid motion can occur. Pursuit of the connection between

[17]See J. G. Armitage and I. E. Farquhar (eds.), *The Helium Liquids*, Academic Press, New York (1975).

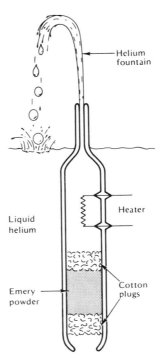

FIGURE 16.10 Schematic illustration of the fountain effect. (From Robert B. Leighton, *Principles of Modern Physics*. McGraw-Hill, New York, 1959. Used by permission.)

particle interactions and the presence of an energy gap is beyond the scope of this book, because of the great variety of interactions that are possible.

The Fountain Effect   This is the most spectacular of the phenomena attributed to superfluidity. To achieve it, a tube is plugged at the bottom with emery powder and cotton and left open at the top. The tube is then immersed in liquid He II, as shown in Figure 16.10. When the heater gives energy to the helium inside the tube, some atoms are raised from the ground state, and superfluid is "destroyed" inside the tube. To maintain a uniform temperature both inside and outside the tube, superfluid rushes into the tube through the plug at the bottom. The plug prevents the escape of normal fluid from the tube (because of the viscosity of normal fluid) so the total amount of fluid in the tube increases. If the tube has a narrow top, a fountain of liquid He eventually shoots out the top of the tube.

ADDITIONAL READING

Charles Kittel, [*Introduction to Solid-State Physics*, 3d edition, Wiley, New York, (1967), Chapters 7–9] has an excellent exposition of electron states in solids, with the free-electron gas model and the effects of the potential of the atomic lattice. A shorter treatment of these topics can be found in *Introduction to Modern Physics* (J. McGervey, op. cit., Chapter 12).

F. London, *Superfluids*, Vol. II, Wiley, New York, (1954) treats Bose–Einstein condensation.

EXERCISES

1. Use the formulas (16.7), (16.8), and (16.10) to compute the values of $W(n_0, n_1, \ldots n_s, \ldots)$ for each distribution in the four-particle example of Section 16.2, and test the agreement of your results with the numbers in Table 16.1.

2. Consider a system whose energy levels are $\varepsilon_s = 0.01s$ (where s is any positive integer), with degeneracy $g_s = 2$ for each level. In this system are seven particles, obeying Fermi–Dirac statistics. The total energy is 0.18 eV.

   (a) Write down all of the possible distributions of the particles in the energy levels.

   (b) Find the probability of occurrence of each distribution.

   (c) Find the average number of particles in each energy level.

   (d) By comparing this distribution with the Fermi–Dirac distribution [Eq. (16.35)] deduce value of ε_f and the temperature of this system.

3. Show that the average energy $\langle \varepsilon \rangle$ of the particles in a Fermi gas is equal to $3\varepsilon_f/5$ at $T = 0$.

4. In the discussion of contact difference of potential (Section 16.3) it was stated that the "number of electrons transferred is minuscule in comparison to the total number of valence electrons." Verify this statement by calculating the total charge transferred when two metals, each of area A and thickness d, are brought together to form a parallel-plate capacitor whose plate separation is also d, and the plates are then joined by a fine wire. Assume that the contact potential difference is 1.0 volt, $d = 0.2$ mm, and $A = 2.0$ cm^2. If one plate is aluminum, find the fractional change in e_f in that plate.

5. Show that the presence of a uniform electric field of magnitude $\mathscr{E}$, directed into the surface of a metal, lowers the effective work function of the metal from W to $W - e^{3/2}\mathscr{E}^{1/2}/(4\pi\varepsilon_0)^{1/2}$.

6. The Debye theory was developed before it was known that a harmonic oscillator has a zero-point energy of $h\nu/2$. Thus Eq. (16.53) does not give the correct total energy of a solid at $T = 0$. Incorporate the zero-point energy into the Debye model to show that the energy at $T = 0$ should be given by $E_0 = 9N_A k\Theta_D/8$.

7. Show that the specific heat of a Bose–Einstein gas is proportional to $T^{3/2}$ for $T < T_c$.

8. Find the value of the Debye specific heat at $T = 3\Theta_D/4$. Use Eq. (16.53) in the form $E = (9RT^4/\Theta_D^3)f(\Theta_D)$, where

$$f(\Theta_D) = \int_0^{\Theta_D/T} \frac{x^3}{e^x - 1}\, dx$$

then write C_v in terms of $f(\Theta_D)$ and $f'(\Theta_D)$, both of which may be found from a graph of $x^3/(e^x - 1)$ vs. x, or by numerical integration.

9. The observed specific heat of metallic nickel (Debye temperature 375 K) is, in arbitrary units (at temperatures in Kelvin degrees):

T	2.0	4.0	6.0	8.0	10.0	12.0	14.0	16.0
C_v	0.4	0.8	1.3	1.8	2.4	3.0	3.8	4.8

(a) Deduce the value of the *electronic* specific heat at 10 K.
(b) Deduce the value of the *total* specific heat at 20 K.

10. At low temperatures, ^{3}He atoms injected into liquid ^{4}He form a monolayer on the liquid surface. We can consider this monolayer to be a *two-dimensional* Fermi gas; we neglect any motion perpendicular to the surface.
(a) Show that the density of states $g(\varepsilon)$ is independent of ε for this gas.
(b) Compute the value of ε_f for an ^{3}He surface density of 6×10^{14} atoms/cm^2.

11. Consider a two-dimensional *Bose–Einstein* gas similar to the two-dimensional Fermi gas of the previous problem. Go through the same steps used to deduce the condensation of a *Bose–Einstein* gas (Section 16.4) and show that condensation does not occur in this case.

APPENDIX **A**

Probability and Statistics

The concepts and language of probability and statistics, which are essential for an understanding of quantum theory, are unknown to many science students. Therefore we present the rudiments of this subject for those students. As a branch of mathematics, probability can be based upon axioms and definitions that are to be accepted without proof. As physicists, however, we need to apply our knowledge to the real world. Therefore, we shall try to make these axioms plausible on physical grounds as we introduce them.

A.1 DEFINITIONS

We define a probability as a number between 0 and 1 which expresses the likelihood that a certain event will occur. Zero probability means that the event will certainly *not* occur; for example, the probability that the moon will turn around and orbit the earth in the opposite direction next month is zero, in our estimation. A probability of 1 means that the event will certainly occur. A probability greater than 1 is meaningless.

Probability, in general, may be thought of as a measure of our degree of ignorance. Two events are equally probable if, in our ignorance, we have no reason to expect one to occur rather than the other. Consider one roll of a single die. We expect each of the six faces to come up with equal probability, because we have not examined that die and have no reason to expect one face to appear more often than any other face. Since there are only six possibilities, and each has the same probability, the probability that any given number will come up on a given roll must be $1/6$, to make the total probability equal to the required value of 1.

A.1.1 A Priori Probability

We can summarize the so-called *a priori* definition of probability in two steps. Given a set of n possible events:

- If we can find no reason why event A should occur more often than event B, then we say that A and B are equally probable.
- If all n events are all equally probable, if only one of them can occur (the events are *mutually exclusive*), and if no other event is possible (the set is *exhaustive*), then the probability P of occurrence of a particular one of these events is $P = 1/n$.

This definition is illustrated by the tossing of two identical coins. We can identify three events that are mutually exclusive and exhaustive: (a) both come up heads; (b) both come up tails; (c) one comes up heads, the other tails. But the probability of each event is *not* $1/3$; there is a good reason why (c) should occur more often than either (a) or (b). The coins are not identical; we can distinguish one from the other, and thus there are *four*, not three, events that are mutually exclusive and exhaustive: (1) both come up heads; (2) both come up tails; (3) the one on the left (coin L) comes up heads, the other (coin R) tails; (4) coin L is tails, and R is heads.

These four events are equally likely because the coins do not communicate with each other. When A is heads, there is no reason that B should be more likely to be either heads (giving event 1) or tails (giving event 3). By the same reasoning, events 2 and 4 are equally probable, and 2 and 3 are also equally probable. Thus the probability of any one of the *four* events is $1/4$, and the probability of the three events labeled (a), (b) and (c) are $1/4$, $1/4$, and $1/2$, respectively, because (c) includes both (2) and (4).

A.1.2 A Posteriori Probability

If this reasoning is not to your liking, you can turn to *a posteriori probability*, determining the probabilities from experience. Simply flip two coins many times, and see how many times each event occurs. The results may or may not agree with the probabilities given above. If they do not agree, the reason may be that the coins are not what you assumed them to be. In that case you must reassess your judgment of the probabilities.

In many situations (as in making weather predictions) this *a posteriori* method is all that you can use, because you have no way of selecting a set of equally likely results as you might with coin tossing.

A.1.3 Compound Probability

We also need to consider the probability that two events, A and B, will occur, when the probability of B is influenced by the occurrence of A. If $p_A(B)$ is the probability that B occurs, *given that A occurs*, then the probability $p(AB)$ that both A and B occur must be

$$p(AB) = p(A)p_A(B) \tag{A.1}$$

If A and B are independent events, then $p_A(B) = p(B)$, and we have

$$p(AB) = p(A)p(B) \tag{A.2}$$

A.2 BINOMIAL DISTRIBUTION

If we flip a normal coin n times, we might expect heads to come up $n/2$ times. Obviously this cannot happen if n is an odd number; in fact it hardly ever happens at all. However, the result is usually quite close to $n/2$. If we are testing the hypothesis that we are flipping a normal coin, it is important to know how much deviation from $n/2$ we should consider to be consistent with this hypothesis.

In general we can consider n occasions (trials). On each trial there is a probability p that event A can occur. Let $q = 1 - p$, the probability that the event will *not* occur on any particular trial. The probability that A will occur *exactly r* times in n trials is denoted as $P_r(n, p)$.

To compute $P_r(n, p)$, we begin with the probability that A will occur on a *specific set* of r trials (and not on the other $n - r$ trials); this compound probability is $p^r q^{n-r}$. There is the same probability that A will occur on any other *specific set* of exactly r trials out of a total of n trials. Since the number of distinct sets of r objects that can be chosen from n objects is equal to $n!/r!(n - r)!$ (Section 16.4), the result is

$$P_r(n, p) = p^r q^{n-r} \frac{n!}{r!(n - r)!} = \frac{n(n + 1) \cdots (n - r + 1)}{r!} \tag{A.3}$$

Graphs of $P_r(n, p)$ for two sets of values of n and p are shown in Figure A.1.

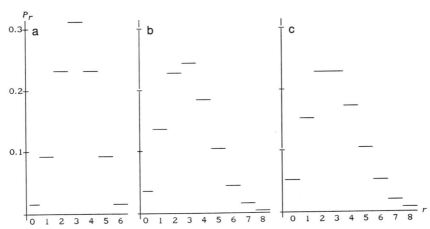

FIGURE A.1 Probability P_r of r occurrences of an event in n trials, if probability of occurrence is p on each trial. (a) Binomial distribution, $n = 6$, $p = 1/2$. (b) Binomial, $n = 18$, $p = 1/6$. (c) Poisson, $np = 3$.

A.2.1 Generating Function

Expression (A.3) appears as the $(r + 1)$th term in the expression for the nth power of the binomial $q + p$:

$$(p + q)^n = q^n + nq^{n-1}p + \frac{n(n-1)}{2} + \cdots$$
$$+ \frac{n(n-1)\cdots(n-r+1)}{r!}q^{n-r}p^r + \cdots$$

The function $(q + ps)^n$ can therefore be used as a generating function for values of $P_r(n, p)$. The *coefficient* of s^r in this function is equal to the probability that event A will occur exactly r times in n trials.

A.3 POISSON DISTRIBUTION

In many situations of interest in physics, n is very large but p is very small. For example, a photon counter might have a resolving time of 1 μs; during each microsecond a photon is either detected or not. Suppose that only one photon per second is detected; that means the probability p of an event during each 1-μs "trial" is only 10^{-6}, but in each second there are

$n = 10^6$ trials. Equation (A.3) is obviously clumsy for such situations, but we can modify the binomial distribution for these cases.

We write the generating function, with $q = 1 - p$, as

$$(1 - p + ps)^n = \left\{ 1 - \frac{m}{n} + \frac{ms}{n} \right\}^n = \left\{ 1 - \frac{m(1 + s)}{n} \right\}^n \quad \text{(A.4)}$$

where m, the mean number of events, is equal to np.

The *Poisson distribution* is found by taking the limit of this function as $n \to \infty$ and $p \to 0$ while m remains constant. In this limit, recalling that $e^x = \lim_{n \to \infty} \{1 + (x/n)\}^n$, we can rewrite Eq. (A.4) in the limit $n \to \infty$, as

$$e^{-m(1-s)} = e^{-m}e^{ms} = e^{-m}\left\{ 1 + ms + \frac{m^2 s^2}{2!} + \cdots \frac{m^r s^r}{r!} + \cdots \right\} \quad \text{(A.5)}$$

This series is the generating function for the Poisson distribution; the coefficient of any given power of s is equal to the probability of s occurrences of the event in question. That is, the probability of r occurrences is, in general

$$P_r(m) = \frac{m^r}{r!}e^{-m} \quad \text{(A.6)}$$

Figure A.1 compares the Poisson distribution with two binomial distributions. The binomial distribution is symmetric for $p = 1/2$, but it must become asymmetric when p becomes smaller, because r cannot be negative. That is, r cannot deviate from the mean by more than m on the negative side, but it can do so on the positive side. The Poisson distribution is thus always asymmetric, but for large values of m it approaches a symmetric curve, as seen in Figure A.2.

A.4 NORMAL DISTRIBUTION

One can, in principle, compute the Poisson probability for any values of m and r, but for large m there is a shortcut, whose result is shown in Figure A.2 for $m = 25$. We see that the Poisson probabilities are matched rather well by the boldface lines showing the "normal distribution," whose values

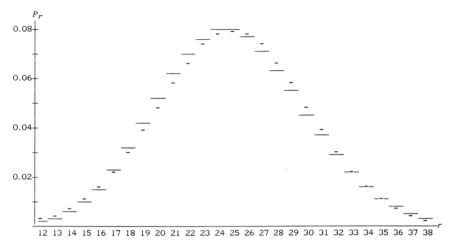

FIGURE A.2 Poisson distribution (thin lines) for $m = 25$. Bold, shorter lines show the normal distribution that most closely fits this Poisson distribution (with $\sigma = \sqrt{m} = 5$).

are obtained from the expression

$$P(r) = \frac{1}{\sigma\sqrt{2\pi}}e^{-(r-m)^2/2\sigma^2} \qquad (A.7)$$

where σ is the *standard deviation*. You can verify that Figure A.2 is consistent with Eq. (A.7) if $\sigma = 5$.

In general, a normal curve is a limiting case of a binomial or a Poisson distribution for large m, with $\sigma = \sqrt{npq}$ or $\sqrt{m}$, respectively.[1]

A short routine for a programmable calculator will give the probability that, for a normal distribution, r will be *greater than* the mean by an amount $x\sigma$ or more. The following code in BASIC will work for a Casio fx-700P:

```
1 INPUT_"X",X:T = 1/(1 + .2316419*X):Q =
  1/SQR_(2*π)*EXP_(X*X/2)
2 A = .31938153:B = −.356563782:C = 1.78147937:D =
  −1.821255978:E = 1.330274429
3 PRINT_"P(X)" = ";Q*(A + T + B*T↑2 + C*T↑3 + D*T↑4
  + E*T↑5):END
```

[1]For more examples and a derivation of the connection between σ and m, refer to *Introduction to Modern Physics* by John D. McGervey, op. cit., Appendix A.

The following table gives the probability that a result will be less than[2] the mean by an amount of $x\sigma$ or more, for selected values of x:

x	0	.25	.50	.75	1.00	1.25	1.50	1.75
P	0.50	.4013	.3085	.2266	.1587	.1056	.06681	.04006

x	2.00	3.00	4.00	5.00
P	.02275	.00135	.00003	3×10^{-7}

A.5 EXAMPLE: STATISTICS IN DETECTING GAMMA RAYS

We sometimes test a photon detecting system by counting. Suppose that we expect to see the same number of photons in each test, but the numbers counted in equal time intervals (using an automated system) are 404, 427, 398, 388, 401, and 298.

The mean of the first five is about 400, suggesting a σ of about 20. The last result is 5σ below the mean; the probability of such a deviation is 3×10^{-7}—not impossible, but less likely than a malfunction of the apparatus. Investigation may show that the plug was pulled during the sixth interval.

[2] Notice the probability of 0.50 for $x = 0$. A number is just as likely to be greater than the mean as it is to be less than the mean. In the limit $n \to \infty$ there is a negligible probability of a result yielding the exact mean value.

APPENDIX **B**

The Boltzmann Factor

In classical statistical mechanics, a system of randomly moving particles has an energy distribution that is proportional to the *Boltzmann factor*, which is $e^{-E/kT}$ for a particle of energy E in a system whose temperature is T. We can illustrate this factor through the law of atmospheres, which shows how the density of a planetary atmosphere depends on altitude and temperature.

The concept of temperature is central to any understanding of the behavior of a collection of particles. Consider a column of gas kept at a uniform temperature. We wish to determine the density of the gas as a function of height. The column is acted on by gravity, so that the pressure P and density ρ in the gas are dependent on height z, although the temperature T is not. Then at height $z + dz$ the pressure is $P + dP$, and dP, the change in pressure between height z and height $z + dz$, is given by

$$dP = -\rho g\, dz \qquad (B.1)$$

because $\rho g\, dz$ is the weight per unit area of a column of gas of height dz, when the acceleration of gravity is g. We also have the ideal gas law, which may be written

$$P = \rho RT/M \qquad (B.2)$$

where M is the mass of one mole of the gas, R is the gas constant, and ρ/M is the number of moles of gas per unit volume.

Division of Eq. (B.1) by Eq. (B.2) eliminates ρ, giving the relation

$$\frac{dP}{P} = -\frac{Mg}{RT}\, dz \qquad (B.3)$$

Integration now gives us the dependence of P on the height z:

$$P = P_0 e^{-Mgz/RT} \tag{B.4}$$

where P_0 is the pressure at $z = 0$. Of course, the number n of molecules per unit volume is proportional to the pressure, so we can also write

$$n = n_0 e^{-Mgz/RT} = n_0 e^{-mgz/kT} \tag{B.5}$$

where we have set R equal to kM/m, m is the mass of a single molecule, and k is the Boltzmann constant (equal to 8.617×10^{-5} eV per degree). Notice that the ratio M/m is equal to Avogadro's number N_A, so k also equals R/N_A. Notice also that mgz is the gravitational potential energy of each particle. Thus the factor $e^{-mgz/kT}$ is the Boltzmann factor.

APPENDIX **C**

Relativistic Dynamics

Here we present the bare bones of the logical basis of special relativity. Information on the crucial experiments leading to acceptance of the theory can be found in many texts [such as R. Resnick and D. Halliday, *Basic Concepts in Relativity & Early Quantum Theory*, 2nd edition, Wiley, New York (1985)].

C.1 THE PRINCIPLE OF RELATIVITY

Galileo stated the principle of relativity as

> No experiment can tell if one is at rest or moving uniformly.

An alternative statement of the principle is

> The laws of nature are the same in all uniformly moving laboratories.

Another name for a uniformly moving laboratory is an *inertial frame of reference*.

C.2 THE SPEED OF LIGHT IN FREE SPACE

Light is a wave, and waves travel in a material medium with a characteristic speed relative to that medium. But light waves, unlike many other waves, also travel in empty space. How is the characteristic speed defined in that case? Einstein, considering the experimental evidence, expanded the principle of relativity to include a second postulate:

Light in empty space always travels with the same definite speed c with respect to any inertial frame of reference, independently of the state of motion of the body emitting the light.

The postulate is in agreement with all known observations of the speed of light. However, the fundamental thrust of the postulate is *speed*, not *light* itself. One could rephrase the postulate to omit any specific reference to light, as follows:

There is a characteristic limiting speed in the universe, such that anything traveling at that speed with respect to one inertial frame must also travel at that speed with respect to any other inertial frame.

Experiments on light all lead to the conclusion that light travels at that characteristic speed.

C.3 INERTIAL FRAMES

A frame of reference may be defined as an infinite set of perfect clocks, each of which can be placed at any point, plus a set of perfectly rigid measuring rods that determine the space coordinates of each point. The rods and clocks then determine the coordinates (x, y, z, and t) of an *event*. Any event could be measured in a different (primed) inertial frame by using clocks and rods that are moving relative to the set used in the first frame. The resulting coordinates for the *same event* are x', y', z', and t'.

Knowing the values of x, y, z, and t for an event, and knowing the velocity v of the second frame relative to the first frame, we should be able to calculate the values of x', y', z', and t' that describe that same event. The formula for finding these values is called a *coordinate transformation*.

C.3.1 Galilean Transformation

Before Einstein, it was universally assumed that for any event $t = t'$. To transform the space coordinates, we can always choose the velocity **v** of the primed frame S', relative to the unprimed frame S, to be parallel to the x axis. This procedure leads to the *Galilean transformation*:

$$x' = x - vt, \qquad y' = y, \qquad z' = z, \qquad t' = t \qquad \text{(C.1)}$$

Figure C.1 suggests that the x' coordinate is related to the x coordinate as in the Galilean transformation. But *this transformation conflicts with Einstein's second postulate.* A flash of light, starting at $x = 0$ when $t = 0$, reaches coordinate x at time t such that $x/t = c$. The same flash of light starts at $x' = 0$ when $t' = 0$, and its speed in the S' frame is, by definition, x'/t'. However, by Eq. (C.1), $x'/t' = (x - vt)/t = c - v$. The speed in the S' frame is thus not c, and the postulate is violated. Therefore we must scrap Eq. (C.1) and look for a better one.

C.3.2 Lorentz Transformation

Einstein's postulate requires that x/t must always be equal to x'/t' for the coordinates of a light flash that goes through the origin of both frames at time $t = t' = 0$. This requirement is satisfied by the Lorentz transformation (for relative motion along the x axis):

$$\boxed{x' = (x - vt)\gamma, \qquad y' = y, \qquad z' = z, \qquad t' = [t - vx/c^2]\gamma} \quad \text{(C.2)}$$

where $\gamma = \sqrt{1 - (v/c)^2}$. You may verify that this transformation gives the required result; divide the expression for x' by the expression for t', set x/t equal to c, and you will see that x'/t' also equals c. Furthermore, for $v \ll c$, the Lorentz transformation becomes equivalent to the Galilean transformation, and Figure C.1 gives a true picture.

C.4 ADDITION OF VELOCITIES

From the Lorentz transformation we can find a formula for the velocity $u'_x = dx'/dt'$ of an object as measured by an observer at rest in frame S', given the velocity $u_x = dx/dt$ of the same object as measured by an observer at rest in frame S. Let us simplify the equations by assuming that

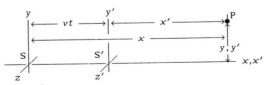

FIGURE C.1 Frame S' moves along x axis with speed v relative to frame S. An event occurs at point P. The origins of S and S' coincide at time $t = t' = 0$.

each velocity is constant. The $u_x = x/t$, $u'_x = x'/t'$, and

$$u'_x = \frac{x'}{t'} = \frac{(x - vt)\gamma}{[t - (vx/c^2)]\gamma} = \frac{(x/t) - v}{1 - (vx/c^2 t)} = \frac{u_x - v}{1 - (u_x v/c^2)} \quad \text{(C.3)}$$

In similar fashion we find

$$u_y = \frac{y'}{t'} = \frac{y}{[t - (vx/c^2)]\gamma} = \frac{u_y}{[1 - (u_x v/c^2)]\gamma} \quad \text{(C.4)}$$

$$u_z = \frac{z'}{t'} = \frac{z}{[t - (vx/c^2)]\gamma} = \frac{u_z}{[1 - (u_x v/c^2)]\gamma} \quad \text{(C.5)}$$

You can verify that, when $u_x = c$, $u'_x = c$ as well. Another example is the case $u_x = 0.6c$ and $v = -0.6c$. The Galilean transformation gives $u'_x = u_x - v = 0.6c - (-0.6c) = 1.2c$.

But nothing can go faster than the speed of light relative to any frame. The Lorentz transformation gives

$$u'_x = \frac{u_x - v}{1 - (u_x v/c^2)} = \frac{1.2c}{1 - (0.6)(-0.6)} = (1.2/1.36)c = (15/17)c$$

$$\text{(C.6)}$$

C.4.1 Synchronization of Clocks

Time is an abstraction that can only be measured by reading a clock. If clocks are to give a true measure of time everywhere, clocks at different places must be synchronized. Two identical clocks at different locations can be synchronized by starting both of them with a flash that originates halfway between them. *If both clocks are at rest* in frame S, both clocks will receive the starting signal simultaneously; being identical, they will remain synchronized from then on. But to an observer in frame S', these clocks are not at rest; one moving toward the light flash will start sooner than the other one, moving away from it. Thus the first clock starts before the other one and stays ahead of it. The conclusion is

> Moving clocks at different locations are, in general, not synchronized.

This conclusion also follows directly from the Lorentz transformation, which shows that the time on a clock that is moving along the x axis is dependent on its x coordinate.

C.4.2 Lorentz Contraction

Obviously, when v approaches c, the length x' in Figure C.1 cannot be the same length x' that appears in the Lorentz transformation equations. The figure shows the distance between the y' axis and point P as determined by a measuring rod *at rest in frame S*. But by definition, x' must be determined by a measuring rod *at rest in frame S'*.

If the length of a rod is L when it is at rest, the Lorentz transformation shows that the rod is contracted to a length $L' = L/\gamma$ when it is moving in the direction parallel to its length. Table C.1 illustrates this point by showing the space and time coordinates of several events. In this table the unit of length is the light-second; thus $c = 1$ in units of light-seconds per second. The relative speed of the two frames is $v = 0.6$ in these units, making γ equal to 1.25. The Lorentz transformation is used to find x' and t', given the values of x, t, and γ.

To measure the length of a rod (or the diameter of a planet), you must locate (by taking a photograph, perhaps) both ends of the object *at the same time*. Events 0 and A, photographs taken at the same time t, show that a certain object has a length L of 4 units.

In frame S' events 0 and B occur at the same time. They show that the ends of the object are separated by 3.2 units in space; therefore the object's length L', *the length seen in S'*, is only 3.2 units. This satisfies the equation

$$L' = L/\gamma \qquad (C.7)$$

because $3.2 = 4/\gamma = 4/1.25$. Observers in both frames *agree on what each photograph shows*, but they still disagree on the length of the object; the observer in S will say that the events 0 and A were at different times, and that the object moved during the time that elapsed between the photographs of those events.

TABLE C.1 Coordinates of Events in Two Reference Frames

Event:	0	A	B	C	D	E	F
$x =$	0	+4	+4	+3	0	+8	0
$t =$	0	+0	+2.4	+5	+4	+8	+8
$x' =$	0	+5	+3.2	0	−3	+4	
$t' =$	0	+3	0	+4	+5	+4	

C.4.3 Time Dilation

As moving rods contract, we find that moving clocks keep time more slowly than identical clocks at rest. The general rule is

> A moving clock runs more slowly than an identical clock at rest.

and the ratio of the clock rates is the same factor γ. Consider a clock that is moving with speed v with respect to frame S and is present at two events (such as 0 and C in Table C.1). If the clock registers a time interval τ between these two events, and the time interval between these two events is t_0 according to clocks in frame S, then

$$\tau = t_0 \gamma \qquad (C.8)$$

You can see that the intervals shown in the table are $\tau = 4$ units and $t_0 = 5$ units, in agreement with the known value of γ for the relative speed of 0.6 between these two frames.

C.4.4 Proper Time

Although observers in different frames may disagree on the time interval between two events, they all agree on the time interval that is recorded by a clock that is present at both events (such as the one that recorded the time of 4 units in the above example). This time interval τ is called the *proper time*. It is given by a formula equivalent to Eq. (C.8):

$$ct = \left[c^2(t_A - t_B)^2 - (x_A - x_B)^2 - (y_A - y_B)^2 - (z_A - z_B)^2 \right]^{1/2} \quad (C.9)$$

where the coordinates apply to any frame of reference whatsoever, and the subscripts A and B refer to their values at events A and B, respectively. For example, the proper time interval between events A and B in Table C.1 above can be computed in frame S to be (with $c = 1$):

$$\tau = \sqrt{(2.4 - 0)^2 - (4 - 4)^2} = 5.76 - 0 = 2.4 \qquad (C.10)$$

and be computed in frame S' to be

$$\tau = \sqrt{(0 - 3)^2 - (3.2 - 5)^2} = \sqrt{9 - 3.24} = 2.4 \qquad (C.11)$$

C.4.5 Space–Time Interval between Events

If we compute the proper time between events C and D, we find that it is imaginary:

$$\tau = \sqrt{(4-5)^2 - (0-3)^2} = \sqrt{-8} = 2.83i \qquad \text{(C.12)}$$

This result is appropriate; a clock that is present at both events must be imaginary! No clock can be present at both events. (To do that, it would have to travel faster than the speed of light between these events). However we can still define τ to be the *space–time interval* between these events. When this interval is imaginary it is called a *spacelike interval*. For such an interval, there must be a frame of reference in which the two events occur at the same time. Conversely, as we have seen, when there is a real proper time between two events, there must be a frame of reference in which these events occur at the same *place* (the place where the clock is located).

The space–time interval between events 0 and E is zero. This set of events illustrates the fact that light travels at the same speed in each frame of reference. In both frames of reference, one must travel at the speed of light to be present at both of these events. Notice that the individual coordinates differ in the two frames, but the speed deduced for a moving object in either frame is simply c.

C.5 CONSERVATION LAWS

The principle of relativity requires that

The laws of conservation of energy and conservation of momentum are valid in all inertial frames of reference.

C.5.1 Momentum

The classical definition of momentum, $p = mu$,[1] can be shown to be incompatible with the above statement when the Lorentz transformation is used. We can find the compatible definition by using the *proper time τ* to determine the velocity. To simplify the form of the following equations, let us consider a particle of constant velocity that goes through the origin at time $t = 0$.

[1] Here we use u instead of v for velocity, to avoid confusion with the use of v for the velocity of the frame of reference S'.

Then the classical momentum components are

$$p_x = mx/t = mu_x, \qquad p_y = my/t = mu_y, \qquad p_z = mz/t = mu_z \quad \text{(C.13)}$$

and the relativistically correct components are:

$$p_x = mx/\tau = mu_x\gamma, \qquad p_y = my/\tau = mu_y\gamma, \qquad p_z = mz/\tau = mu_z\gamma$$
$$\text{(C.14)}$$

We see that Eqs. (C.14) reduce to (C.13) in classical physics, where $v \ll c$ and therefore $\gamma \to 1$. However, we must also allow for a *fourth* component of the momentum, because the Lorentz transformation is four-dimensional, involving time as well as space. We can write this component, analogous to the first three, as

$$p_t = mt/\tau = m\gamma \qquad \text{(C.15)}$$

C.5.2 Energy

In any interaction, all four components of the total momentum of the system must be conserved. Therefore p_t as well as the other components must be conserved. We also have a law of conservation of energy, and we can now assert that p_t should be proportional to energy, so that the law of conservation of energy becomes a part of the law of conservation of momentum. Let us therefore multiply both sides of (C.15) by c^2 to obtain a quantity with the dimensions of energy, and test this assertion. Calling this quantity E, we have

$$c^2 p_t = mc^2\gamma = E \qquad \text{(C.16)}$$

If this is indeed energy, this expression should agree with the classical definition of kinetic energy for $u \ll c$. The binomial theorem gives

$$mc^2\gamma = mc^2\left[1 - (u/c)^2\right]^{-1/2} = mc^2[1 + u^2/2c^2 + 3u^4/8c^4 + \cdots] \qquad \text{(C.17)}$$

If $u \ll c$ we use only the first two terms, and the presumed energy E becomes

$$E \cong mc^2 + mu^2/2 \qquad \text{(C.18)}$$

C.5.3 Definition of Kinetic Energy

The second term in Eq. (C.18) is the *classical* kinetic energy. The first term is the energy of the particle when it is at rest. The *relativistic* kinetic energy K is given by $E - mc^2$, with $E = mc^2\gamma$.

Thus the kinetic energy may be written

$$K = mc^2\gamma - mc^2 = mc^2(\gamma - 1) \tag{C.19}$$

If u were to equal the speed of light c, then γ would be infinite. This shows how the speed of light is a limiting speed.

C.5.4 Energy–Momentum Relation

By squaring and adding both sides of each of the Eqs. (C.14) and (C.16), we obtain the relation

$$mc^2 = \left[E^2 - c^2\left(p_x^2 + p_y^2 + p_z^2 \right) \right]^{1/2} \tag{C.19}$$

or

$$\left(mc^2\right)^2 = E^2 - c^2 p^2 \tag{C.20}$$

where $p^2 \equiv p_x^2 + p_x^2 + p_z^2$.

Equation (C.20) makes it possible to find the mass m of a particle by measuring its energy and momentum in any frame of reference; mass, like proper time, is independent of frame of reference. If the particle is a photon we have found that $E = cp$; we conclude that the photon has zero mass.

C.5.5 Inelastic Collisions

An inelastic collision is defined as one in which kinetic energy is converted into some other form of energy, such as heat, while the total energy is always conserved. With relativistic mechanics, all such forms of energy are included in the value of E. Figure C.2 shows what happens in a particular elastic collision, as seen in two frames of reference. In frame S, body B is at rest; in frame S′, the center of mass of the two bodies is at rest and the total momentum is zero.

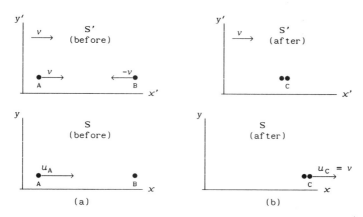

FIGURE C.2 Inelastic collision of two equal-mass objects as seen in two reference frames, S and S', (a) before the objects collide and (b) after they collide. Frame S' is moving with constant velocity v relative to frame S.

Let us define M to be the rest mass of each separate body. In frame S' each mass has the same speed v, and the total initial energy is

$$E_i' = 2Mc^2[1 - (v^2/c^2)]^{-1/2} \qquad (C.21)$$

Let us assume that A and B stick together to form C. Then:

In frame S', energy is conserved, but the speed of the resulting particle is zero. Since the kinetic energy has gone to zero without a change in total energy E', *the total mass must have increased.* Suppose that $v = 0.6$. Then A and B each begin with an energy of $Mc^2\gamma_i'$, and $\gamma_i' = 1.25$, so the total energy is to 2.5 Mc^2. After the collision, the composite body C must have 2.5 Mc^2 of *rest energy only.* Therefore the mass of C must be 2.5 M.

Mass is not conserved; in this case it has increased.[2] Energy is conserved, but it does not have same value in all frames of reference.

In frame S, the total mass has also increased from $2M$ to $2.5M$, because mass is independent of reference frame. However there is kinetic energy after the collision. The speed of the combined particle C must be equal to the speed of frame S' relative to frame S (because C is at rest in frame S'); thus $u_C = v$.

[2] Here we always use the word "mass" to denote an invariant quantity. This quantity is sometimes called "rest mass" to distinguish it from so-called "relativistic mass." The use of the latter term has often led to confusion.

Energy is also conserved in frame S. The final energy E_f in frame S is that of the single object C of mass 2.5 M moving at speed v, so

$$E_f = 2.5Mc^2\gamma_C, \quad \text{where } \gamma_C = [1 - (v^2/c^2)]^{-1/2} \qquad \text{(C.22)}$$

The initial energy E_i in frame S is the sum of the energy of A and the energy of B, or

$$E_i = E_A + E_B = Mc^2\gamma_A + Mc^2 \quad \text{where } \gamma_A = \left[1 - (u_A^2/c^2)\right]^{-1/2} \quad \text{(C.23)}$$

To show that E_i is equal to E_f as given in Eq. (C.23), we must use Eq. (C.3) to find the value of u_A in terms of v. Solving that equation for the unprimed x component of u_A gives us

$$u_A = \frac{v + v}{1 + (v/c)^2} \qquad \text{(C.24)}$$

The results *for frame S* are:

$$u_A = \frac{0.6 + 0.6}{1 + 0.36} = 15/17c, \qquad \gamma_A = [1 - (15/17)]^{-1/2} = 17/8$$

$$\gamma_C = [1 - (v^2/c^2)]^{-1/2} = [1 - (3/5)]^{-1/2} = 5/4$$

so that the initial energy and momentum are

$$E_i = E_A + E_B = (17/8)Mc^2 + Mc^2 = \underline{(25/8)Mc^2}$$

and

$$p_i = Mu_A\gamma_A = \underline{(15/8)Mc}$$

and the final energy and momentum are

$$E_f = E_c = 2.5Mc^2\gamma_C = \underline{(25/8)Mc^2} \quad \text{and} \quad p_f = 2.5Mv\gamma_C = \underline{(15/8)Mc}$$
$$\text{(C.25)}$$

Thus energy and momentum are conserved in both frames, although each of these variables has a different value in S than it has in S'.

In each frame *kinetic* energy is converted to *rest* energy. As you can see, in S', kinetic energy decreased from $0.5Mc^2$ to zero; in S, it decreased *by the same amount*: from $(9/8)Mc^2$ to $(5/8)Mc^2$. In classical mechanics we say that the kinetic energy lost in an inelastic collision has been converted to *heat*. It has; creation of mass is always accompanied by an increase in temperature.

We conclude that rest mass increases with temperature, and therefore the temperature of a body, like its rest mass, is invariant; it is the same in all frames of reference. For example, if an object melts or is vaporized, all observers agree that this phase change has occurred.

The reaction depicted in Figure C.2 can also proceed in the opposite direction. To analyze the reverse reaction, we simply exchange the labels E_i and E_f and reverse the signs of all velocities. With the data given above, we have an increase in kinetic energy and loss of rest mass. *Nuclear fission* is an example of this sort of reaction.

C.5.6 Relating p and E to u

In general, $pc/E = u/c$ (where $p = \sqrt{p_x^2 + p_y^2 + p_z^2}$). This useful relation is derived by eliminating γ from Eqs. (C.14) and (C.16).

Derivation of the Eigenfunctions of the L^2 Operator

In Chapter 6 it was shown that the wave function $\Psi(r, \theta, \phi)$ could be written as a product of functions $R(r)\Theta(\theta)\Phi(\phi)$, and that the three-dimensional Schrödinger equation could be separated into three differential equations. When this was done, the equation for the function $\Theta(\theta)$ was shown to be (here we let $m = 0$; we can then generate the solutions for the case $m \neq 0$)

$$\frac{\sin \theta}{\Theta(\theta)} \frac{d}{d\theta} \left\{ \sin \theta \frac{d}{d\theta} \right\} \Theta(\theta) = \alpha \hbar^2 \sin^2 \theta \qquad (6.49)$$

It was shown by a statistical argument that the constant α must be equal to $l(l + 1)$. This can also be shown by solving Eq. (6.49), as follows.

We further simplify the form of the equation by changing the variable to $\xi = \cos \theta$, so that $\sin^2 \theta = 1 - \xi^2$. We define $F(\xi) = \Theta(\theta)$, and we eliminate $d/d\theta$ by the equation

$$d\Theta/d\theta = (dF/d\xi)(d\xi/d\theta) = -\sin^2 \theta \, (dF/d\xi) = (\xi^2 - 1)(dF/d\xi) \qquad (D.1)$$

Substitution into Eq. (6.49), with $m = 0$, then yields (after some algebra) *Legendre's equation*:

$$\frac{d}{d\xi} \left\{ (1 - \xi^2) \frac{dF}{d\xi} \right\} + \alpha F = 0 \qquad (D.2)$$

Let us seek a solution to Eq. (D.2) in the form of a power series in ξ:

$$F(\xi) = \sum_{s=0}^{\infty} a_s \xi^s = a_0 + a_1 \xi + a_2 \xi^2 + \cdots + a_k \xi^k + \cdots \quad \text{(D.3)}$$

whose derivative may be written

$$\frac{dF}{d\xi} = a_1 + 2a_2 \xi + \cdots + ka_k \xi^{(k-1)} + (k+1)a_{k+1} \xi^k + \cdots \quad \text{(D.4)}$$

If we now write D.2 in the form

$$\frac{d^2 F}{d\xi^2} - \xi^2 \frac{d^2 F}{d\xi^2} - 2\xi \frac{dF}{d\xi} + \alpha F = 0 \quad \text{(D.5)}$$

the individual terms may be expanded to obtain

$$\frac{d^2 F}{d\xi^2} = 2a_2 + \cdots + (k+2)(k+1)a_{k+2} \xi^k + \cdots$$

$$-\xi^2 \frac{dF}{d\xi^2} = -6a_2 \xi^2 - \cdots - k(k-1)a_k \xi^k - \cdots$$

$$-2\xi \frac{dF}{d\xi} = -4a_2 \xi^2 - \cdots - 2ka_k \xi^k - \cdots$$

$$\alpha F = \alpha a_0 + \alpha a_1 \xi + \cdots + \alpha a_k \xi^k + \cdots$$

The sum of all the terms on the left must be zero *for all values of* ξ, if Eq. (D.5) is to be satisfied. For example, if we set $\xi = 0$, we must have $0 = 2a_2 + \alpha a_0$. In general, the sum of the right-hand sides of these expressions can be zero only if the sum of the coefficients of *each power* of ξ is equal to zero. Thus, from the coefficients of the general term ξ^k, we obtain

$$(k+2)(k+1)a_{k+2} - k(k-1)a_k - 2ka_k + \alpha a_k = 0 \quad \text{(D.6)}$$

Rearranging terms gives us

$$(k+2)(k+1)a_{k+2} = [k(k-1) + 2k - \alpha]a_k \quad \text{(D.7)}$$

and a_{k+2} as a multiple of a_k is

$$a_{k+2} = \frac{k(k+1) - \alpha}{(k+2)(k+1)} \alpha_k \tag{D.8}$$

Equation (D.8) provides solutions of Eq. (D.5) consisting of either a series of odd powers of ξ or a series of even powers of ξ. The odd and even powers cannot be mixed, because of the requirement that the wave function must have a definite parity. When $a_0 = 0$, all even powers of ξ are missing, and when $a_1 = 0$, all odd powers of ξ are missing.

Not all solutions of Eq. (D.5) lead to acceptable wave functions, because the wave function must be continuous and it must be *single valued*, as we saw in Chapter 6 in the solution for the function $\Phi(\phi)$. To satisfy these conditions we require that the series representing $F(\xi)$ have a finite number of terms. This requirement is satisfied if the numerator in Eq. (D.8) is equal to zero for some value of k. This value of k is *defined* as l. We see that $\alpha = l(l+1)$.

The solutions of Eq. (D.5) for $l < 4$ are (using l as a subscript)[1]

$$F_0 = 1, \quad F_1 = \xi, \quad F_2 = (3\xi^2 - 1)/2, \quad F_3 = (5\xi^3 - 3\xi)/2 \tag{D.9}$$

Notice that all of these are normalized so that $F_l(1) = 1$ for any value of l. The general solution is, with this normalization,

$$F_l(\xi) = \frac{1}{2^l l!} \frac{d^l}{d\xi^l} (\xi^2 - 1)^l \tag{D.10}$$

The solutions of Eq. (6.49) are now found by returning to the definition $\xi = \cos\theta$. They are, therefore,

$$P_l(\cos\theta) = \frac{1}{2^l l!} \frac{d^l}{d(\cos\theta)^l} (\cos^2\theta - 1)^l \tag{D.11}$$

These are called *Legendre polynomials*. The first four are given in Eq. (6.51).

[1] In terms of θ, the first three of these are also given in Eq. (6.51), and they can be seen as factors in the list of spherical harmonics below Eq. (6.52).

Solutions when $m \neq 0$. The solutions of Eq. (6.46) [which reduces to Eq. (6.49) when $m = 0$) are called *associated Legendre functions* and denoted by the symbol $P_l^{|m|}(\cos \theta)$. These functions are listed in Chapter 6 for $l < 3$. They can all be generated from the Legendre polynomials by using the stepping operator method (used in Section 10.3 for generating eigenfunctions of J_z).

APPENDIX E

Solution of the Radial Equation for the Hydrogen Atom

We wish to derive bound-state solutions of Eq. (9.11) for the single-electron atom or ion with $V(r) = -Ze^2/4\pi\varepsilon_0 r$ (Z being the atomic number). Thus

$$\frac{d^2}{dr^2}[rR_{nl}] = -\frac{2m_r}{\hbar^2}\left\{E_n + \frac{Ze^2}{4\pi\varepsilon_0 r} - \frac{l(l+1)\hbar^2}{2m_r r^2}\right\}rR_{nl} \qquad (E.1)$$

where n is the radial quantum number and $l(l+1)\hbar^2/2m_r r^2$ the "centrifugal potential." To simplify the subsequent equations we will introduce the symbols $\alpha_n = \sqrt{-2m_r E_n/\hbar^2}$ and $C = 2m_r Ze^2/4\pi\varepsilon_0\hbar^2$. (Because E_n is negative, α_n is real.)

With these substitutions, Eq. (E.1) becomes

$$\frac{d^2}{dr^2}[rR_{nl}] = \left\{\alpha_n^2 - \frac{C}{r} + \frac{l(l+1)}{r^2}\right\}rR_{nl} \qquad (E.2)$$

To solve such an equation it is useful to begin with the limit as $r \to \infty$. In that limit we have

$$\frac{d^2}{dr^2}[rR_{nl}] = \alpha_n^2 rR_{nl} \qquad (E.3)$$

One solution of this equation is $rR_{nl} = e^{-\alpha_n r}$. We now show that any function of the form $r^q e^{-\alpha_n r}$ is also a solution of Eq. (E.3). We write

388

$rR = r^q e^{-\alpha_n r}$, so

$$\frac{d}{dr}(rR) = -\alpha_n r^q e^{-\alpha_n r} + qr^{q-1}e^{-\alpha_n r} \rightarrow -\alpha_n r^q e^{-\alpha_n r} \quad \text{as } r \rightarrow \infty$$

and

$$\frac{d^2}{dr^2}(rR) \rightarrow \alpha_n^2 r^q e^{-\alpha_n r} = \alpha_n^2(rR) \quad \text{as } r \rightarrow \infty \tag{E.4}$$

which is precisely Eq. (E.3) without the subscripts on the function $R(r)$. Therefore we try the product of $e^{-\alpha_n r}$ and a *power series* in r, as a solution for all values of r, and we write

$$rR(r) = g(r)e^{-\alpha_n r} \tag{E.5}$$

where $g(r)$ is a power series whose form we now seek. Substitution into Eq. (E.2) yields (after dividing out the factor $e^{-\alpha_n r}$)

$$\frac{d^2g}{dr^2} - 2\alpha_n \frac{dg}{dr} + \frac{Cg}{r} - \frac{l(l+1)g}{r^2} = 0 \tag{E.6}$$

Substitution of the power series $g(r) = r^s \sum_{q=0}^{\infty} b_q r^q$ into each term of Eq. (E.6) gives the following four series:

$$\frac{d^2g}{dr^2} = s(s-1)b_0 r^{s-2} + (s+1)sb_1 r^{s-1} + \cdots$$

$$+ (s+q+1)(s+q)b_{q+1} r^{s+q-1} + \cdots$$

$$-2\alpha_n \frac{dg}{dr} = -2\alpha_n \Big[sb_0 r^{s-1} + (s+1)b_1 r^s + \cdots$$

$$+ (s+q)b_q r^{s+q-1} + \cdots \Big]$$

$$\frac{Cg}{r} = C\Big[b_0 r^{s-1} + b_1 r^s + b_2 r^{s+1} + \cdots + b_q r^{s+q-1} + \cdots \Big]$$

$$-\frac{l(l+1)g}{r^2} = -l(l+1)\Big[b_0 r^{s-2} + b_1 r^{s-1} + \cdots + b_{q+1} r^{s+q-1} + \cdots \Big]$$

To satisfy Eq. (E.6), the sum of these four series must equal zero for *all values of* r. This can be true only if the sum of the coefficients of each power of r is equal to zero. The lowest power, with exponent $s - 2$, appears only twice, in the first series and in the last; the sum of the coefficients of r^{s-2} is thus

$$s(s - 1)b_0 - l(l - 1)b_0 = 0$$

The constant b_0 is nonzero by definition, because it is the coefficient of the smallest *nonzero* power of r in the solution. Thus

$$s(s - 1) = l(l - 1)$$

There are two values of s in terms of l for this "indicial equation." We rule out the negative value of s, because that would make $R(r)$ infinite at $r = 0$, and we find that $s = l + 1$.

Setting the coefficients of r^{s+q-1} equal to zero gives us a relation between the coefficients b_q and b_{q+1}:

$$b_{q+1} = b_q \left\{ \frac{2\alpha_n(s + q) - C}{(s + q + 1)(s + q) - l(l + 1)} \right\} \tag{E.7}$$

or, with $s = l + 1$,

$$b_{q+1} = b_q \left\{ \frac{2\alpha_n(l + 1 + q) - C}{ql + (q + 2)(l + q + 1)} \right\} \tag{E.8}$$

The value of b_0 is determined by normalization of the wave function. We can then determine any of the values b_q from Eq. (E.8). However, the resulting wave function must satisfy the boundary condition that $R(r) \to 0$ as $r \to \infty$, and it can be shown that an *infinite* series with these coefficients goes to infinity as $r \to \infty$.

We therefore need to find a value of E_n (or α_n) for which this series has a finite number of terms. This will happen if one coefficient, b_{q+1}, is zero; Eq. (E.8) will then ensure that all succeeding coefficients will be zero as well. When this occurs (and $b_q \neq 0$), Eq. (E.8) shows that

$$2\alpha_n(l + 1 + q) - C = 0 \tag{E.9}$$

and therefore

$$\alpha_n = \frac{C}{2(l + 1 + q)} \tag{E.10}$$

From the definitions of α_n and C we now find the energy E_n, in agreement with Eq. (9.12):

$$E_n = \frac{-m_r Z^2 e^4}{2\hbar^2 n^2} \cong -13.60 \frac{Z^2}{n^2} \text{ eV} \tag{E.11}$$

where n is the radial quantum number, equal to $l + 1 + q$. The possible values of n are therefore $l + 1, l + 2, \ldots, l + t$, where t is the number of terms in the radial function $R(r)$.

Equation (E.8) can be used to write each radial function $R_{nl}(r)$ explicitly. For $n = l + 1$ we have only one term, and with the help of Eq. (E.5) we have

$$rR_{nl}(r) = g(r)e^{-\alpha_n r} = b_0 r^s e^{-\alpha_n r}$$

or

$$R_{nl}(r) = b_0 r^l e^{-\alpha_n r} \quad (\text{if } n = l + 1) \tag{E.12}$$

where the value of b_0 is determined by normalization.

When $n = l + 2$, $R_{nl}(r)$ has two terms, with nonzero coefficients b_0 and b_1, respectively, and all other coefficients equal to zero. Using Eq. (E.8), first with $q = 0$ and again with $q = 1$, we have

$$b_1 = \frac{2\alpha_n(l + 1) - C}{2(l + 1)} b_0 \tag{E.13}$$

$$b_2 = \frac{2\alpha_n(l + 2) - C}{l + 3(l + 2)} b_1 = 0 \tag{E.14}$$

and so, from (E.14),

$$2\alpha_n = \frac{C}{l + 2} \tag{E.15}$$

and Eq. (E.13) then becomes

$$b_1 = -\frac{C}{2(l + 1)(l + 2)} b_0 \tag{E.16}$$

and therefore

$$R_{nl}(r) = b_0 r^l \left\{ 1 - \frac{Cr}{2(l+1)(l+2)} \right\} e^{-\alpha_n r} \quad (\text{if } n = l + 2) \quad (E.17)$$

where again b_0 is determined by normalization. The polynomials in brackets in Eq. (E.17), as well as analogous polynomials for $n = l + 3$, $n = l + 4$, etc., are called *Laguerre polynomials*.

The functions tabulated in Table E.1 are written in terms of the "reduced Bohr radius" a_0', obtained from the first Bohr radius [Eq. (1.14)] by replacing the electron mass by the reduced mass m_r. In terms of a_0', $C = 2Z/a_0'$ and $\alpha_n = C/2n$. Physically, a_0' is the distance between the positive and negative charges in the ground state of a Bohr atom.

TABLE E.1 Some Radial Wave Functions

$n = l + 1$	$n = l + 2$
$R_{10} = b_0 e^{-Zr/a_0'}$	$R_{20} = b_0 \left\{ 1 - \dfrac{Zr}{2a_0'} \right\} e^{-Zr/2a_0'}$
$R_{21} = b_0 r e^{-Zr/2a_0'}$	$R_{31} = b_0 \left\{ 1 - \dfrac{Zr}{6a_0'} \right\} e^{-Zr/3a_0'}$
$R_{32} = b_0 r^2 e^{-Zr/3a_0'}$	$R_{42} = b_0 \left\{ 1 - \dfrac{Zr}{12a_0'} \right\} e^{-Zr/4a_0'}$

Numerical Solution
of the Schrödinger
Equation

By a numerical integration routine obtainable with this book, the user can compute and observe one-dimensional probability amplitudes and probability densities for a single particle in a variety of potential-energy fields, and observe the results of wave-function calculations for energies that are not allowed. This MS/DOS program uses the Schrödinger equation to find the value of d^2u/dx^2. Then it finds the values of u and du/dx at the point $x = \delta x$ from the equations

$$u(x + \delta x) = u(x) + \delta x(du/dx) \tag{F.1}$$

and

$$[du/dx]_{x+\delta x} = [du/dx]_x + \delta x(d^2u/dx^2)_{x+\delta x} \tag{F.2}$$

where δx is one step in an iteration. This program makes eight iterations between one plotted point (one pixel) and the next. It can be used to solve problems in Chapters 3, 4, 8, and 9.

F.1 STARTING THE PROGRAM

At the opening menu you choose one of the following potentials:

For solving problems involving:	Enter:
Square Wells (including multiple wells)	1
Linear Potentials $V = kz$ (the quantum bouncer)	2
Hydrogen Atom or One-Electron Ion	3
Spherical Square Well (bound states and scattering)	4
Simple Harmonic Oscillator	5
Single Potential Step	6

After the initial choice (made by pressing the desired number and then pressing ⟨Enter⟩), you enter the parameters of the particle, described as follows for each choice of problem. The wave function is then computed from a point-by-point numerical integration of the Schrödinger equation.

The function $u(x)$ is normalized so its maximum value does not exceed the limits of the screen. When the curve appears you can plot the probability, which appears directly above the graph of $u(x)$. You can also change the value(s) of quantum numbers n and/or l, or change the value of E by a small amount to observe a function that is *not* an eigenfunction. (This energy option is useful in searching for eigenfunctions that have not been loaded into this program.) The procedures for each initial option are described below. *Pressing ⟨Esc⟩, then ⟨Enter⟩ at any time brings back the opening menu.*

F.1.1 Square Wells

This option lets you plot the wavefunction and probability density in one dimension for a particle of energy E in an infinitely deep potential well of width W. Outside this well the potential energy is infinite. Within the well you choose the potential energy V of a series of finite potential wells having a constant separation between one well and the next. Each of these wells has width WW, symmetrically placed within the infinite well.

You will be able to choose the values of the following well parameters:

- The number N of potential steps within the well (default: $N = 1$).
- For $N = 1$ or $N = 2$, the width WS of the individual finite wells. (For $N = 2$, the choice is limited to 1 nm or 0.5 nm.) For $N > 2$, the value of WS is determined by the value of N. The overall width W is determined by the values of N and WS.

- The potential energy V in each finite well (default: $V = -4$ eV).
- The energy E or quantum number n of the quantum state.

Example: Choosing all default values will bring up the wave function for the lowest energy level of an electron in a single finite well of width one nm and depth 4 eV (Fig. 3.1a). E will be set at -3.7366233, the lowest allowed energy. (The *precision* must be that great to make the wavefunction go to zero at the right side of the well; however the *accuracy* is not that great.) To see the next higher level (Fig. 3.1b), enter $\langle 1 \rangle$, then select $n = 2$.

To see an allowed function for another value of n, just enter that value (within the range shown on the screen), and the eigenfunction will be displayed. To see a non-allowed function, you can change the value of E. In this case you will see that the absolute value of the wave function rises exponentially as it approaches the right-hand side of the well.

The value of the wavefunction $u(x)$ will always be zero at the left wall. If your choice of energy is an allowed value for the well parameters that you choose, then the wavefunction will be zero at the right wall as well. In almost every case, this point-by-point numerical integration yields a function that is automatically symmetric or antisymmetric when it goes to zero at the right wall.

The function $u(x)$ is always normalized to permit its largest value to be displayed on the screen. *Caution*: If you choose an energy that is too far from an allowed energy level, $u(x)$ will blow up, and the last plotted value may be so large that most of the other plotted points will not be visible. You then can experiment with larger or smaller values of energy until you see these points. To be sure of seeing an allowed function you can select a value of the quantum number n and the computer will display a wavefunction for the appropriate preset energy eigenvalue. You can then change the value of E if you wish to see a non-allowed function.

You can set N, the number of wells inside the infinitely deep well, equal to any of the following values: 1, 2, 3, 4, 7, 10, and 22. Choosing N to be greater than 1 allows you to see the effect of multiple steps on the energy eigenvalues. For $N = 1, 2, 3, 4, 7$, and 10, with $V = -4$ eV, the energies for a number of allowed levels have been stored in the program. The functions for these levels can be plotted simply by entering a value for the quantum number n. With $N = 7$, for example, you can choose any of the lowest 14 allowed levels (all of which have $E < 0$), and you will see that these split into two bands of allowed energies, separated by a wide gap. The difference between the wavefunction for $n = 7$ and the wavefunction for $n = 8$ is striking.

For $N = 10$ or $N = 22$, only the lowest level is displaced automatically, but it does not take long to find additional levels by trial and error.

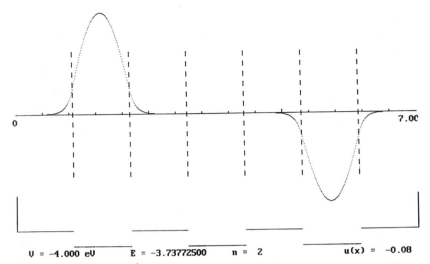

FIGURE F.1   Graph of wavefunction for the $n = 2$ level in a set of three equally-spaced potential wells. These wells are shown below the graph. The potential energy V in each well is also shown below, as well as the energy level E and the value of n. Notice that there is only one node, in the middle. The wave decays exponentially toward the center well in both directions; if the vertical scale were greatly expanded, you could see that it goes to zero only at the center of the graph. There is a tiny nonzero probability of finding the electron in the middle well.

When both E and V are negative, and $E > V$, there will be alternately allowed and forbidden regions, the allowed regions being where the potential energy has the value that you chose, and the forbidden regions being in between, where the potential energy is zero. In these forbidden regions the wavefunction will decay instead of oscillating. (Figure F.1 shows an example, for $N = 3$.) If you choose a value of E for which $E - V$ is positive everywhere, the wavefunction will oscillate in all regions.

You can change parameters and draw a new graph by following the instructions at the top of the screen. Options are

- Enter $\langle 1 \rangle$ to choose a new value for n only.
- Enter $\langle 2 \rangle$ to change the energy by a small amount δE.
- Enter $\langle 3 \rangle$ to display the probability density $P(x)$.

F.1.2 Linear Potential

This option shows wavefunctions and probability densities for the linear potential $V = kz$, where k can have any positive or negative value. The particle mass m can also be chosen. The Schrödinger equation can be solved analytically for a linear potential, but this is seldom done in quantum mechanics texts.

Eq. (4.43) shows from the correspondence principle that the energy levels for a particle of mass m are proportional to $k^{2/3}m^{-1/3}(n + n_0)^{2/3}$, where n is any positive integer. [See F. C. Crawford, *Am. J. Phys.*, **57**, 621 (1989)]. The correspondence principle does not provide the value of n, which can be found by solving the Schrödinger equation. However, the value of n can also be found by numerical integration, as is done in this program.

When you enter a value of $n \leq 13$, the program will generate acceptable wavefunctions for that value. You will see the value of E for each n that you choose. You can also verify that the wavefunction has a point of inflection where $E = kz$. You can restart this potential by pressing ⟨Enter⟩; this allows you to test the theoretical dependence of the energy levels on k and m.

The linear-potential option provides the opportunity of changing the horizontal scale by entering ⟨3⟩. [You display $P(x)$ by entering ⟨4⟩.] Changing the scale is useful in searching for higher levels ($n > 13$).

F.1.3 Hydrogen Atom or One-Electron Ion

This option lets you plot the radial part of the probability amplitude (the product of r and the radial factor $R(r)$ in the wavefunction) for a hydrogen atom, a one-electron ion, or an "exotic" atom such as muonium. It gives you a choice of the values of the atomic number Z, the mass m of the negative particle—electron, muon ($m = 207$ electron masses), etc.—in units of the electron mass, and the quantum numbers $n(\leq 100)$ and l. The probability amplitude $rR(r)$ will be displayed, on the appropriate horizontal scale.

You can change this scale as well as other parameters by following instructions at the top of the screen. Pressing ⟨Enter⟩ restarts this option with a new set of values for the parameters. Other choices are as above, except that choice ⟨1⟩ lets you choose a new value for l as well as for n.

For the hydrogen atom you make $Z = 1$ and $m = 1$ (units are electron masses). Making $m = 207$ gives muonic hydrogen (proton plus muon). Making Z an integer greater than 1 yields eigenfunctions for a one-electron ion (e.g., Li^{++}).

F.1.4 Spherical Well

This routine (4 on the opening menu) provides two options, as follows:

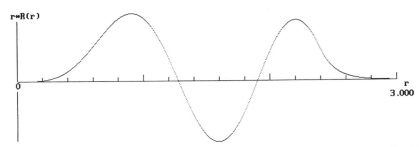

FIGURE F.2 Graph of wavefunction for an electron confined to a spherical well of radius 2.4 nm. The electron is in the state for $n = 3$, $l = 3$.

Option 1   You may plot the radial probability amplitude $rR(r)$ for an electron in a spherically symmetric potential well of finite depth. In this well the potential energy is zero for $r < 2.4$ nm, then suddenly it increases to $V = 4$ eV. The maximum plotted value of r is initially set at 3.0 nm. Figure F.2 shows the result for $n = 3$, $l = 3$; notice the point of inflection at $r = 2.4$ nm; also notice the forbidden region produced by the centrifugal potential near the origin.

You may choose any value up to 19 for the angular quantum number l, then you may choose a value of $n < 9$. There may be no solution if n is too large for your value of l; a prompt will tell you which values of n are allowed for your choice of l. If you do enter a value of n outside the allowed range, the value of n will be set at 1.

It is interesting to see how the wave is pushed out to higher values of r as l increases with fixed n. This happens because the width and depth of the effective potential energy are reduced as l increases. You can see the radial probability density as a function of r in each case.

Option 2   You may plot the wavefunction for an *unbound* state in a spherical well of the same depth and radius as the well of option 1. In this case you select the energy E rather than the quantum number n, because the levels are not discrete. With this exception, the options after the graph is plotted are the same as those for the hydrogen atom.

F.1.5 Harmonic Oscillator

Entering 5 at the opening menu lets you plot the wavefunction and the probability density for a particle of mass m with potential energy $V = Kx^2/2$, where K is the spring constant. You can choose the values of K (in units of eV/nm^2), m (as a multiple of the electron mass), and the quantum number n. In this case, the numerical approximation that is used becomes noticeably inaccurate for values of n greater than 7, making the wavefunction asymmetric for these values of n.

Again, you can see that the point of inflection occurs where $E - V = 0$, you can show that the energy levels have the right dependence on K and m, and you can see the variation of the peaks in the probability density.

F.1.6 Single Potential Step

This option shows both bound and unbound levels for an electron in a one-dimensional well in which V is finite for $0 < x < 2$ nm, and is infinite outside that range. For $0 < x < 1$ nm, $V = 0$, and for 1 nm $< x < 2$ nm, V has a value chosen by you (default is $V = +4$ eV). You may select the following:

- The value of V (≥ 0). Entering zero gives the default—the previous value of V. To make $V = 0$ over the full 2 nm, enter .000001.
- The value of E (> 0).
- The value of n (≤ 7). This graphs eigenfunctions for the case $V = +4$ eV. To see graphs for $n > 7$, find E by trial and error. Observation of the peaks in the probability curve shows that, where $E > V$ and $V > 0$, the electron is more likely to be found in the region where its wavelength is greater.

Appendix G Table of "Elementary" Particles

Data are taken from Particle Data Group, Review of Particle Properties, *Physical Review D*, **45** (1992), Sec. II.1.

Class	Symbol	Spin parity	Mass (MeV/c^2)	Mean lifetime (seconds)	Common decay modes (percent)	
Photon	γ		0			
Lepton	ν_e	1/2	< 0.0000073			
	ν_μ	1/2	< 0.57			
	$e^\pm$	1/2	0.51099906 ±0.00000015			
	$\mu^\pm$	1/2	105.658389 ±0.000034	2.19703 ± .00004 ×10^{-6}	$e\nu\nu$	(100)
	$\tau^\pm$	1/2	$1784^{+2.7}_{-3.6}$	3.05 ± .06 ×10^{-13}	$\mu\nu\nu$ $e\nu\nu$ hadrons*	(18) (18) (64)
Meson	$\pi^\pm$	0^-	139.5679 ±0.0007	2.6030 ± .0023 ×10^{-8}	$\mu\nu$	(100)
	π^0	0^-	134.9743 ±0.0005	0.848 ± .0067 ×10^{-16}	$\gamma\gamma$ γe^+e^-	(98.8) (1.2)
	$K^\pm$	0^-	493.646 ±0.009	1.2371 ± .0029 ×10^{-8}	$\mu\nu$ $\pi\pi^0$ $\pi\pi\pi$ $\mu\pi\nu$ $e\pi\nu$	(63.5) (21.2) (7.3) (3.2) (4.8)
	K^0	0^-	497.671 ±0.031			
	$K_1(K_S)$			0.8922 ± .0020 ×10^{-10}	$\pi^+\pi^-$ $\pi^0\pi^0$	(68.6) (31.4)
	$K_2(K_L)$			5.173 ± .040 ×10^{-8}	$\pi^0\pi^0\pi^0$ $\pi^+\pi^-\pi^0$ $\pi\mu\nu$ $\pi e\nu$	(21.6) (12.4) (27.0) (38.7)

*Hadron is a generic term for baryons and mesons. This decay leads to many possible combinations.

Appendix G (continued)

Class	Symbol	Spin parity	Mass (MeV/c^2)	Mean lifetime (seconds)	Common decay modes (percent)	
Baryon	p	1/2	938.27231 ±.00028	$> 5 \times 10^{32}$		
	n	1/2	939.56563 ±.00028	889.1 ± 2.1	$pe^-\,\nu$	(100)
	Λ	1/2	1115.63 ± .05	2.632 ± 0.02 $\times 10^{-10}$	$p\pi^-$	(64.1)
					$n\pi^0$	(35.7)
	Σ^+	1/2	1189.37 ± .07	0.799 ± 0.004 $\times 10^{-10}$	$p\pi^0$	(51.6)
					$n\pi^+$	(48.3)
	Σ^0	1/2	1192.55 ± .10	7.4 ± 0.7 $\times 10^{-20}$	$\Lambda\gamma$	(100)
	Σ^-	1/2	1197.43 ± .06	1.479 ± 0.11 $\times 10^{-10}$	$n\pi^-$	(99.8)
	Ξ^0	1/2	1314.9 ± 0.6	2.90 ± .09 $\times 10^{-10}$	$\Lambda\pi^0$	(100)
	Ξ^-	1/2	1321.32 ± .13	1.639 ± 0.015 $\times 10^{-10}$	$\Lambda\pi^-$	(98.8)
	Ω^-	3/2	1672.43 ± .32	0.822 ± .012 $\times 10^{-8}$	ΛK^-	(67.8)
					$\Xi^0\pi^-$	(23.6)
					$\Xi^-\pi^0$	(8.6)

Appendix H Table of Physical Constants

Values are taken from Particle Data Group, *Physical Review D*, **45** (1992), Sec. III.1.
Numbers in parentheses are standard deviations in the last digits of the quoted values,
determined on the basis of consistency of fits to all relevant experiments.

Quantity	Symbol	Value and Units
Speed of light in vacuo	c	2.99792458×10^8 m/s (exact)
Avogadro's number	N_A	$6.0221367(31) \times 10^{23}$ mole^{-1}
Electron charge	e	$1.60217733(49) \times 10^{-19}$ C
Planck's constant	h	$4.1356692(27) \times 10^{-15}$ eV-s
Fine-structure constant	α	$1/137.0359895(61)$
Atomic mass unit	u	$931.49432(28)$ MeV$/c^2$
Electron mass	m_e	$0.51199906(15)$ MeV$/c^2$
Proton mass	M_p	$938.27231(28)$ MeV$/c^2$
		$= 1836.152701(37)\ m_e$
		$= 1.007276470(12)\ u$
Classical electron radius	r_e	$2.81794092(38) \times 10^{-15}$ m
First Bohr radius	a_0	$0.529177249(24) \times 10^{-10}$ m
Compton wavelength of electron	λ_c	$2.42631058(22) \times 10^{-12}$ m
Rydberg constant for infinite-mass nucleus	R_∞	$10973731.45(32)$ m^{-1}
Bohr magneton	μ_B	$5.78838263(52) \times 10^{-11}$ MeV/T
Electron magnetic moment	μ_e	$1.00115965241(20)\ \mu_B$
Nuclear magneton	μ_N	$3.15245166(28) \times 10^{-14}$ MeV/T
Proton magnetic moment	μ_p	$0.001521032209(16)\ \mu_B$
Boltzmann's constant	k	$8.617385(73) \times 10^{-5}$ eV/K
Gravitational constant	G	$6.67259(85) \times 10^{-11}$ m^3/(kg-s^2)
Wavelength of 1-eV photon		$1.23984244(37)\ \mu$m
Temperature for 1 eV/particle		$11604.45(10)$ K

INDEX

ISBN 0-12-483545-7